prefiguring cyberculture

AN INTELLECTUAL HISTORY

prefiguring cyberculture

AN INTELLECTUAL HISTORY

SENIOR EDITOR
Darren Tofts

EDITORS
Annemarie Jonson and Alessio Cavallaro

The MIT Press
Cambridge, Massachusetts
London, England

Power Publications
Sydney

This book was set in Charlotte Book and Univers Ultra Condensed by Diagram.

Printed and bound in Australia.

Library of Congress Cataloging-in-Publication Data
Prefiguring cyberculture: an intellectual history / edited by Darren Tofts, Annemarie Jonson, and Alessio Cavallaro
   p. cm.
   Includes index.
   ISBN 0-262-20145-3 (hc.:alk. paper)
  1. Technological innovations—History. 2. Information technology—Social aspects.
3. Technology and civilization I. Tofts, Darren. II. Jonson, Annemarie. III. Cavallaro, Alessio.

T173.8 P688 2003
303.48´34M—dc21   2002029375

CONTENTS

SECTION TWO

VIRTUALITY : WEBWORLDS AND CYBERSPACES

SECTION THREE

VISIBLE UNREALITIES : ARTISTS' STATEMENTS

SECTION FOUR

FUTUROPOLIS : POSTMILLENNIAL SPECULATIONS

CODA

ACKNOWLEDGMENTS

In 1998, the idea for this book – a volume that would illuminate the rich intellectual history of cyberculture by probing "framing texts" drawn from fiction, science and philosophy – was originated by Alessio Cavallaro. Since then, it has been a long, at times tortuous and always complex process. We are indebted to the many people and organizations who have contributed so generously their personal and institutional support to its realization.

Julian Pefanis, then General Editor of Power Publications, recognized the value of the concept, and, on the basis of an inchoate outline, offered his unreserved support and the backing of Power Publications for the project. Professor Terry Smith, until recently Director of the Power Institute Foundation for Art and Visual Culture at the University of Sydney, was instrumental in overseeing the development of the project through its middle and crucial later stages. His belief in this project and his expert advocacy on its behalf were decisive in securing co-publication with MIT Press and MIT's mentoring of the project. Former and current Managing Editors of Power Publications, Liz Schwaiger and Greg Shapley, gave invaluable assistance in steering the project through the multiple hoops of project management.

Several Australian funding bodies committed financial assistance to the book: Cinemedia (now the Australian Center for the Moving Image), the Australia Council and the Australian Film Commission. In particular, we would like to thank the people from these organizations whose support was key in our successful applications for funding. They are Ross Gibson, Lisa Colley, Julie Regan and Kate Ingham.

We are very grateful to Roger Conover, executive editor of MIT Press, for his staunch commitment to the book and his consistent and unstinting application of editorial rigor to the text. His critical contribution has made *Prefiguring Cyberculture* a much stronger, more cohesive book than it otherwise might have been. Thanks also to Lisa Reeve, Janet Timmerman, Claude Lee and Ann Rae Jonas from MIT Press.

We are indebted, also, to our institutional bases and colleagues at The School of Social and Behavioural Sciences, Swinburne University of Technology, the Arts Informatics program at the University of Sydney, the School of Design at the University of Western Sydney and the Australian Center for the Moving Image (formerly Cinemedia). It is against the backdrop of these institutions' intellectual culture and the support of and discursive engagement with our colleagues, that the project took shape.

A number of people have provided vital assistance at key moments in the project's development and implementation. Cathy Gray did a sterling job copyediting the long and complex manuscript and we are grateful for her expertise and attention to detail. Christopher Waller of Diagram took the text and interpreted it into a design which we believe beautifully complements the book's contents and thematic. Thanks also to Anne Mason and Greer Sansom for their work in the design process, and also to Jane Ellery and Lisa Carlon for their contribution to this process. Robin Appleton commenced the index, which was completed by Nicholas Strobbe, who also proofread the manuscript. Robert Hassan, Rowan Wilken, Stephen Johnson and Rex Butler are also owed thanks. And for their personal support throughout this and all our other projects, professional and otherwise, we thank Lisa Gye, Peter Lowe and Antoanetta Ivanova.

Finally, a great debt of gratitude is owed to the writers and artists for their patience and responsiveness over the last five years and, of course, for their outstanding contributions. It is on the basis of their efforts that *Prefiguring Cyberculture* makes its mark, we hope, as a collection which, in N. Katherine Hayles' words, helps to define the moment to which we have come.

Darren Tofts, Annemarie Jonson and Alessio Cavallaro

FOREWORD

N. Katherine Hayles

The appearance of *Prefiguring Cyberculture* indicates that new media studies is no longer in its infancy, for it bespeaks a field mature enough to wonder about its ancestors, eager to explore family connections and claim relations. The essays, richly diverse, converge in arguing that technology alone is not sufficient to understand the evolution of cyberculture. Historical context shapes how the technology will be conceived, understood and implemented. The child is born not into a vacuum but a social and cultural matrix whose lines of force stretch back to Plato and Descartes, arc forward through *Frankenstein* and *Erewhon*, and connect with such breathtakingly different 20th century visionaries as Alan Turing, Alvin Toffler, Donna Haraway and Vernor Vinge.

There are of course recursive loops in play here. This project of recovering antecedents and evaluating, with the benefit of hindsight, their assumptions, achievements and predictions, will soon enough itself become another point along the trajectory of the past, to be sized up by more recent arrivals. As this project prepares to take its place in the slipstream of time, one might feel tempted to wonder how its evaluations will fare. At the risk of implying that its retrodictions are already outdated before they are officially published (as if this collection were to suffer the same fate as the New York TWA terminal that Mark Dery analyzes in his fine essay "Memories of the Future"), I suggest that the concerns voiced here are very much of the moment, a remarkable index to the historical junction between the past millennium and the new one we are learning to call home.

Where can we see this mark of the present? Consider Erik Davis' brilliant essay on the persistence – or rather, apotheosis – of the Cartesian subject in the cultural formations identified with cyberspace. Although Davis does not draw the connection, there is a strong link between his argument and Margaret Wertheim's astute analysis of More's and Bacon's utopias in "Internet Dreaming." Davis and Wertheim both see a tincture of metaphysical and religious concerns swirled in with cyberspace rhetoric. Yet the Cartesian division between body and mind that Davis addresses is more characteristic of cyberspace as it was conceptualized in the 1980s than now. The 80s construction of cyberspace as an immaterial space where mind reigned supreme, captured in John Perry Barlow's quip that "my everything was amputated," is giving way to what Marcos Novak has called "eversion" and others, such as artist Troy Innocent, call "mixed reality;" the seamless mixing of virtual and real spaces through such technologies as embedded sensors, pervasive computing and eyeglasses superimposing simulated displays on real landscapes. In electronic art, liquid architecture, remote surgery, distributed agent simulations and a host of other areas, such fusions incorporate computation with real-world dynamics, simulation with real life. As my colleague Tim Lenoir has remarked, our children will not call such phenomena virtual reality; they will simply call it reality.

Of course, the materiality that made virtual reality anything but a purely mental phenomenon was always available as a resource to understand it as a real world practice. Indeed, future generations may well be struck with astonishment that people wearing the unwieldy Polhemus helmets of the 1980s could for one instant ever have believed that they had somehow left their physical bodies behind. The claim is testimony to the malleability of perception, the ability of cultural presuppositions to shape experience. In part the movement toward mixed reality can be attributed to the evolving technology as it moves toward ubiquity and invisibility. Also contributing is a changed understanding of humans as participants in distributed cognitive systems that include multiple nonhuman cognizers, a perspective diametrically opposed to the Cartesian cogito constructing itself out its own thoughts. In pointing to the importance of what Andy Clark has called "extended mind," John Sutton, in "Porous Memory and the Cognitive Life of Things," remarks that this view brings the interface into focus as a crucial component, for it is through the interface that the experiences, capabilities and actions of the human are structured in relation to the distributed system in which she participates. Understanding human cognition will increasingly mean analyzing the affordances that suture us into the flows of information as we are incorporated into systems that are at once material and conceptual, virtual and real.

In a sense, extended mind can be seen as a down-home version of mixed reality, for it does not necessarily depend on high tech appliances to position the human as part of a distributed cognitive system. Such everyday objects as pencils, compasses and room arrangements are seen as objects humans enrol in their cognitive systems to enable them to perform sophisticated computational tasks. What emerges is a richer, fuller picture of human cognition as an embodied activity that takes place in environments embedded with a host of objects that create flexible, creative systems of thought. In retrospect we may wonder why the Cartesian cogito could ever have been convincing as a model of thought, for humans have always used their environments, natural and artificial, as resources to extend their cognitive grasp.

The desire to give a fuller account of the relation of cognition and subjectivity to cultural, social, technological and economic contexts forms another strong thread through these essays. In re-visiting Donna Haraway's famous essay "Manifesto for Cyborgs," Zoë Sofoulis remarks upon the difficulty of naming what Haraway was after, for what she wanted was what artificial life researchers call "the whole iguana," a representation that reaches toward holistic interconnection. In her case, the object of desire was a rhetoric and mode of analysis adequate to depict the complex webs between capitalism, informatics, gender, racial politics and the construction of scientific knowledge. No wonder her famous essay employs a notoriously difficult style of writing, for it is crafted to evoke ironic skepticism, critical distance, and utopian yearnings simultaneously, forcing these disparate affective and cognitive modes to occupy the same space at the same time.

Also concerned with using the past as a resource to give a richer account of present possibilities is Elizabeth Wilson's essay on the importance of affect and emotions in Alan Turing's work. She uses her analysis to invest his "childishness" with a different valence than it had for his detractors (and for his mother), suggesting that his work was centrally concerned with ways to incorporate in intelligent machines the childish capability for surprise, creativity, and emotional spontaneity. The impetus she finds in Turing's work could be seen as a trajectory that arcs forward toward the creation of intelligent machines capable of emotional responses.

In a different way, Evelyn Fox Keller's deeply researched essay is also a commentary on the whole iguana. She explores the decline of teleology as a biological idea as the schism deepened between molecular biology, with its reductive emphasis on the gene as the primary actor in reproduction, and embryology, with its search for more holistic explanations that could account for an organism's power to self-organize. Tracing the rise of cybernetics and its interest in circular causality, she documents the budding alliance between cyberneticians and developmental biologists as they sought to resurrect discredited ideas of self-organization, complexity, and emergence. Although this alliance came to "not very much," in her words, the arrival of more sophisticated computers powered a new turn of events when connectionist ideas could actually be implemented in hardware. Developmental biology became one of the inspirations for computer simulations seeking to capture evolutionary processes in artificial media. Although she ends her account here, the story continues as research into artificial life constructs it as both parallel to and a competitor of biological life. If we splice this narrative together with John Sutton's account of the cognitive life of things, we could arrive at the techno-enthusiast breathlessness of Damien Broderick's "Racing Toward the Spike," envisioning a "singularity" in which the conditions of human life, inextricably entwined with intelligent machines and artificial organisms, will be utterly transformed in ways that we cannot now possibly imagine.

Or we could remember that such predictions, even predictions that seek to prolong their shelf life by shrouding the future in mystery (if not mysticism), prove reliably inadequate to account for the complex interplays between technology, culture and cognition that are central to this fascinating collection of essays. I agree with Broderick in this, if nothing else: whatever the future brings, we can be certain it will be different from how we imagine it now. All the more reason, then, to look to the past to understand in fuller, richer terms the possibilities of the present. *Prefiguring Cyberculture* is a collection whose time has come; or rather, a collection that, through its insightful and compelling interventions, helps to define the moment to which we have come.

prefiguring cyberculture

AN INTELLECTUAL HISTORY

ON MUTABILITY

Darren Tofts

This is a book about technology and change. Or, more specifically, it is about mutability, the tendency towards change and alteration. *Prefiguring Cyberculture* addresses the matrix of themes to do with the integration of human life and technology. These themes gesture to profound and, in many cases, disturbing, incomprehensible change, such as the concept of artificial life, or disembodied virtual space, or the prospect of a future in which intelligent machines, rather than human beings, dominate life on earth. These are contemporary themes, the stuff of popular culture, technoscientific research and cultural theory alike. While of the moment, of our time, they also resonate with the traces of earlier, much older encounters with change, of apprehension for the transformative impact of technology. *Prefiguring Cyberculture* is concerned with exploring particular historical traces of technological change that, in retrospect, seem prescient, foreshadowing the lineaments of our contemporary moment. In doing so it reveals that mutability is not simply about change, but is rather an ongoing inclination to change, a constancy in human thinking on matters of technology – a constancy that can be characterized by the idea of becoming.

The essays collected in *Prefiguring Cyberculture* are responsive to a phenomenon that has primarily been associated with the final decades of the 20th century – the emergence, no less, of a new conception of human life, a redefined ontology that goes by a number of different, yet complementary names: posthuman, cyborg, informatic. While the inflections of these terms suggest important differences, they are nonetheless cultural indices of change and variation, ways of thinking about and defining what we are becoming. This ontology emerges out of the changes in the way people live, communicate and comprehend themselves in a world of technological complexity. The vast social apparatus of the computer network has aligned people with technology in dramatic and, it is argued, unprecedented ways. The intimacy of the human-computer interface, as intuitive to quotidian experience as speaking and writing, has meant that it is largely irrelevant to distinguish technology from the social and cultural business of being human. At a deeper, metaphysical level, this intimacy with the virtual space of the network, which extends our ability to be present elsewhere, to be here *and* there at the same time, has altered some of the defining parameters of human nature. In the most literal sense of the term, such extension is a mutation, a demonstrable alteration of what we are and what we are capable of doing in the name of being human. The figure of the cyborg has been invented to describe this advanced state of augmented potential. However, as we shall see, the cyborg may have assumed many guises in the past, all embodying contemporary articulations of mutation.

Cyberculture is the broader, epochal name that has been given to this process of becoming through technological means. Cyberculture, though, is not, in advance, the most apposite concept for thinking about extreme ideas to do with technological transitions in and abstractions of human life. The etymology of *cyber*culture only partially denotes computers, yet despite this the term has nonetheless stuck as a connotative marker of our engagement with computer technology. More specifically for our purposes, the prefix cyber has a more specialized and specific denotation to do with the control and flow of information; or, to be even more specific again, with information as the common unit or element of organization within organic and inorganic systems. Norbert Wiener defined information as "the content of what is exchanged with the outer world as we adjust to it, and make our adjustments felt upon it" (Wiener 1968, 19). "We," in Wiener's conception of cybernetics, referred to human beings as much as steam engines, guided missiles or thermostats. Cybernetics postulated that to understand human nature one had to understand information and to envisage the human as just another closed system like a machine, adapting and adjusting to its environment on the basis of the flow and control of information.

This cybernetic turn in the middle of the 20th century was not so much anti-humanist, as an articulation of what we now understand by the term posthuman. The posthuman is a revisionary conception of the category "human," a coupling of the human and the technological, in which it is "no longer possible to distinguish meaningfully between the biological organism and the informational circuits in which it is enmeshed" (Hayles 1993, 80). Cyberculture, as it is being invoked in the title of this book, is to be understood within this context of a conception of the human that has gone beyond – hence post – the organic, a-technological vision of "man" of classical antiquity, the Renaissance and the Enlightenment. Similarly, the concept of becoming informatic, used throughout this book as a framework for understanding historical and contemporary mutability, also encapsulates this recognition of the human as a special kind of information-processing device (Haraway 1991, 164). It is a category of emergence that recognizes the fundamental importance of the information circuit to the ontology of autonomous, self-regulating systems.

The tendency, then, to see the human in informatic or posthuman terms has clearly been predominant since the 1940s with the advent of cybernetics. However, the tendency to change, mutability, is historical and its traces can be found in the history of ideas, the rich legacy of philosophical, literary and scientific thought that underpins the western intellectual tradition, of which cybernetics, itself, is a product. The notion of prefiguring cyberculture is concerned with identifying historical encounters with mutability, with intimations of transformation, variations in thinking about the division, indeed, the indivisibility, of human life and technology. By no means teleological – the view that cyberculture was a state of being towards which we have been irrevocably moving – the prefiguring described here attests to a less causal, elliptical trajectory of change. That is, a recognition that what we call cyberculture is an instance of an ongoing tendency to alteration, a re-configuration of what it means to be human in the context of technology.

The essays in *Prefiguring Cyberculture* are structured around three thematic modalities: subjectivity (Section One: *I, Robot*), spatiality (Section Two: *Virtuality*) and temporality (Section Four: *Futuropolis*). While these sections are concentrated around discrete themes, the ideas discussed within them are also skeined throughout the book as a whole, resonating the historical prevalence of ideas that occupy our attention today. Within each section, writers engage with a framing text or concept to tease out and discuss the ways in which it has in some way prefigured or forecast some aspect of the becoming informatic. The framing texts have, by and large, been selected on the basis of their recognized importance in debates to do with cybercultural discourse, such as Alan Turing's essay on computing machinery and intelligence, or William Gibson's *Neuromancer*. But we have also selected less orthodox texts that cast surprising light on cybercultural matters, such as Samuel Butler's novel *Erewhon* and Thomas More's *Utopia*. There are notable differences in the manner in which different writers engage with the framing texts. Some, such as McKenzie Wark in his discussion of Ray Bradbury's short story "The Veldt," offer close, detailed readings that are singularly focussed in their pursuit of a concept, such as virtuality as a post-representational space; others, such as Gregory Ulmer's treatment of Plato's simile of the cave, diverge dramatically away from the text, using it as a springboard to make new ideas from it, such as a model for interface design. These differences are indicative of the vitality and richness of an intellectual legacy of ideas that either overtly suggest issues of technological change, or obliquely gesture towards tendencies that, only now, seem very recognizable. The writers contributing to this volume have stamped their particular "take" on the becoming informatic in the ways in which they either analyze or parlay with their framing texts.

There are differences, too, in the conclusions the writers draw regarding the human implications of technological change. Most writers share the view that cyberculture involves a stage of advancement in the human interface with technology, though not all agree on the degree or significance of this change – indeed, Richard Slaughter argues that cyberculture is not even a particularly useful concept for thinking about the future and technological change. There are unavoidable assessments of the positive or negative implications of becoming informatic. Cyborg subjectivity, for example, is seen as the reconfiguration of a new unity in Zoë Sofoulis' essay on Donna Haraway, while Erik Davis sees it as the arrival of the Cartesian split that hybridity purports to reconcile. Apart from these binary extremes of response, there are also profound expressions of ambivalence, of the inability to fully comprehend or predict what technological complexity portends for concepts such as life, let alone human life, as in Damien Broderick's discussion of Vernor Vinge. Such differences are highlighted in the section introductions, which set out to contextualize the principal themes and the writers' interpretations of them.

The issues and themes discussed in these essays are also explored in a selection of artists' statements and accompanying images (Section Three: *Visible Unrealities*). Ten internationally regarded new media artists articulate their practice in the context of the broad cybercultural themes implicit in the essays. Art has always been a reliable barometer of change and the work sampled in this section attests to the social, cultural and technological implications of mutability. Artists, too, have historically been technological innovators, exploring the aesthetic possibilities of new media. Specifically, these artists dramatize the ways in which the computer, as medium and metaphor, has impacted upon aesthetic practices in the last decades of the 20th century. While the term "electronic arts" has come to represent a specific epoch, these artists suggest that their pre-occupations – to do with nature, mind, the body and the interface between humans and technology – are also part of an ongoing historical inquiry into what it means to be human in the face of technological change.

Overall, then, this book presents a range of responses to a common theme, that what we call cyberculture has been a long time coming. *Prefiguring Cyberculture* is not concerned with identifying how we "became posthuman," but rather demonstrating how becoming informatic is an ongoing process of the continuous present tense.

REFERENCES

Haraway, D. 1991. *Simians, cyborgs and women. The reinvention of nature.* New York: Routledge.

Hayles, N. K. 1993. Virtual bodies and flickering signifiers. *October* 66 (fall).

Wiener, N. 1968. *The human use of human beings. Cybernetics in society.* London: Sphere Books.

SECTION ONE

I, ROBOT:
AI, ALIFE AND CYBORGS

"It's not an easy thing to meet your maker"
BLADE RUNNER

Annemarie Jonson and Darren Tofts

What does it mean to be human? This question, perhaps it is the eternal question, has been asked many times throughout history. It is fundamental to being human and, arguably, is the defining issue of cyberculture, since the process of becoming informatic involves the transformation into something other. The figure of the cyborg has emerged as a kind of totem for this emerging state of otherness, a state that presages the ascendancy of a dramatic new phylum – the posthuman. The posthuman is a category or classification of life that goes beyond essentialist thinking and the traditional binary oppositions by which the human has historically been defined. Its paragon, the cyborg, is a hybrid rather than pure being, for whom the technological is elemental rather than optional. Less a terminal event (the end of the human), the cyborg has evolved, theoretically, as a transitional evocation of what the human is becoming in the informatic age.

The technological developments and theoretical explorations that are most readily associated with cyborg subjectivity – artificial intelligence, artificial life, genetic engineering, robotics, advanced prosthetics – exert a conspicuous presence in cybercultural literature as 20th century phenomena. However, as the essays in this section suggest, they are actually contemporary manifestations of a historical continuum of metaphysical and speculative thinking to do with questions such as what is life, what is mind, can machines think, can they be self-aware? These are ideas that humans have been grappling with for a long time, which suggests that the becoming informatic has an impressive, though perhaps unrecognized, longevity.

A common strand examined within these essays in this context is the continued relevance of dualistic or binary thinking for conceptualizing the posthuman. If hybridity entails non-oppositional thinking, what paradigms do we have to define the ontology of cyborg subjectivity, to understand its strange, ambivalent mode of being? How can we understand what it is, when we can no longer define it in terms of what it is not? To think outside binary oppositions is the challenge posed by such developments as artificial intelligence and the deepening intimacy of our interface with computer technology. However, the more urgent and disturbing challenge may be in having to face what such developments inadvertently disclose – the proposition that humans have never been purely human to begin with, that human nature is more technologized than we might like to admit.

Erik Davis, in his discussion of the Cogito, "Synthetic Meditations: Cogito in the Matrix," fires a spirited sortie at the proponents of post-dualism and argues that cyberculture is underpinned by a "profound if often unconscious Cartesianism." While he is against a dualistic split in terms of subjectivity, Davis notes that the disembodiment associated with cyberspace, virtual reality and artificial intelligence, are reinforcements of what he calls "a splinter of the cogito." For Davis, what we have been calling cyborg subjectivity is predicated not on the breaking down or imploding of binary oppositions between matter/spirit, mind/body, but is rather a "new picture of man (sic) as an instrumental agent of his own incorporeal will." In discussing the Cartesian skepticism that led to the modern concept of the subject, Davis offers an inventive and timely reading of the film *The Matrix* in the context of Descartes' first Meditation, with its dark scenario of "an evil genie" responsible for creating powerful illusions that the enlightenment philosopher of doubt unwittingly takes to be reality. The "false reality" narrative, of which *The Matrix* is an example, further evidences the pervasiveness of Cartesianism for Davis. In this he anticipates discussions in Section Two about virtual space and simulation and the fate of the real under such conditions.

The dualism identified by Davis concerns an historical split between the controlling mind and the responsive body. Catherine Waldby sees in Mary Shelley's novel *Frankenstein* a different kind of dualism, a classifying distinction between the natural and unnatural body, between a pure or essential humanity and monstrosity. In "The Instruments of Life: Frankenstein and Cyber-culture," Waldby shifts the emphasis away from the apocalyptic connotations of the novel – the creature as abject chimera roaming the countryside seeking revenge on its maker – and instead reads it in the context of 19th century scientific explorations into the concept of vitality, the fascination with the idea that life can be assembled ("technogenesis") and need not originate in nature. But as *"un corps en morceau,"* an amalgam of heterogeneous bits and pieces that cohere into neither one thing or another, Frankenstein's monster represents the "open-ended nature of the body's becoming" in the technoscientific age. Unlike Davis, Waldby finds support for this post-dualistic approach to human nature in 17th century discussions of the "economy of man," which portray the body in terms of separable, material components that work together to form an "organic machine." For Waldby, this suggests that human nature is already machinic and that what constitutes life is less a matter of matter, than of classification, which is always open to change.

The historical tendency to see *Frankenstein* in dualistic terms is seen by Waldby as a distraction from its portrait of a possible ontology of the machinic organism, that both inscribes and pre-cedes it. Elizabeth Wilson, in her account of Alan Turing's discussion of machine intelligence, "Imaginable Computers: Affects and Intelligence in Alan Turing," proffers a similar impatience with the dualism that characterized Turing's approach to thinking machines. Turing's description of his "electronic brain," controlling its roving mechanized body from a distance – itself an uncanny echo of Frankenstein's monster – represents for Wilson a moribund, unsustainable philosophy of disembodiment. However, she is more impatient with critical assessments of Turing's work that are preoccupied with this dualism, arguing that they neglect the inventive, heterogeneous thinking that went into his work on machine intelligence. Wilson argues that Turing was less interested in creating a machine that could think than encouraging "serious consideration of the possibility of machine intelligence." As Wilson suggests, the distinction is an important one, for the possibility of machine intelligence had little to do with computational power and more to do with speculative force. In this, Turing's contribution to cyborg subjectivity resides not in his famous, eponymous test, but rather in his delineation of the inventive space in which machine intelligence can be hypothesized. Within this space of immanence, Turing postulated an affective rather than programmed approach to intelligence, in which an artificial mind developed and learned in a similar way to that of a child, entertaining such human factors as surprise, excitement and repetition.

Turing's privileging of childishness, as one of the necessary components in any intelligent machine, foregrounds the notion of vitality so central to discussions of artificial organisms. The question of artificial life, like artificial intelligence, begins with a discussion of first principles, such as what is life, what conditions make it possible, what is the source of living order? As Evelyn Fox Keller argues in "Marrying the Premodern to the Postmodern: Computers and Organisms after WWII," discussions regarding the nature of vitality have a turbulent prehistory. We witness the issue of this prehistory in the mid-20th century convergence of the advent of the computer, the work of Norbert Wiener and the early cyberneticists and the new meanings of mechanism they theorized. In centralizing notions of organization and feedback as the defining characteristics of an organism – whether it be a person, a thermostat or a steam engine – cybernetics bypassed over 100 years of molecular biology in favor of a discredited, 18th century notion that identified the interaction between parts in a whole as the defining characteristic of life. Ironically, this convergence occurred around the same time that the materialist model of physio-chemical order reached its apotheosis in the discovery of DNA and the genetic program. But in the biological concept of organization, Wiener found the conceptual basis for putting "the principles of Life to work in the world of the non-living." In so doing, he developed the "ultramodern" organic theory that we associate with cyborg subjectivity, complex systems and artificial life.

While predicated on the notion of organization, complexity is profoundly unstable. Cybernetic organisms are one thing, but it is another thing altogether to think of the weather or the behavior of stock markets as "living" systems. Within the frameworks of understanding available to us, how do we make sense of such phenomena in this context? Furthermore, what kind of mind develops such theories as complexity? In seeking out the conceptual terrain of complexity, Samuel Umland and Karl Wessel find that it is derived from the reductive rather than holistic mind, from inward states of intense fixation akin to autism (a state of mind identified in two of the 20th century's most profound agents of technological change, Alan Turing and Albert Einstein). In "Cassandra Among the Cyborgs," Umland and Wessel pursue the relationship between mind and complexity through a "misreading" of Philip K. Dick's 1976 essay, "Man, Android and Machine." Treating this essay as a kind of palimpsest, they weave fragments of Dick's writings throughout their essay, enacting, in an elliptical and suggestive manner, the questions about consciousness raised in his work, of which "Man, Android and Machine" was a single snapshot. Tackling the problem of unitary subjectivity common to this section, Umland and Wessel go beyond Dick's well-known ambivalence towards ontological distinctions between man and machine, focussing instead on his interest in the idea of dissociated consciousness. In pursuing Dick's interest in the divided self, the idea of multiple minds within the one body, Umland and Wessel find that neural abnormalities, such as alien hand syndrome (in which the hand moves independently), evidence an extreme case of the body developing a mind of its own. Observing the "law of unexpected consequences," this phenomenon is an example of the links between aberration and complexity, in which any notion of an organic whole, as we have seen so often throughout this section, is but an illusion.

In the increasingly hyperconnected world of cyberculture, there may be surprises in store for us that we can't possibly begin to anticipate. In a world of increasing technological complexity, of multiple online personae and the notional distancing of our minds from our bodies, can we remain in control of ourselves? An apt metaphor of divided cyborg subjectivity, of the technologized body doing things we didn't have in mind, is Stelarc's 1995 performance piece, *Stimbod*, which, through the use of a touch screen interface, allowed users to stimulate the movement of limbs independently of the artist's central nervous system (a later 1997 work, *PingBody*, took this indeterminacy even further by actuating the movement of the body not through the promptings of another body, but through the flow of internet activity). What does it mean, then, to become cyborg? In "Cyberquake: Haraway's Manifesto," Zoë Sofoulis responds to this question in her discussion of Donna Haraway's landmark "Cyborg Manifesto," and in the process crystallizes a number of the themes developed in this section. Sofoulis observes that "the cyborg figure cannot be traced back to a pure origin or natural essence." An "assemblage of heterogeneous parts," it is the talisman for the semantic, political and metaphysical disruption of dualistic thinking that (pace Erik Davis) cyberculture promises. By treating terms in a binary opposition – such as man/woman, natural/artificial – as negotiable and open boundaries, to be crossed and traversed in the context of particular situations, Haraway re-defined, rather than transcended, dualistic thinking. This hybridity, the fusion of categorical difference, provided Haraway with the philosophical foundation for her figure of the cyborg. As such, the cyborg, as Sofoulis suggests, was an effective, defamiliarizing way of imaging the emergent forms of experience and subjectivity associated with our prolonged interactions with computers and other technoscientific realities.

Haraway's claim that the cyborg "gives us our ontology" is suggestive of the contemporary outcome of a historical legacy of, in Erik Davis' words, "reconceiving the world of bodies and nature under the sign of the machine." This ontology, whether we call it "postmodern," as Keller does, or "posthuman," as N. Katherine Hayles does, involves a complex matrix of challenges to modes of classifying life, human nature, intelligence and organicism. These essays prompt consideration of an ancient question, from the point of view of an age dominated by the prefix "artificial." We should perhaps no longer ask, "what is the meaning of life?" but rather, "what does it mean to be alive?"

SYNTHETIC MEDITATIONS :
COGITO IN THE MATRIX

"Find what Descartes wanted, what it was possible for him to want,
what he coveted, if only half consciously"
PAUL VALÉRY

"The only thing real is waking and rubbing your eyes"
THE FALL

Erik Davis

INTRODUCTION : TECHNO COGITO

Of all the lumbering giants of the Western philosophical tradition, none now resembles a punching bag as much as René Descartes. He gets it from all sides: cognitive scientists and phenomenologists, post-structuralists and deep ecologists, lefty science critics and New Age holists. The main beef, of course, is the stark divide that Descartes drew between mind and body, a dualism that, by its very claim of rationality, now appears even more obscene than the religious dualisms that stretch back to Zarathustra. Nearly across the board, contemporary thought calls us to defend and affirm the body that Descartes rendered a machine, a soulless automaton under our spiritual thumb. It doesn't really matter that the body so affirmed is itself multiple and even contradictory: the materialist object of biology, the phenomenological bed of Being, a feminist site of anti-patriarchal critique, the New Age animal immersed in Gaia's enchanted web. Regardless of the framework, the song remains the same: we are bodyminds deeply embedded in the world. For many thinkers now, the sort of abstract, disengaged soul-pilot pictured by Descartes – the "I" immortalized in the famous *Cogito ergo sum* – is not only bad thinking but, ideologically speaking, bad news.

In many ways I share this urge to trace the networks that embed consciousness in phenomenal reality, and to insist on the extraordinary (though not exclusive) value of causal explanations rooted in the history of matter. But I am no absolutist. The fact that Descartes keeps popping up like a Jack-in-the-box suggests that a splinter of the cogito remains in our minds, some fragmentary intuition or glimpse that we cannot accommodate and so wall off in order to reject. I am not interested in philosophically defending the cogito, or at least the metaphysical cogito we are familiar with: the rational disengaged instrumentalist manipulating the empty machinery of matter. But I am interesting in probing for that splinter, which I suspect is lodged somewhere in the apparently yawning gap between self-conscious awareness and the phenomenal world – a gap that, despite some hearty attacks from nondualists East and West, continues to inform subjectivity.

One zone that magnifies this gap is technoculture. Cyberspace and its allies (AI, VR, robotics) are shot through, on sociocultural, methodological, and philosophical planes, with a profound if often unconscious Cartesianism. First and foremost, this Cartesianism is what one might call "technical": the operating assumption that the mathematical recoding of reality is the golden road to the mastery of nature. But this assumption has powerful and various socio-cultural ramifications as well. As we'll see, some archetypal technopop fantasies – downloaded minds, manipulative technological demiurges, the breakdown between VR and real life – derive in part from the Cartesian imagination.

One field of technoculture particularly marked by Cartesian assumptions is artificial intelligence. Classical AI conceives the mind as a disembodied symbolic processor manipulating representations and information in order to reason about the world. Perception, sensation, and behavior are seen as inputs and outputs of an essentially logical machine, a machine whose essential activity is, to take an example fetishized by the AI community, expressed in chess. Though starkly reductive when compared to humanist or existential conceptions of consciousness, classical AI has the peculiar characteristic of reinforcing the familiar "Christian" priority of mind over matter.[1] The ultimate fantasized outcome of this line of thought, famously characterized in chilling detail by the Carnegie Mellon roboticist Hans Moravec, is the ability to upload the mind into silicon – effectively immortalizing the subject. After all, since there is nothing magical about the processes that coax the mind from our neural flesh, then nothing in theory should prevent a computer from simulating an individual brain to such a degree that the self originally booted up by the physical brain couldn't re-emerge inside the simulacrum.

In light of the pivotal role that absolute doubt plays in Descartes' *Meditations* – the doubt that calls into question the existence of the world presented by our senses – it is important to underscore how thoroughly the uploading scenario depends on erasing the material distinction between reality and copy. In essence, the argument goes, we *already* live inside a virtual reality; sights, sounds, textures, and flavors are all ghosts in the brain, woven out of pre-configured cognitive patterns and the incoming signals we receive from senses that shape those signals on the fly. These signals do not carry the things themselves, but only information about those things. In this view, I am not tied to the world. "I" am a kind of foam that forms atop a swirling stew of memory, perception, and various cognitive recursion loops staged in the virtual operations of the brain. However, the flipside of this rather contingent if not degrading view of subjectivity is that the self that might one day find itself a computer would be, for all intents and purposes, me. The difference between the material brain and the simulated brain does not affect the ontological status of the mind that arises from the formal operations of both organic and synthetic neural networks.

Unfortunately, classical AI wasn't able to make much practical headway over the decades, and this failure created room for rival theories and strategies to arise. In the 1980s, the MIT roboticist Rodney Brooks helped revolutionize his field with ideas that challenged the symbolic and Cartesian assumptions of AI. Instead of the classical approach to automata, which attempts to program them with complex centralized symbolic representations of the world around them, Brooks imagined robots who learned about their environment by exploring it according to simple behaviors distributed throughout the mechanism. The results of these simple interactions are

---

1 Elsewhere (Davis 1998, 121-128), I have argued that this dualism is really more gnostic than Christian.

then subsumed into higher global behaviors – a "bottom up" rather than "top down" approach. Tellingly, the inspiration for Brooks' first robots were not chess-playing automata, but insects.

Even from Brooks' own practical perspective, his ideas were always more than mere design strategies. Turning away from the Cartesian premises of classical AI, Brooks held that cognition emerges from the history and memory of the organism's interactions with the world around it, interactions which are thoroughly distributed throughout the body. In human beings, the increasingly complex behaviors emerging from lower-order processes ultimately lead to consciousness, but at no point does some distinct, underlying, and potentially self-sustaining formal symbolic language of representation pop up. To be conscious is to be engaged in a world that embeds and defines the subject.

One can overplay the conflict between symbolic and behaviorist AI – the "society of mind" model championed by Marvin Minksy, a towering figure in classical AI, shares a number of important characteristics with many of the more "bottom up" theories of human consciousness. But for most cultural theorists who have waded into the field, the distinction is key. For many critics, the rationalist enlightenment ideals that undergird classical AI are just as ripe for attack as the rest of the enlightenment project, whereas the behaviorist AI model can be seen to affirm pet concepts like contingency, relativity, and situated embodiment. In *How We Became Posthuman*, for example, N. Katherine Hayles (1999) has offered, in the name of a sophisticated account of embodiment, a historically rich critique of the rhetoric of disembodiment found in much AI and cybernetics. She shows how the apparent incorporeality of information – an incorporeality essential for the uploading model – is itself the product of ideological forces and institutional practices which serve to obscure the social and material bases that circulate and produce information. In this latest transform of historical materialism, then, the tension between Brooks and Minksy involved a distinctly moral dimension. As noted by Michael Mateas, a creator of a number of AI-based artworks, "[behaviorist AI] is associated with freedom and human rights and [classical AI] with oppression and subjugation" (Mateas 2000).

Readers of cultural theory should be familiar with the various associations and lines of thought that would lead to the denigration of symbolic AI, so open is the science to critiques of patriarchy, logocentricity, and the white privilege of disembodiment. It may also be the case that the Cartesian project will contribute little to the task of constructing mobile machine minds (the jury is still out). But the philosophical and even psychological underpinnings of Cartesianism are not so easily written off, let alone banished. As Slavoj Zizek notes in the introduction to *The Ticklish Subject* (1999), academia continues to be haunted by the specter of the Cartesian cogito. In other words, we have by no means sealed up the mad void out of which the cogito first arose – a void which in some sense founds modernity. So whatever happens to the vast edifice of rationalist procedures derived from Cartesian science and mathematics, the splinter of Descartes' true cross – the cogito – will continue to swell within the increasingly posthuman spaces of technoculture. In fact, I take Zizek at his cryptic word when he claims that Cartesian subjectivity is not only alive and kicking, but that only now, in the age of the Internet, are we truly arriving at it.

I. THE EVIL GENIE

With his otherworldly skepticism, Descartes cracked open the ontologically consistent universe of the premodern mind. He split the "great chain of being," and that split became the subject, a creature he came to identify as a rational and individual soul fundamentally divorced from the world of extension. How did Descartes, through his own philosophical unfolding, open up this revolutionary split? As he explains in the *Meditations*, he begins by undermining his conventional habits of thought and perception through the operation of hyperbolic doubt. Sitting robed at his fire, holding a piece of paper not so different than the one you're now reading, Descartes subjects himself to a series of "what if?" scenarios, soberly swallowing the conceivable possibility that he might be insane, or dreaming, or that an evil genie, "exceedingly potent and deceitful," might be conjuring up the illusions that he takes to be reality.

The next stage of the story is well-known: having plumbed the pit of doubt, Descartes realizes that even if reality is an elaborate deception engineered by an evil demon, there remains *someone* who is being deceived. To put it another way, even as Descartes strives to think everything false, "he" is still there, a something that thinks, and which therefore participates in existence. With this move, Descartes chiseled his keystone, reifying the subject who doubts into a metaphysical foundation. And though the cogito itself winds up resting on the even more fundamental foundation of God – a story we will leave by the wayside – the subject remains the first move in Descartes' pivotal game. "Observing that this truth 'I am thinking, therefore I exist' was so solid and secure that the most extravagant suppositions of the skeptics could not overthrow it, I judged that I need not scruple to accept it as the first principle of philosophy that I was seeking" (Descartes 1975, 32).

Despite the likelihood that few readers find the cogito mantra very solid and secure at this stage of the history of thought, I cannot resist taking a few pot shots. Turning within and recognizing that thinking is going on gives one no warrant to assume that an "I" exists whose predicate is thought. There is simply thinking. Admittedly, this move only shifts the problem, because there is still the "one" who recognizes that thinking is going on, the one who is tempted to assume the mantle of an "I who thinks." But even if we grant that this "one" and "I" truly exist, we have not healed the gap. The one who is aware that thinking is going on does not become transparent to itself by positing an I that thinks, because there is no reason, except for habits of speech, to identify the I that thinks with the one who is aware. In other words, I am not (the) one. Or, if you prefer, one does not think. Rather, as Zizek characterizes the situation, it is the "thing that thinks" (Zizek 1993). To this a philosopher stung by the Buddhist bug might add that there is no compelling intuitive reason to move from "thinking is going on" to "some *thing* is thinking." Why reify the process in the first place? The whole shadow play of substance and identity may be nothing more than conceptual imputation, a whirlpool of linguistic reflexivity arising in the foundationless stream of mental activity, boundless and unclear. The one who is aware may not be a one at all. There is simply the mind's intrinsic mirror-like capacity to reflect phenomena that arise.

I mention these final concerns because a great deal of Buddhist philosophy and practice is explicitly designed to undermine the precise act of introspective reification which founds the cogito – the act of hardening William James' "stream of consciousness" into a substantial self. But the invocation of Buddhism also lets us recognize an aspect of Descartes' method that is generally overlooked. His first meditation, wherein he imagines the evil genie, is not simply a skeptical argument; it is also a *procedure*, an introspective experiment that erodes the cognitive ground that Descartes (thinks he) stands upon. In this sense, his meditation is a meditation, one not altogether unlike the more analytic meditations found in, say, the Gelugpa school of Tibetan Buddhism. Throughout their career, Gelugpa monks will engage in contemplative practices which take the explicit form of dialectically interrogating the conceptual assumptions which structure their own consciousness. Winging it without a lama, Descartes found his own way of pulling the rug out from under his mundane convictions, a practice he clearly hopes the reader will try at home. The recipe: seriously take on the possibility of the evil genie, and see what remains. Don't slip back into your familiar habits. Risk the dark.

The distinction between the *Meditations* as the record of a conceptual experiment and the *Meditations* as a philosophical system is mirrored in the fact that Descartes is really talking about two cogitos. On the one hand, there is the epistemological void of doubt that conditions and expresses the first "I think." On the other hand, there is the *res cogitans* that Descartes subsequently constructs: a substantial and rational locus of thought and will, a self-transparent representation in a series of representations ultimately and necessarily established by God. Derrida and Zizek have both drawn attention to the cleft between these two cogitos. Derrida makes a distinction between Descartes' initial ahistorical passage through the madness of hyperbolic doubt, and the shelter the philosopher immediately takes inside the historical structure of reason and representation (Derrida 1978, 45-63). Zizek in turn brings up the Lacanian distinction between the subject of the enunciation and the subject that is enunciated. As we will see in more detail later, the former is an empty, logical variable devoid of the fantasies and representations that materialize personality, whereas the latter, in this case the *res cogitans*, is the conceptual "stuff" that fills in that void.

Descartes himself papered over this difference, believing that the "I think" ineluctably implied a rational person transparently aware of his own status as a thinking thing. In a sense, though, Descartes simply displaced the split between the two cogitos onto the grosser division between mind and body, a division that, in the *Discourse* anyway, is the first conclusion that follows the discovery of the solid and secure cogito: "From this [the cogito] I recognized that I was a substance whose whole essence or nature is to be conscious and whose being requires no place and depends on no material thing. Thus this self … is entirely distinct from the body … and even if the body were not there at all, the soul would be just what it is" (Descartes 1975, 32).

Today this line of thinking smells like religion – specifically, a kind of rationalist gnosticism. Descartes, of course, remained a believing Catholic throughout his life. Moreover, there is no Cartesianism without God, because God guarantees the order of representations that vanquishes the evil genie. On the other hand, while Descartes was convinced that his account of the cogito supported Church doctrine, believers in Descartes' day were by no means settled on the issue of whether corporeal existence would eventually follow us into the afterlife. Cartesian disembodiment seems to arise at least as much from the "gnostic" tendencies inherent in the reification of rational interiority as from the structures of 17th century belief.

Nonetheless, the Christian life certainly carried with it a tradition of disciplinary detachment from, if not outright loathing of the body. This basic distrust of carnal reality can be largely chalked up to Augustine, who, perhaps under the lingering influence of the Manichaean dualism he imbibed as a youth, reconceived the body as a perverse and untrustworthy product of Adam's sin. In his eyes we are torn between the "two loves" of body and soul. For Augustine, the desires and dispositions of the flesh are no longer natural expressions of an ordered world but our own inner demons, idiotically and destructively repeating their endless fall away from God.

This is harsh stuff, bemoaned by everyone today from hedonic New Agers to critical historians of thought. But Augustine's rejection of the body also went hand-in-hand with his revolutionary interiority, an intensification of inwardness that, as Charles Taylor explains in *Sources of the Self* (1989), was transformed by Descartes into the cogito, the seed of modern subjectivity. Augustine did not look to God primarily as the ordering principle of the cosmos that surrounds us – a view you could characterize, risking a certain simplicity, as the Platonic legacy. Instead, Augustine turned away from the world and conceived of God as the basis for our own knowing activity. By shifting the location of what Taylor calls "moral sources," Augustine thereby pried open a space of radical reflexivity within awareness. Suddenly our own experience of ourselves as subjects peels back from embodied experience, becoming the separate space of an internal order illuminated with an inner light. "Do not go outward; return within yourself. In the inward man dwells truth" (quoted in Taylor 1989, 129).

Descartes rationalized this spiritual withdrawal into the skeptical questioning that opens the *Meditations*. Descartes also transformed Augustine's two loves into two substances, one of which he neatly renders void. In other words, once Descartes identifies the soul as an immaterial consciousness, he reduces the remaining material world, including the body, into a hollow coordinate space of extension utterly devoid of the occult forces that animated premodern matter. But he does so not simply to render the material world a fit object for mathematical analysis. As Taylor astutely argues, the striking withdrawal of spirit from the material world enables Descartes to maintain the adamantine form of the rational soul he had crystallized as the *res cogitans*. Compared to the Platonic soul, which realizes its eternal nature by becoming absorbed in the supersensible, "the Cartesian discovers and affirms his immaterial nature by objectifying the bodily" (Taylor 1989, 146).

This, of course, is what mechanism is all about. By reconceiving the world of bodies and nature under the sign of the machine, one also constructs a new picture of man as an instrumental agent of his own incorporeal will. But where to draw the line in the bodymind? For Descartes, the human being is basically an automaton that moves according to the disposition of its limbs and organs – a doll with advanced plumbing. Given his lingering commitment to the soul, which he lodged in the ajna chakra (a.k.a. the pineal gland), Descartes' radical mechanism was not yet absolute – that would have to wait a century, until La Mettrie's *L'homme – Machine*. Nonetheless, as John Cottingham notes (1992), Descartes characterized many activities that we would consider "psychological" as blind functions of the animal machine. Memory, internal passions, the imprinting of sensation on the imagination – none of these demands the intervention of the soul. However, where mental attention is needed, Descartes posits a separate rational agent, a conscious spirit capable of diverting the flows of the body into various channels.

Descartes avoided a lot of grief by simply identifying agency with consciousness (which I will generally refer to as awareness, defined as the phenomenal activity of attention). In the world of making dinner and paying cable bills, we also adopt this identification: we become aware of a need or desire, and seemingly choose to act and plan accordingly. But what happens when there is a split between awareness and agency, at least in theory? What happens when I take on board the consideration that I am not actually thinking and doing, but that "the Thing" is thinking and doing? In some sense, this split between awareness and agency defines the anxiety of post-Romantic, increasingly cybernetic subjectivity. The mechanistic philosophy that Descartes birthed is now thoroughly undermining – at least in scientific terms – the notion of a single incorporeal point of awareness, rationality, and control. Today, we are anxious because we do not and cannot know who or what is pulling the strings of the subject. Throughout elite and mass culture, we argue and wonder about where the pivot of control lies: with corporate cabals or strands of DNA, with brainwashing advertisers or karmic forces, with historical forces or the structure of language, with the unconscious or the market's invisible hand. We wonder if our own sense of agency is actually blind causation in disguise, nothing more than a negative feedback loop in a cyborganic system of memes and genes. We wonder to what degree we are "programmed" – by media or social regimes, introjected concepts, or neural pathways laid down in infancy. Or we project the anxiety into the technological field: are machines becoming conscious; are they going to run the show; are they *already* running the show?

These doubts reach their most audacious limit in the techno-fantasies of paranoid schizophrenics, but they also lurk in cultural phenomena like conspiracy theory and *X-Files* fandom. They even exist to some degree in the popular discourse surrounding evolutionary psychology, which finds Cro-Magnon subroutines lurking beneath every sorrow and lust. The paradox is that these doubts place us back in front of Descartes' fire, with a bathrobe on and a book in our hands, pulling the rug out from under the world. Today the void is not epistemological – we no longer care particularly about how it is we seem to know things. The void we face is the self – how or why (or even if) we perceive ourselves as conscious agents in the first place. This, I believe, is why it is only now that we arrive at the cogito.

If now is the time, then where is the place? According to the Lacanian from Ljubljana, the answer is cyberspace, the supreme techno-fantastic implementation of illusion and control. "Only in cyberspace do we approach what Cartesian subjectivity is all about," Zizek claims, noting that virtual space is simply the materialization of the evil genie's deceptive powers (Zizek 1998a). We all wonder about reality now, how it is constructed, the claims of space and time. So it is hard to avoid occasionally slipping into giddy cyber-doubt: "What if everything is just digitally constructed, what if there is no reality to begin with?"[2] These are obvious questions, of course, the kind of thing that intrigues drug fans or 14-year-olds. But the "naiveté" of these questions is simply a sign of their universality, and it shouldn't prevent one from taking them seriously. As adults, we learn to not ask "What is reality?" or "Who am I?" because we know there are no answers, and so either develop more complex questions or drop the whole line of inquiry. But these interrogations aren't really questions; they are *devices*. If you sit with them without trying to find an answer, they can eat away at certainty and resistance, taking you to the point of bafflement, disassociation, insight. And somewhere, a stage along this path, lies the pure cogito, the void of the subject that is "our" homeless home.

II. THE LABYRINTH

In *Neuromancer*, the *Odyssey* of cyberlit, William Gibson delineated the Cartesian fantasy of cyberspace with the precision of a nanotechnologist. With its "lines of light ranged in the non space of the mind," Gibsonian cyberspace unfolds as an abstract, disembodied realm of geometry in motion, splayed across a three-dimensional coordinate system devoid of all secondary qualities but color (Gibson 1984, 51). In essence, the fantasy-reality of cyberspace, of virtual reality, is an analog of Descartes' view of matter: a zone of spatial extension under the rule of causality and essentially identical "to what the geometers call quantity" (quoted in Cottingham 1992, 14). Even today's budding 3D Internet and game consoles achieve, or at least suggest, Descartes' abstract virtualization of the material world into infinite mechanized extension.

Gibson also hit the Cartesian nail on the head when he characterized his hero Case's banishment from cyberspace as a fall into "the prison of his flesh." The dualistic denial of the body encouraged by virtual technologies is so often lamented today that neither it nor its supposed Cartesian origins need repeating. Obviously, virtual technologies encourage a distinct shift of

---

2   http://www.heise.de/tp/english/inhalt/co/2492/1.html

identification away from our phenomenal embeddedness in the material world where we eat, defecate, and die. In *How We Became Posthuman*, Hayles characterizes this shift in epochal terms: a movement away from the embodied dialectic of presence and absence, and towards an informational dialectic of pattern and randomness. Given this it's not surprising that the embrace of pattern has enabled some computer scientists to dust off the dualism that, I would argue, has always been implicit in the Cartesian foundations of the modern engineer.

Cyberspace is Cartesian in an epistemological sense as well, because the growth of the Internet as a medium of knowledge raises deeply Cartesian questions about the status of the external world – say, for example, the snoozing hippos or bubbling coffee pots we see through supposedly "live" webcams. In his article "Telepistemology: Descartes' Last Stand," Hubert Dreyfus (2000) argues that Descartes' original skeptical turn was itself partly inspired by the appearance of new perceptual media. The telescope and microscope both extended perception while simultaneously opening up doubts about the reliability of those perceptions. At the same time, sense organs were also increasingly imagined as transducers bringing information to the brain – senses that, as in Descartes' example of the phantom limb, could not always be trusted. Similarly, today's new media, as well as the new models of the nervous system they breed, have re-invoked the evil genie. "New tele-technologies such as cellular phones, teleconferencing, telecommuting, home shopping, telerobotics, and Internet web cameras are resurrecting Descartes' epistemological doubts" (Dreyfus 2000, 54).

Dreyfus notes ironically that most professional philosophers are no longer very interested in these epistemological questions. The problem is that the sophomores who slouch into today's philosophy classes (or ignore them altogether) often live in a world defined by virtual technologies, cyborg entertainments, and the popular fictions – sonic as well as narrative – that construct those emerging technocultural spaces and the shifting subjectivities they imply. These kids are already down with the evil genie. At the very least, they've seen *The Matrix*, the phenomenally successful 1999 Wachowski brothers film that imagined a vast simulation lorded over by evil computers and populated by hundreds of thousands of duped human beings.

The claim that so-called "consensus reality" is an elaborate construct that enslaves perception and occludes our "true" condition is hardly original. A staple of science fiction, where it was deployed with greatest sublimity by Philip K. Dick, the "false reality" set-up has become an increasingly common theme in Hollywood, from *The Truman Show* to *Dark City*. But I would also like to suggest that this set-up attempts to narrate a fundamental split in consciousness between consensus reality – or in Lacanian terms, the Symbolic – and the capacity of the human mind to disengage from the immediate claims of that reality. Skepticism can open up such doubts of course, but so will the ancient, non-philosophical evidence of dreams, drugs, or altered states of consciousness. This is why we find false realities popping up everywhere, from Indian dream fables to gnostic myths of cosmic prisons to Zhuangzi's famous question: "How do I know I am a man dreaming he was a butterfly, and not a butterfly dreaming he is a man?" The fundamental accessibility of the false reality scenario also accounts for its cheesy, adolescent character, a comic-book quality that makes sophisticated intellects cringe. And yet, if Descartes's *Meditations* did indeed help spawn the modern subject, then that subject – who is, in some sense, "us" – emerges from the shadow of such pulp musings.

Besides being a rite of passage for any budding cogito, the "false reality" question becomes especially unavoidable in the age of virtual technologies. These technologies constantly narrate their own totalizing dreams of "building worlds" and "providing experience," and produce

– consciously or not – the corresponding "gnostic" desire to escape the prison of manufactured dreams. I'd like to think both these factors help explain the immense popularity of *The Matrix*, especially among younger viewers. Alongside the video-game fight scenes and the nifty FX, *The Matrix* presents a narrative that articulated the seductive disassociation one feels as a subject of the popular digital spectacle, as well as the yearning for the cracks in the symbolic surface that offer the possibility of escape – an ultimately spiritual transcendence that, in the film's basic twist, is actually embodiment.

So we too are in that decrepit hotel room with Lawrence Fishburn's Morpheus, who is really speaking to us when he addresses Neo, the ever-wooden Keanu Reeves:

> You know something. What you know you can't explain, but you feel it. You've felt it your whole life, felt that something is wrong with the world. You don't know what, but it's there like a splinter in your mind, driving you mad.

Establishing the itch – which I suppose most of us share, however we interpret it – Morpheus offers to scratch it. He will give Neo "nothing more" than knowledge of the truth (that is, no solution to the problems posed by the truth). Moreover, this knowledge comes wrapped in the package of immediate experience. "No one can be told what the Matrix is," says Morpheus. "You have to see it for yourself." This lends it an explicitly gnostic character – not only did the Gnostics of antiquity believe that we were immortal sparks slumbering in an illusory cosmos manufactured by an evil or ignorant demiurge, but they also held that escape occurs through knowledge of our condition, a knowledge that is necessarily non-ordinary and experiential.

So like the serpent in the Garden of Eden, which the Nag Hammadi codex *The Apocryphon of John* claims was a liberating Christ in disguise, Morpheus offers Neo a pill. Neo, of course, swallows the molecular package, which is really the most heroic act in the film. For Neo must then face his own Cartesian "passage through madness," melting into a mirror that alludes not only to Lewis Carroll but to the mystic-psychotic collapse and disappearance of the externalized ego that stabilizes our inner void. As Neo phases out of the Matrix, he opens up, however briefly, the fractured bardo that is the secret thrill of every fan of the "false reality" genre: the moment when baseline reality dissolves but no new world has yet emerged in its pixelating wake. This is the most radical moment of the cogito, but it's tough to sustain. In *The Matrix*, the flux quickly crystallizes into what Morpheus, sampling Baudrillard, calls the "desert of the real": a ruined planet dominated by evil AIs who keep humanity mentally imprisoned inside the computer-generated Matrix. At this point, *The Matrix* stages an orthodox reversal of gnosticism's dualistic undermining of the world. Just as Irenaeus affirmed the reality of Christ's material body against the Docetist claim that God merely simulated human flesh, so do Morpheus and crew affirm the reality of the suffering material body against the mundane dream of the Matrix. Moreover, they do so in the name of the One who will come, a One that organizes the reality of their struggle the way that God provides the ultimate foundation for Descartes' metaphysical vertigo.[3]

The body is an understandable object of nostalgia in virtual fiction, though rarely in a pop film is the real we are rooting for so grimly depicted. At the same time, *The Matrix* subtly undermines the apparently "solid and secure" foundation of the flesh. Consider two intercut scenes focused on food. While the crew of Morpheus' ship, the Nebuchadnezzar, eat yucky nutritious slop ("everything the body needs") in a parody of communion, the Judas-like Cypher dines on

20

---

3   Thanks to Carlos Seligo, Ph.D., for this point.

steak inside the Matrix. Cypher agrees to betray Morpheus in exchange for blissful ignorance: to wake up rich and happy in the Matrix, with all memories of the desert of the real removed. Meanwhile, back on the ship, the young Mouse brags about having designed a sexy virtual character that Neo had earlier encountered in a training simulation. Mouse offers to arrange a sexual (pornographic?) encounter with the woman for Neo; when the other crew members give him grief, Mouse calls them hypocrites: "To deny our own impulses is to deny the very thing that makes us human." Here Mouse recognizes one paradox of desire – that the body's carnal impulses are fused with "virtual" fantasy – but he mis-states the case: what makes us human is the gap between impulses and the alienated awareness that both the object of those impulses and the body that wants them are in some sense virtual.

The Matrix also undercuts any simple valorization of carnality in its portrayal of the "virtual bodies" that play such an important role in the guerrilla war Morpheus wages within the Matrix, where he struggles against the all-powerful evil agents (sentient programs disguised as human beings). In this struggle, the knowledge that the Matrix is unreal is not sufficient to bend its rules; the freedom fighters must train their false Matrix bodies in order to leap through the air, bend spoons, and, ultimately, slow time. In other words, "the body" becomes a virtual field of affect and extension that resists what they already know, a resistance that gives way not through further knowledge but though *practice*. Here the film is even more "eastern" than the debt its fight scenes owe to Hong Kong cinema and Japanese video games would suggest. As in yoga, T'ai chi, and other martial arts, the mind awakens through the disciplined and devotional unfolding of the capacities and energies of the body. Of course, the bodies trained for the Matrix are composed of code, no more fleshy than the brutes and ninjas in *Mortal Kombat*. But that misses the point: the "magical" body – a body immortalized in Chinese and Japanese popular cinema, as well as the half-Hollywood hit *Crouching Tiger, Hidden Dragon* – arises through a practice that constructs a liminal phenomenological vehicle between body and mind, a vehicle simultaneously virtual and carnal.[4] Similarly, though the "bodies" that players of first-person computer games like *Quake* and *Doom* control are not actual, they are certainly phenomenological.[5]

Manex, the company behind *The Matrix*'s excellent special FX, placed a strong emphasis on the phenomenological or subjective dimension of such virtual bodies. In popular film, most digital FX depict the "objective" world of extension – either new macroscopic worlds (*Star Wars: The Phantom Menace*), natural or supernatural phenomenon (*Twister, Spawn*), or microscopic scales of perception (*Heavenly Creatures*). These images present a publicly accessible "real" space. But verisimilitude, fantasized or otherwise, ultimately limits FX, which have nothing intrinsically to do with representation or reality and everything to do with mobilizing new phenomenological openings and synesthetic becomings. FX are not really about *what* we see; in fact, they are not "about" anything at all. They reconfigure how we see, and how that subjective seeing mutates into often ambiguous and explosive feelings and relations. That's what makes them so hard to talk about – "pure" effects are much more like roller-coasters or the space-time distortions of drugs than they are like signs or icons.[6]

---

4 Perhaps the "energetic" body diagrams found in Taoism and Tantra, with their chakras, nadis, and meridian lines, depict traditional formulations of this liminal bodymind.

5 See Canny and Paulos 2000.

6 This probing of nameless affects and desire explains why the subjective rhetoric of speeds and slownesses, including the bullet-time photography mentioned below, more often appear in advertisements for sports utility vehicles and McDonalds than in mainstream cinema.

What makes *The Matrix* such a great FX movie is that the film maps its "false reality" theme onto the objective/subjective divide that underpins the visual rhetoric of Hollywood FX. The Matrix as such characterizes the imprisonment of FX by verisimilitude – FX as illusion, as secular fairyland, as the seamless artificial product of what Disney calls "imagineering." But when Neo reaches the peak of his power, FX become an expression of his own subjective mastery of speeds and slownesses. The most notable FX device here is the bullet-time photography featured, most memorably, in the scene where the leather-clad Neo confronts an agent on the roof of a building and manages to slow down time enough to lean away from the agent's oncoming bullets. Using an array of multiple still cameras whose images are subsequently treated like animation cells, the technique creates the effect of a single camera sweeping in a long arc around a static or very brief slice in time. Time appears to slow, and yet the movement of the (virtual) camera keeps things up to speed. So Trinity, who watches Neo dodge the gunfire, comments on how fast he moved, as fast as an agent. But for the viewer, as, significantly, for Neo, the action moves like molasses.

The affirmation of slowness is remarkable enough, especially given the usual strategy of overwhelming the audience at a peak moment with quick cuts and superfast images. Slowness is the phenomenological effect that Neo must master in order to detach himself from the logic of the Matrix while remaining inside its narrative framework – a slowness that is manifested in both mind (this is Keanu Reeves after all) and body. In the final action sequence, Neo is apparently killed by an agent inside the Matrix. Then a kiss from Trinity, monitoring Neo on the Nebuchadnezzar, revives the hero in the material world. With this carnal affirmation, Neo returns to the Matrix, where he stops a barrage of bullets in mid-air, slowing down time to the point of stasis. It is only then, when he fully inhabits the gap he has opened in virtual time, that he "sees into" the Matrix. The hallway before him melts into rushing streams of green computer code – the "real" beneath the Matrix's symbolic fantasy. When the head agent subsequently engages him in hand-to-hand combat, Neo's movements are cool, slow, meditative, almost bored. He has seen through the fantasy in the midst of the fantasy, a seeing which is the equivalent of dying. He becomes the One.

But this gnostic-Christian resolution is not for us, or most of us anyway, for we have no access to such singular foundations, Cartesian or otherwise. For us there is no One, no *deus ex machina* who can found the order of true representations that describe the mechanisms driving the production of the phenomenal world. The digital figures that Neo glimpses, after all, are representations of electrons flip-flopping through material circuitry, and at that point neither the pattern of bits nor the electro-dynamic substrate can claim ontological priority. The moment of subjective transformation that interests us is much earlier, before Neo even hears that Morpheus thinks he's the One. It is the moment when Neo swallows a pill in a seedy room, and becomes, for a spell, no-one at all.

III. A CRACK IN THE SKY

In the great eighth chapter of the *Confessions*, Augustine (1961) describes his endless difficulties cleaving to God, at one point comparing his situation to a sleeping man. Though he knows that Jesus Christ is for him, the call of the world and the lusts of the body weigh on him like slumber, and he feels like a fellow who, though he knows it is time to get out of bed, keeps hitting the snooze button. "Just a little bit longer," he keeps telling God, "let me sleep a little more." Though he partly blames the body, Augustine identifies sleep less with carnal lust than with "the force of habit, by which the mind is swept along and held fast even against its own will" (165).

Besides underscoring how fundamental the natural analogy of awakening is to both religious and philosophical discourse, this passage provides an angle on the somewhat peculiar paragraph that closes Descartes' first meditation. Earlier, Descartes had convinced himself that only by embracing hyperbolic doubt – hypostasized as the evil genie – could he undermine the habitual force of his "old and customary opinions." As he closes the meditation, however, Descartes admits how difficult it is to keep these habits at bay, acknowledging that "a certain indolence" continually creeps in, drawing him back to his ordinary perceptions of life. Taking Augustine's analogy a step further, Descartes compares his state to a prisoner dreaming of his liberty, a captive who, when sensing that the moment of awakening is at hand, "conspires with the agreeable illusions that the deception may be prolonged" (Descartes 1966, 133). Descartes then admits a fear that does not trouble Augustine: that even if he does awaken, he will not be able to see his way out of the darkness unleashed by the genie.

Here we taste something of the frightening vertigo opened up on the way to the cogito. Despite the rational and theological foundations that soon come, Descartes' initial movement has nothing intrinsically to do with philosophical concepts – the evil genie as a "possible world" – and everything to do with the phenomenological process of emptying oneself by turning that self inside-out through doubt. Descartes decoupled his internal awareness as much as possible from the contents of consciousness, effectively declaring "I am not in this dressing gown, not before this fire, not holding a piece of paper." Like a shaman offering his body to the ferocious spirits of the underworld, Descartes submitted himself to the genie, who tore away the certainties that stabilize the ordinary non-skeptical self in its sleep of habit. On the other hand, Descartes did not even have the ontological stability of the shaman's premodern cosmos to rely on, for the void that he opened up was precisely the void that separates the modern mind from the great chain of being.

For Descartes, this was a passage through madness, a madness that subsequently founds the modern sense of disjunction from tradition and the enchanted world. The paradox is that even the acknowledgment of such madness affirms the certainty that, for Descartes, grounds the cogito. As Derrida explains, "the Cogito escapes madness only because at its own moment, under its own authority, it is valid even if I am mad, even if my thoughts are completely mad" (Derrida 1978, 55). In other words, the cogito stabilizes itself in the gap that opens up between the madness of thought and the I whose thoughts are mad. One might even say that the cogito is on the far side of madness, a cool and impersonal witness, utterly untethered from the objects that arise in thought and perception. "This is why it is not human," says Derrida, "but rather metaphysical and demonic" (55). Descartes then draws back from this "zero point" into factual historical structures of thought, and it is these structures – at least the metaphysical ones – that are now almost ritualistically vilified. The Descartes we love to hate knows where he stands. But as Derrida states, "Nothing is less reassuring than the Cogito at its proper and inaugural moment" (56).

Even the conceptual condensation of the cogito that follows Descartes' passage through madness is none too comfy. In mapping his dualistic divide between mind and body, Descartes separates the pure modes of consciousness that characterize the incorporeal *res cogitans*, such as intellection and volition, from those mixed modes that also depend upon the body, such as imagination and sensation. As John Cottingham notes (1992, 241), this division leads to a rather creepy state of affairs: after death, "the soul will be devoid of all particularity," condemned to an eternity of chewing over abstract and general ideas. (241). Later Christian Cartesians had to jump through hoops explaining how any sort of personality could survive this distillation –

indeed, how such impersonal souls could even be distinguished from one another at all. In other words, the cogito is essentially inhuman, at least in the sense that it does not participate in the order of habits, memories, images, and symbolic identifications that structure embodied personality and the perceptual stream of ordinary life.

The first time that Neo returns to the Matrix after joining Morpheus' crew, he passes one of his favorite restaurants. "They have really good noodles," he recalls, his words trailing off as he realizes that the dispositions and memories that structured his personality are, at least from the perspective of his new reality, utterly false. Realizing that he can no longer sustain, or desire, his normal round of identifications, he asks Trinity what it all means. "That the Matrix cannot tell you who you are," she responds. If you hit the pause button right there, before the film fills in this space of not-knowing with Neo's emerging identity as a Christ hero, then you are at the empty heart of the subject.

This picture of the cogito differs significantly from the now-classic postmodern portrait of the "decentered subject." That story basically claims that the crusty old idea of the individual – the self-aware "Cartesian" locus of will and understanding – has been decentered in the light of its fundamental multiplicity and the myriad elements that make up the construction of identity – floating signifiers, ideological forces, historically constituted forms. But as Zizek explains, what really decenters the subject is the fact that the subject that enunciates is not the subject of the enunciation. The subject that enunciates is a logical void, a kind of empty place holder – $ in Lacanese – for the material that, loosely speaking, congeals into the personality, i.e., the subject of the enunciation. This material is largely determined by the already established network of the Symbolic (a.k.a. the Matrix). The fact that the symbolic identifications that attempt to found the subject of the enunciation are themselves constructed and drifting without foundation is almost beside the point; what is decentered is the point of speaking (or knowing) itself; i.e., the cogito.

In this account, the cogito does not arise from the Symbolic. Instead, it emerges "at the very moment when the individual loses its support in the network of tradition; it coincides with the void that remains after the framework of symbolic memory is suspended" (Zizek 1993, 42). Zizek's most forcefully futuristic account of this void appears in his discussion of the paradox posed by *Blade Runner*: the subject who knows she is a replicant. "Where is the cogito, the place of my self-consciousness, when everything that I actually am is an artifact – not only my body, my eyes, but even my most intimate memories and fantasies?" (40). Here Zizek takes one of Descartes' more paranoid musings to its logical conclusion. In the second meditation, Descartes asks himself, observing a street below, "What do I see from the window beyond hats and cloaks that might cover artificial machines, whose motions might be determined by springs?" (Descartes 1966 139). This is not simply a mercilessly skeptical spin on the perennial problem of "other minds"; it is also, mutatis mutandis, an inquiry into the (replicant) self within. How deep does your automaton go? Zizek's paradoxical and beautiful conclusion is that *Blade Runner*'s replicants become, in recognizing their own artificial nature, "pure subjects." As far as the subject of the enunciation goes, they know they are replicants, not human beings, which is why Rachel weeps when Deckard (Descartes?) tells her the truth. But it is precisely at that moment, when her confusion over whether she is human or not melts into nostalgia for a lost humanity, that Rachel is most like us – that is, most human.

Zizek concludes that "I am a replicant" is the statement of the subject at its purest. But we might just as easily say "I am an avatar," or simply "I am online." For as *The Matrix* suggests, cyberspace – the technologized space of virtuality – increasingly constitutes the Symbolic as such, and thus begins to infect and dominate the material of subjectivity. As Zizek explains, cyberspace externalizes us, translating the contents of subjectivity into an objective space of technical operations. So on the one hand we have the endless play of virtual identity, in which we lend "reality" to stray fragments of the psyche by externalizing them into a field of technologically sustained symbolic intersubjectivity. On the other hand, we enter a paranoid dystopia, where our every move is tracked, controlled, and manipulated by an increasingly intelligent virtual environment. In either case, there is a deprivation of sorts, although this deprivation comes with a twist. "What you are deprived of are only your positive properties, your personality in the sense of your personal features, your psychological properties. But only when you are deprived of all your positive content, can one truly see what remains, namely the Cartesian subject" (Zizek 1998a).

The ferocity of this deprivation will only increase as e-commerce intensifies its marketing technologies. The dream of e-commerce could be dubbed "molecular marketing": the thoroughly targeted individual whose unique desires and dispositions have been data-mined, tracked, extrapolated, commodified, and, most importantly, fed back to the target in a personalized, even obscenely intimate form. In this process, the statistical generalities that govern demographics are brought down to the scale of the individual without losing their abstract and utterly impersonal instrumentality. The new goal is to anticipate and nudge the precise and singular unfolding of subjectivity in its encounter with information and commodities. Perhaps in the future our own shifting moods, interests, and needs will be so sensitively monitored that we will be able to read our own state of mind by the variations in the incoming streams of newsfeeds, advertisements, and animated spiels. So if one night our lover whispers about her childhood fantasies of Arabia before we sleep and dream of sinking in quicksand, the next morning our virtual homeroom will be infested with travel agents on flying carpets and pop-up adbots offering the latest anti-anxiety neuro-cocktails specifically designed to generate the proper degree of subservient enthusiasm. One day we may reach the point when our needs and desires are fully externalized as semi-autonomous avatars, so that we hardly need to intervene in order to "satisfy" the identifications that structure the subject of the enunciation.

Similar problems arise with the great dream of virtual reality, which, in its fantasized image at least, at once fulfills the contents of consciousness and subtly alienates the subject from those contents. In the standard account, VR and other designer realities create a plastic playground of the self, allowing us to explore and experience the hidden "real me" lurking beneath that mask of socially constrained subject positions and the ever-present resistance of the Real. But even if we accept this naïve account of the self, the very engine of virtual production undermines the "fullness" of the simulated experience. McLuhan described the evolution of technologies as a progressive amputation of human capabilities; with virtual reality, or the similar plasticity of material reality achieved through nanotechnology, we amputate the drives and desires that structure the subject by fully externalizing them and feeding them back to the subject. It's the problem of the hedonist: the self that manipulates and refines techniques of pleasure is not the same self that luxuriates in those experiences, and this anxious gap yawns ever wider the more rounds we make on the technical pleasure circuit. (The appeal of S&M partly derives from apparently splitting these two functions between two individuals).

So as designer realities radically fulfill the contents of fantasy, the existential remainder – that modern spark which voids or demythologizes all fantasy – becomes ever more refined and impossible to avoid. Then it will be even more obvious that we are not our avatars – that the Matrix cannot tell us who we are. We still won't know who we are, of course, because that quest for equivelence itself is a mode of the Symbolic, a way to "resolve" the ambivalent emptiness of the pure subject by injecting it back into the round of identifications.

But we will know that, like the sages in the Upanishads or Descartes before the genie's fire, we are *neti, neti* – not that, not that. We are not just contingent historical agents embedded in a finite horizon of meaning, but nor are we the solid and secure foundation of the *res cogitans*. And though we emerge from the process of embodiment, we are not "the body," if by the body we mean a fixed chunk of space-time or a founding representation or a neurobiological object of science.

Though I have no room to explore my argument here, I believe the kind of *via negativa* suggested here describes the "native" spirituality of the post-Romantic modern subject. In his 1928 essay, "Freedom Without Hope," René Daumal – Gurdjieffean pataphysician, Sanskrit scholar, and author of *Mount Analogue*, one of the 20th century's few masterworks of spiritual literature – described this rather astringent path in terms reminiscent at once of surrealist manifestos and the traditionalist rants of René Guénon:

> The essence of renunciation is to accept everything while denying everything. Nothing that has a form is me; but the determining factors of my individuality are thrown back on the world … The soul refuses to model itself on the image of the body, of desires, of reason; actions become natural phenomena; and man acts the way lightning strikes. In whatever form I find myself, I must say: 'that is not me.' By this negation, I throw all form back to created Nature [or cyberspace], and make it appear as object. I want to leave whatever tends to limit me – body, temperament, desires, beliefs, memories – to the sprawling world, and at the same time to the past, for this act of negation creates both consciousness and the present; it is a single and eternal act of the instant. Consciousness is perpetual suicide (Daumal 1991, 4).

Authentic consciousness, for Daumal, is simply the pure subject constantly re-awakening to itself. And in an utterly un-Cartesian move, this vast impersonal awareness is reached only through the negation of individual autonomy. Freedom – for this is what Daumal is talking about – has nothing to do with the Cartesian image of an operator lodged in the theater of the mind. That supposedly free agent is just an avatar roving around, slurping noodles, getting and spending, running on auto-pilot.

Zizek seems to waver on whether this pure subject is accessible to us through the ascesis of dis-identification, or whether it remains the subject of the unconscious alone, available only in theory or the cracks of language. In his essay on Daniel Dennett, he asks "What if the ultimate paradox of consciousness is that consciousness – the very organ of 'awareness' – can only occur insofar as it is unaware of its own conditions?" (Zizek 1998b, 269). But this implies that the site of consciousness is fixed. In other words, even if the paradox Zizek describes holds, the site of consciousness could nonetheless shift as more and more of its structuring conditions are brought into the circuit of consciousness. This is one way of characterizing the sort of psychological self-observation and self-programming whose various permutations infest the cybernetic world of self-help. Here the claim is that certain conditions that structure consciousness can be known, recognized, and managed. At the same time, this process shifts the seat of consciousness into another frame, maintained by another set of unknown structures.

The pure subject is a void, a not-knowing, a suicide. But this void *moves*, an empty roaring stream we enter without resolution or understanding. For just as we cannot know what a body can do, neither can we know what consciousness can do – especially when it is becoming-empty, which if the Nyingmapas are right is equivalent with becoming-radiant. So I'll leave you with the challenge the Sixth Ch'an Patriarch threw at his students: Show me your original face. What original face? The face you had before your parents were born. That is, before you tried to find yourself in the Symbolic matrix of identification and signification, a "before" that does not lie in some foundational past but in the bottomless pit of the passing present.

REFERENCES

Apocryphon of John, The. Trans. Frederick Wisse. Nag Hammadi Library. See http://www.gnosis.org/naghamm/apocjn.html (accessed Jan. 26, 2001).

Augustine. 1961. *Confessions*. Trans. R. S. Pine-Coffin. New York: Penguin.

Canny, J. and E. Paulos. 2000. Tele-embodiment and shattered presence. In *The robot in the garden*, ed. K. Goldberg. 277-294. Cambridge, MA: MIT Press.

Cottingham, J. 1992. Introduction; and Cartesian dualism. In *The Cambridge Companion to Descartes*, ed. J. Cottingham. Cambridge: Cambridge University Press.

Daumal, René. 1991. *The powers of the word*. Ed. and trans. M. Polizzotti. San Francisco: City Lights.

Davis, E. 1998. *TechGnosis*. New York: Harmony.

Derrida, Jacques. 1978. Cogito and the history of madness. In *Writing and difference*. Trans. A. Bass. Chicago: University of Chicago Press.

Descartes, René. 1966. Meditations. In *The philosophy of the 16th and 17th centuries*, ed. R. Popkin. New York: The Free Press.

————— 1975. *Philosophical writings*. Trans. and ed. E. Anscombe and P. T. Geach. London: Nelson's University Paperbacks.

Dreyfus, H. 2000. Telepistemology: Descartes' last stand. In *The robot in the garden*, ed. K. Goldberg. Cambridge, MA: MIT Press.

Gibson, W. 1984. *Neuromancer*. New York: Bantam.

Hayles, N. Katherine. 1999. *How we became posthuman*. Chicago: University of Chicago Press.

Mateas, M. 2000. Expressive AI. In *Electronic art and animation catalog*. Art and Culture Papers, SIGGRAPH 2000. New Orleans, LA. Also downloadable from pw1.netcom.com/~apstern/interactivestory.net/papers/MateasSiggraph00.pdf (accessed Feb. 5, 2001).

Taylor, C. 1989. *Sources of the self*. Cambridge, MA: Harvard University Press.

Zizek, S. 1993. I or he or it (the Thing) which thinks. In *Tarrying with the negative*. 9-44. Durham, NC: Duke University Press.

————— 1998a. Hysteria and cyberspace: Interview with Slavoj Zizek by Urik Gutmair and Chris Flor. *Telepolis* Oct. 7. At http://www.heise.de/tp/english/inhalt/co/2492/1.html (accessed Jan. 26, 2000).

————— 1998b. Cartesian subject versus Cartesian theater. In *Cogito and the unconscious*, ed. S. Zizek. Durham, NC: Duke University Press.

————— 1999. Introduction. In *The ticklish subject*. 1-6. New York: Verso.

THE INSTRUMENTS OF LIFE:
FRANKENSTEIN AND CYBERCULTURE

"Man makes man in his own image"
NORBERT WIENER

Catherine Waldby

Mary Shelley's *Frankenstein, or the Modern Prometheus* has been utilized as a stock narrative for technology out of control since its publication, a use given impetus by the James Whale movie versions of the 1930s. We are all familiar with the images of the lumbering artificial man, bent on the destruction of his maker. Successive waves of cinema have taken up the motif of the human-like creation that hates its creator; the neurotic AI HAL in *2001: A Space Odyssey*, which expels its human handlers into deep space; the *Terminator* films, in which a series of cyborgs are sent back in time from a machine-controlled future to eradicate potential human resistance; the more nuanced *Blade Runner*, in which four replicants seek revenge on the creator who has programmed them to "terminate" at a predefined date. Each of these films, and many many others,[1] rework the apocalyptic reading of *Frankenstein*, finding dramatic potential in the capacity of tools to exceed their passive status and take on a malevolent autonomy.

In an equally apocalyptic vein, many developments in the biosciences and biomedicine are habitually discussed in relation to the Frankenstein story. A surf through Medline (www.medline.com) produced 30 recent science articles with "Frankenstein" in the title.[2] The word is used as an instantly recognizable allusion to what are considered the destructive potentials of recom-

---

1 The films dealing with some variation of the Frankenstein theme are far too numerous to be discussed here. See for example *The Stepford Wives, Demon Seed, Metropolis, Weird Science, Making Mr. Right, Eve of Destruction, Species I* and *II*, and the *Robocop* films, as well as various remakes of *Frankenstein* itself.

2 The following are a random selection: R. Sikorski, R. Peters. 1998. Cryopreservation. A Frankenstein experiment. *Science* 281(5380, Aug. 21):1163-4.
H. Have. 1995. Letters to Dr Frankenstein? Ethics and the new reproductive technologies [editorial]. *Social Science & Medicine.* 40(2, Jan.):141-6.
B. E. Rollin. 1986. 'The Frankenstein thing': the moral impact of genetic engineering of agricultural animals on society and future science. *Basic Life Sciences.* 37:285-297.
H. J. Morowitz. 1979. Frankenstein and recombinant DNA. *Hospital Practice.* 14(1, Jan.):175-176.

binant DNA technologies, in vitro fertilization techniques, cryopreservation practices, xeno-transplantation – any scientific practice that threatens to change the human body in new and unfamiliar ways, and is indifferent to notions of human purity.

These apocalyptic or phobic readings of *Frankenstein* are well worn, and I do not want to pursue them further here. Rather, I would like to consider what the novel might still have to contribute to an understanding of the consequences of a certain technics of vitality: the scientific ability to engineer living systems. Shelley's novel was inspired by 19th century scientific experimentation with vitality – efforts to technologize the trajectory of living entities through galvanization. It is one of the earliest attempts to deal with the malleability of life and the possibilities of human "technogenesis," the loss of an origin securely located in nature. In creating her monster, Shelley takes the first fictional step towards understanding what artificial life and artificial intelligence might mean, and how they might be lived. *Frankenstein* has become the archetypal techno-monster story in part because of this prefigurative power, its anticipation of the potentials of science to create new forms of life. As Margaret Shildrick argues, monsters are always speculative, implying a potential future. "Monsters signify … the otherness of possible worlds, or possible versions of ourselves, not yet realized" (Shildrick 1996, 8). The monstrosity of today is the transformation of tomorrow, disturbing precisely because it suggests the open-ended nature of the body's becoming. Victor Frankenstein's creature is of interest to us today, nearly 200 years after the novel was written, because it is the first fictional text to speculate about possible transformations of organic life in response to technoscientific innovations. *Frankenstein, or the Modern Prometheus* takes the potentials of machinic life seriously, and explores some of its consequences.

Now, at the beginning of the third millennium, at least some of the possibilities for artificial life investigated in the novel have become our everyday reality, as the engineering of new animal species, the cloning of existing ones, and the generation of vital information (artificial life forms in silico) have become commonplace. The possible application of these techniques to human beings is a topic of constant journalistic speculation and bioethical condemnation, and has produced anxious legislation to try to control the distinction between human and non-human technogenesis. While sheep and monkeys may be technically reproduced, the possibility that humans might be cloned has proved too confronting for an idea of the human still invested in notions of unique identity and exceptional status. *Frankenstein* suggests the kind of radical re-examination of this status that these developments demand. The novel both describes and prefigures a world in which the human body and human sociality owe a greater and greater debt to technoscientific and machinic systems of production and reproduction, and are less and less able to be thought outside of those systems. It suggests that the ethical and intellectual problems of these new forms of technogenesis are old problems, and that reference to a natural humanity is always anachronistic. Victor, the inventor, seems so contemporary to us because he is confronted with the spectacle of his potential resemblance to an invention. The creature seems so sympathetic because his subjectivity, the subjectivity of the scientific object, does not seem so far from our own. His body's complex entanglement with the anatomical and galvanic technoscience of the early 19th century rehearses our own entanglements with informational and bioinformational systems and suggests the possibility of infinite regress in the search for a natural origin. The monstrosity of the creature's body implies the monstrosity of our own.

Frankenstein belongs to a long lineage of mythic creation stories – Pygmalion, Prometheus, der Golem – in which a human creator usurps the power of the divine and imbues a creature with life. While Victor Frankenstein shares the hubris of these human creators, what distinguishes Frankenstein from these earlier narratives is that the life conferred on the creature is generated through technical instruments and scientific procedures. In the creation stories that precede Frankenstein, vitality is conveyed through a supernatural or magical process that reaffirms the divine origins of life, even if this divinity is usurped. The Golem, for example, is fashioned from clay and brought to life by uttering the name of God.

Unlike his predecessors, Victor Frankenstein treats life as a material quality, one that can be understood analytically as mechanism rather than as transcendental essence. This materialist approach to living entities is what locates the story in modern scientific practice. The mechanistic philosophy of living entities, the understanding of life that has become associated with scientific biology and biomedicine, treats vitality as a separable force, which exceeds its location in any particular body. Like the forces of magnetism, gravity, and entropy found in physics, the force of life as mechanism can be abstracted from any particular body that it might animate. Hence life can be isolated as a capacity, analyzed, quantified, and controlled, irrespective of the form of embodiment in which it is temporarily located. Life of this kind is open to instrumentation.

Georges Canguilhem (1992), in his studies of the relationships between machines and organisms, locates the possibility for conceiving life as abstractable force in the development of automata. He suggests that it only becomes possible to conceptualize life as specifiable force – rather than transcendental and unquantifiable essence – when it is possible to build models that demonstrate self-motivating power:

> For a long time, kinematic mechanisms were powered by humans or animals. During this stage, it was an obvious tautology to compare the movement of bodies to the movement of a machine, when the machine itself depended on humans or animals to run it. Consequently it has been shown that mechanistic theory [of biology] depended, historically, on the assumption that it is possible to construct an automaton, meaning a mechanism that is miraculous in and of itself, and does not rely on human or animal muscle power (Canguilhem 1992, 47).

In other words, the idea of living entities and the quality of life itself as machinic is historically associated with the development of mechanical sources of energy and a certain level of dynamic technical complexity. If life is machinic, this is because machines can sustain a certain level of vivacity. Canguilhem's assertion suggests that Victor's creature has a strong family relationship with those chess-playing, letter-writing, eating and drinking clockwork and electrical automata that populated the scientific academies and wealthy parlors of 18th century Europe.[3] Like these automata, the creature's process of construction involves the building of a mechanical device, which is then animated by the application of a technical force. In *Frankenstein*'s case, the mechanism to be constructed is an *organic* machine. While it is to have the self-organizing and self-sustaining capacities that define the organism, its body will not be produced through a process of conception, gestation, birth, and growth, but rather through a mechanical process of design and production. It will be a living body not born but assembled.

---

3   Robert Malone (1978) claims that one particular 19th century clockwork automaton, a human figure that could write a short dictated message, was the inspiration for *Frankenstein*.

Frankenstein makes the body of his monster by a reverse anatomization. The scientific dissection of corpses, a commonplace practice by the early 19th century,[4] was one of the earliest assertions of the fundamentally mechanical nature of the human body and the materialism of the natural world (Sawday 1995). It suggests that the vitality of living things might arise from their material organization rather than from divine inspiration, and hence can be understood through the study of that organization. Anatomization analyzes the body according to its macro-level machine logics. It treats the dead body as "organ-ism," an assemblage of tools, whose value lies in its capacity as a useful machine, with potential relationships with other orders of instrumentation. It was (and still is) commonplace for anatomists to itemize parts of the body through comparison to various tools. One 17th century doctor writes, for example:

> Examine carefully the physical economy of man: What do you find? The jaws are armed with teeth, which are no more than pincers. This stomach is nothing but a retort, or heat chamber; the veins, the arteries and indeed the entire vascular system are simply hydraulic tubes; the heart, a pump; the viscera nothing but filters and sieves (cited in Canguilhem 1992, 47).

Classical anatomy treats the body as a set of interlinking yet separable parts, organs as in-themselves technologies that nevertheless work together to form a living body. This is the assumption behind Victor's creation. In his "workshop of filthy creation" (Shelley [1831] 1980, 55) Victor builds the creature's body by piecing together the organs of a number of other, dead bodies as components in an organic machine, a kind of primitive recombinant technology. He goes to graveyards and medical schools to collect the fragments of corpses and anatomized bodies:

> Who shall conceive the horrors of my secret toil, as I dabbled in the unhallowed damps of the grave, or tortured the living animal to animate the lifeless clay? … The dissecting room and the slaughter-house furnished many of my materials (54-55).

The creature's body is assembled piece by piece. Here we find one of the earliest fictional speculations about that new space – the laboratory – where powerful relationships are created between organic and technical capacities. While the laboratory appears as a sober, rational space of inquiry in early scientific papers,[5] Shelley proposes the laboratory as a site of unregulated and passionate production. The laboratory is, she implies, not a place for the rational analysis of natural laws but rather a place where new kinds of nature are brought forth into the world, with unknowable consequences. Having created the synthetic body of his being, Victor must then find the motive force, the life force to animate the body, so that, unlike a conventional machine its vitality is self-renewing and self-organizing. His experimentation with the creation of life is conducted through a close examination of dead bodies, and through attempts to manipulate the threshold that separates the living from the dead. "Life and death," Victor states, "appeared to me ideal bounds, which I should first break through, and pour a torrent of light into our dark world" (54). Throughout the novel, Victor's search for the secret of life is never pursued as an end in itself but is always driven by a general desire for reanimation of the dead. Victor's feverish experimentation is precipitated by the death of his mother. "I thought," he muses, "that if I could bestow animation upon lifeless matter, I might in process of time renew life where death had apparently devoted the body to corruption" (54). If life is mechanism for Victor, it is appar-

---

4   Anatomical dissections for scientific purposes took place in Italy as early as the 14 century (Cazort 1996).

5   See Shapin & Schaffer (1985) on Boyle's gentlemanly laboratory method and its pivotal position in establishing laboratory culture.

ently a linear mechanism, able to be function both forwards and backwards. As one commentator puts it, "readers are expected to accept the idea that the same forces are at work in the dissolution, as in the formation, of … bodies" (James and Field 1994, 5). By studying dead bodies, Victor hopes to discover how to reanimate and reformulate the body, restoring the life that had departed:

> Whence, I often asked myself, did the principle of life proceed? … To examine the causes of life, we must first have recourse to death. I became acquainted with the science of anatomy: but this was not sufficient; I must also observe the natural decay and corruption of the human body … I paused, examining and analysing all the minutiae of causation, as exemplified in the change from life to death, and death to life, until from the midst of the darkness a sudden light broke in upon me … After days and nights of incredible labour and fatigue, I succeeded in discovering the cause of generation and life; nay, more, I became capable of bestowing animation upon lifeless matter (Shelley [1831] 1980, 51-52).

The force of life that Frankenstein develops is galvanic: the animation of organic matter through the application of electricity. Shelley, in her 1831 introduction, refers to the potentials of electricity to reverse the state of death and confer life. She reports that she, Percy Shelley and Byron frequently discussed,

> the nature of the principle of life, and whether there was any probability of it ever being discovered and communicated … Perhaps a corpse would be reanimated; galvanism had given token of such things: perhaps the component parts of a creature might be manufactured, brought together, and endued with vital warmth (9).

Here Shelley is referring to a particular kind of spectacular 19th century automaton. During the 1790s the ability to send electrical impulses through organic matter had been developed by the Italian physician Luigi Galvani, who made the leg muscles of dead frogs twitch by jolting them with a spark from an electrostatic machine. This ability to reanimate dead matter had also been used on dead human bodies, both as part of serious scientific experiment and as a popular spectacle of the time. The galvanization of corpses was both a serious scientific experiment and a popular spectacle of the time. Galvanism seemed to represent the force that was subtracted from the dead, a kind of excess of vivacity that might be so powerful as to reverse the process of dissolution and carry the body back across the threshold of death into the realm of the living. Shelley's story creates an interface between electrical life and bodily systems, experimenting with their possible relationships and outcomes. In doing so it lays out the logic of life that informs subsequent and current experimentation with vitality: that is, life is calibrated according to the technical trajectories and capacities of particular machines. Life in Shelley's novel is wattage, electrical force, and power that can be intensified and managed by electrical machines. Life today is information, molecular or neuronal data, and it circulates freely between bodies and computational systems.

Hence Victor's discovery allows him to "bestow animation upon lifeless matter" (52) – to galvanize the dead. In Victor's laboratory, the creature is finally sparked into life:

> With an anxiety that almost amounted to agony, I collected the instruments of life around me, that I might infuse a spark of being into the lifeless thing that lay at my feet. It was already one in the morning; the rain pattered dismally against the panes, and my candle was nearly burnt out, when, by the glimmer of the half-extinguished light, I saw the dull yellow eye of the creature open; it breathed hard, and a convulsive motion agitated its limbs (57).

The being Victor creates is not, however, a singular being that has been brought back from death to life in an act of symmetrical reversal. Rather it is an assemblage of mismatched parts, all with different histories and different origins, taken from the charnel house, the slaughter pen and the dissection room. Some parts, this implies, are animal, some human, some belonging to criminals, some to women, some to men. Even more alarming for Shelley's readership, the creature is a mixture of the organic and the artificial – once living matter that has been technically induced to live again. From the point of view of any conventional notion of human identity, the creature is utterly heterogeneous, a being at odds with the notion of identity as such. Moreover, he is visibly marked by his unnatural origin; it is clear from his appearance that he is an awkward assemblage of reanimated body parts:

> His yellow skin scarcely covered the work of muscles and arteries beneath; his hair was of a lustrous black, and flowing; his teeth of pearly whiteness; but these luxuriances only formed a more horrid contrast with his watery eyes, that seemed almost the same colour as the dun white sockets in which they were set, his shriveled complexion and straight black lips (57).

Victor is repelled by his creature's appearance once he has been brought to life. "Unable to endure the aspect of the being I had created," he runs from the room, abandoning his creature to fend for itself.

THE SUBJECTIVITY OF THE OBJECT

The rest of the novel is, among other things, a detailed investigation of the subjectivity of the abandoned creature, presented in the first person from the creature's point of view. Here again Shelley's novel displays its prefigurative reach. This is the first investigation of the ontology of the technoscientific object, its conditions of being in the world. How, the novel asks, do the conditions of artificial creation and technically conferred life generate certain possibilities for being? What does it mean to be embodied, when the body cannot claim the status of nature? How does a variation on human being, not born of woman but rather manufactured through scientific procedures, stand to the human social order? How can an artificial life situate itself in the world, and what kind of world does it make for itself?

Reading *Frankenstein* in the present recalls that other, recent, investigation of the ontology of machinic organisms, Donna Haraway's "Cyborg Manifesto," discussed at length by Zoë Sofoulis in her essay in this volume. Haraway asks a similar set of questions in relation to contemporary innovations in technoscience, particularly the convergence of biological and informational systems that she characterizes as "the translation of the world into a problem of coding" (Haraway 1991, 164). What forms of life are emerging from new entanglements of organic and informatic matter, and how is the human to be situated among these forms? Like Shelley, she authors a speculative creature, the cyborg, "a hybrid of machine and organism," social reality and fiction, through which to think these questions, and asks what kind of world such a creature would inhabit. Ultimately Haraway's experiment is a more optimistic interpretation of the possibilities for human-machine relations than Shelley's. It is an optimism made possible by the dynamic and highly public nature of new millennium technoscience and the general inability to deny the proliferation of hybrid life that it produces. Haraway lays out possible terms for the recognition of human-machine exchanges, while Shelley's nameless creature, in an earlier era, finds itself utterly excluded from such exchanges.

Initially Victor imagines his creation as occupying a social position of dependent kinship with himself, a social harmony of creature and creator in which he occupies the position of both father and God. As the creature's point of origin, Frankenstein expects to know his creation's life and subjectivity comprehensively, and expects it to act as a tool for his narcissism and ambition:

> A new species would bless me as its creator and source; many happy and excellent natures would owe their being to me. No father could claim the gratitude of his child so completely as I should deserve theirs (Shelley [1831] 1980, 54).

As a result of this self-gratifying fantasy, Frankenstein is utterly unprepared for the autonomy or particularity of the creature he manufactures – the fact that it is not simply an extension or benign reflection of himself. He seems incapable of recognizing its separateness, and hence can only interpret this as an affront to his own being. Hence his violent rejection of the creature and his disgust at its appearance.

After Frankenstein abandons it the creature wanders into a nearby forest, where it must slowly put together a world for itself. At first, it is in a state of fragmented confusion, and cannot interpret the stimuli that bombard it from all sides. Like the Lacanian infant, its body is *un corps en morceau*, a body in bits and pieces. Awkwardly assembled by its creator and abandoned without instruction, it must find ways to integrate the disorganized universe of phenomenal impressions that its disparate organs apprehend. Despite the apparently robotic nature of its manufacture, its creation as a machine assembled from discrete parts, the bodily and subjective capacities that emerge as the creature develops are not at all *mechanical* in the strict sense. As Canguilhem (1992) points out, a mechanical system (as distinct from an informational or electromagnetic system for example) is highly stable and non-dynamic. It consists of a set of movable parts that work together and periodically return to a set relation with respect to each other. The dynamics of a mechanical system are preset, and unlike an informational system, are not capable of self-organization or non-predictable forms of behavior. They do not learn, nor can they grow more complex over time. Their interactions with their environment are limited and fixed. Victor's expectations about his creature's being seems to be based on this kind of model of machine life – one in which his inputs would have predictable outputs, and his creature's life would belong to its inventor.

However, the creature proves very capable of dynamic forms of perception, thought, and action. Here we can see Shelley speculating about the same questions that drive artificial intelligence engineering. How does a perceptual system become a learning system? How can disparate perceptions and perceptual organs be integrated? How does information processing relate to learning, and how does learning relate to memory? The creature's machinic life proves to be of a complex non-stable kind, cumulative rather than additive. Slowly, phenomenal coordination emerges from its interactions with its environment and the creature is able to create multiple relationships with the world:

> A strange multiplicity of sensations seized me, and I saw, felt, heard, and smelt, at the same time; and it was indeed, a long time before I learned to distinguish between the operations of my various senses. By degrees, I remember, a stronger light pressed upon my nerves, so that I was obliged to shut my eyes. Darkness then came over me, and troubled me … Several changes of day and night passed, and the orb of night had greatly lessened, when I began to distinguish my sensations from each other (Shelley [1831] 1980, 102-103).

The creature slowly learns how to gather food, make fire, and find its way. Its body proves extremely strong, agile, and hardy. It can remember nothing of its origins, and cannot account for its own existence or its singularity. It finds, to its distress, that it is without kinship:

> Where were my friends and relations? No father had watched my infant days, no mother had blessed me with smiles and caresses … From my earliest remembrance I had been as I then was in height and proportion. I had never yet seen a being resembling me, or who claimed any intercourse with me. What was I? (121).

Its social education begins when it first encounters humans, a family of cottagers who live in the forest. The creature conceals itself, and by observing the family, learns to speak (French), and to read (*Paradise Lost*, *Plutarch's Lives* and the *Sufferings of Young Werther*). By observing the interactions between the cottagers, it learns of the nature of kinship and social bonds, and passionately desires some relationship, some form of social inclusion. Its attempts to make contact with the peasants provoke fear and panic, and the creature is again violently rejected. When it learns of its "unhallowed origin" and its prior rejection by its creator, it realizes that it is excluded from the human social order. It comes to understand itself in terms of *Paradise Lost*, interpreting its technogenesis as a horrible parody of Genesis, the narrative that, for a pre-Darwinian scientific culture, still offered a guarantee of human exceptionalism and foundational status. As a perversion of Genesis, the creature understands that it will never be included, that it will always remain alien. The creature, in its rage against the irresponsibility of its creator, says to Victor:

> Remember, that I am thy creature; I ought to be thy Adam; but I am rather the fallen angel, whom thou drivest from joy for no misdeed. Everywhere I see bliss from which I alone am irrevocably excluded (100).

Victor dismisses the creature's pleas for recognition and states that they will be forever locked in antagonistic hatred of each other. "Begone! I will not hear you. There can be no community between you and me; we are enemies" (100). Yet the novel makes clear that Victor and his creature are mirror images of each other, and that Victor's loathing is a loathing of the same. The creature reminds Victor that he too has become an outcast because, as a kind of precursor to Darwin and his *On the Origin of Species*, he has utterly disrupted the Genesis narrative's foundational power. By his experiments he has demonstrated that the force of life is not divine, but is rather susceptible to technical manipulation. Moreover by creating a copy of a human being, he has cast doubt on the foundational status of the human. To copy the human is to demonstrate that the originality of the original is questionable, and that a natural origin is potentially exchangeable for a technological one. The monster fully understands that his pariah status derives from being a copy and, necessarily, a bad copy. "My form," he tells Victor, "is a filthy type of yours, more horrid even from the very resemblance" (130). The monster is a mirror-image of the human, but one in which the human debt to technical systems is exposed rather than concealed. It suggests that the human body has been mis-classified as natural all along and threatens to expose this category of error for what it is. This is the monster's "ugliness," the reason for his constant repudiation and rejection by humans who refuse to acknowledge their own debt, who do not want to see how they all resemble the monster.

In response to his exclusion from the human social world and his rejection by his creator, the creature kills Frankenstein's family and friends, one by one. Frankenstein eventually dies in pursuit of his creation, exhausted by the creature's superior endurance and intelligence. In most popular readings of the tale, the creature's destructiveness is treated with a certain inevitability. It is taken for granted that any bad copy of the human would be so driven by *resentiment* that it would naturally desire to destroy humankind. The robot or android that longs for humanity is, of course, the flip side of this *resentiment* – its commonly proffered alternative, thematized in much bad science fiction. See, for example, the glutinous movie *Bicentennial Man* (1999), in which Robin Williams plays a robot who desperately desires to be human. Another persistent example of this desire is Data in *Star Trek*, an android who seems permanently nostalgic for a human status he never quite attains. It is as if all machine life must be caught between a fixed set of human or anti-human values, good and bad uses, and that it utterly lacks specificity of its own – its own trajectories and logics.

A more attentive reading would suggest otherwise. Frankenstein is killed by his creature because he has utterly neglected the ethical dimension of his project. He has failed to consider the ethics involved in the machinic production of life, or the consequences of the way that such life changes what life and human being *means*. Shelley's detailed consideration of the creature's experience, its affective and imaginative life, suggests that each possible creation represents a possible world of experience, feeling, embodiment, and desire. Each possible creation introduces new possibilities for relationship, and shifts the terms of human ontology in unexpected ways. Hence each possible creation demands and contributes to an ethical engagement with its specificities. Victor fails to engage with his creature in any way. Instead he descends into a mire of antagonism, posing his human identity in direct opposition to that of the creature and vowing to destroy it.

The ethics of machinic life is another of the threads in *Frankenstein* that Haraway picks up in the "Cyborg Manifesto."[6] The figure of the cyborg is used as monster in Shildrick's sense, as a way to speculate about potential modes of becoming, about possible worlds of life. Frankenstein's monster and Haraway's cyborg are both ways to think about human becoming. To reject them is to reject possible human futures, to refuse to engage with the consequences of shifting modes of embodiment, reproduction, and living process. Both monsters are ways to think about how major transformations in technical systems (from clockwork to electricity, from heat engines to information) upset the putative stability of the category "human" and send its naturalized modes of embodiment into disarray. I have argued at length elsewhere (Waldby 2000) that any shift in the logic of technical systems changes the material terms in which the human takes place. Such shifts open human being up to unforeseen possibilities for new modes of embodiment and translation, extension, supplementation, and loss. They suggest new forms of entanglement between human, technical, and animal life, new forms of enablement and disablement. As Haraway puts it, "we are all chimeras, theorized and fabricated hybrids of machine and organism" (Haraway 1991, 150). If transformations in technical systems always implicate human bodies, then such developments cannot be dealt with adequately through a simple reassertion of the inherent stability and naturalness of the human category. As Shelley's tale instructs us, such a reassertion only produces violence.

---

6   I am not trying to suggest that the "Manifesto" is a direct reply to *Frankenstein*. Nevertheless, at several points Haraway clearly draws on the Frankenstein narrative, often in order to move beyond its various impasses and "stand-offs" between creator and creation.

Learning Shelley's lesson, Haraway's "Manifesto" is an attempt to move beyond the antagonism between inventor and invention that characterizes the story, and beyond the anti-technological ways it has been interpreted. It argues for non-oppositional ways of understanding the relationship between humans and technologies, and calls for an end to the privileging of a putatively natural origin. Such a privileging merely perpetuates unserviceable ways of thinking that are unable to deal with the ethical complexity of human indebtedness to technogenic processes and the imprecision of the distinction between organic and technologic life. The word *organ* is, after all, derived from the Greek word for *tool*, suggesting the futility of a search for a pure organic origin prior to the machinic. The "Manifesto" articulates the terms of a new social contract, in which all the actants – human, animal, and machine – can recognize and acknowledge their mutual implication and the non-exclusive nature of their categories. It calls for a recognition of the cultural nature of nature, a recognition that all actants "populate worlds ambiguously natural and crafted" (Haraway 1991, 149) and that responsibility must be exercised in both living in and making such a world:

> A cyborg world might be about living social and bodily realities in which people are not afraid of their joint kinship with animals and machines, not afraid of permanently partial identities and contradictory standpoints (149).

Nearly 200 years after Frankenstein's creature is excluded from the human social contract for exposing the weaknesses in its foundational narratives, the "Manifesto" examines the costs of a social contract that still retains vestiges of these narratives. While technoscientific innovations multiply ways to hybridize humans, animals, and machines in ever more complex internetworks, orthodox social scientific and political discourse remains unable to incorporate such developments, and continues to assume the coherence of each of these categories. Both Shelley's *Frankenstein* and Haraway's "Manifesto" offer imaginative resources for those who would move beyond these entrenched positions, creating possible worlds in which human beings cannot repress their indebted relationship to technogenic processes. Rather than repudiating monstrosity, both argue, along with Latour (1993), for the centrality of monstrosity to any social contract adequate to the articulation of a technoscientific social order. It is only when the artifice and hybridity of "human nature" is acknowledged that it becomes possible to think through the terms of human-nonhuman collectives. Cast in this light, the monster's desire for kinship is not so much a desire for organic origin as it is for inclusion in a world where monstrosity is an essential dynamic, a valued moment of transformation mediating the unstable relations of humans and machines.

REFERENCES

Canguilhem, G. 1992. Machine and organism. In *Incorporations*, ed. J. Crary and S. Kwinter. New York: Zone Books.

Cazort, M. 1996. The theatre of the body. In *The ingenious machine of nature: Four centuries of art and anatomy*, ed. M. Cazort, M.Kornell and K.B. Roberts. Ottawa: National Gallery of Canada.

Haraway, D. 1991. *Simians, cyborgs, and women: The reinvention of nature*. New York: Routledge.

James, F. and J. Field. 1994. Frankenstein and the spark of being. *History Today* 44(9):47-54.

Latour, B. 1993. *We have never been modern*. Cambridge, MA: Harvard University Press.

Malone, R. 1978. *The robot book*. New York: Jove Publications.

Sawday, J. 1995. *The body emblazoned: Dissection and the human body in renaissance culture*. London and New York: Routledge.

Shapin, S. and S. Schaffer, S. 1985. *Leviathan and the air pump: Hobbes, Boyle and the experimental life*. Princeton, NJ: Princeton University Press.

Shelley, Mary [1831] 1980. *Frankenstein, or the modern Prometheus*. Oxford: Oxford University Press.

Shildrick, M. 1996. Posthumanism and the monstrous body. *Body and Society* 2(1):1-15.

Waldby, C. 2000. *The visible human project: Informatic bodies and posthuman medicine*. London and New York: Routledge.

Wiener, N. [1950] 1968. *The human use of human beings: Cybernetics and society*. London: Sphere Books.

IMAGINABLE COMPUTERS :
AFFECTS AND INTELLIGENCE IN ALAN TURING

"The interrelationships between the affect of interest and the function of thought and memory are so extensive that absence of the affective support of interest would jeopardize intellectual development no less than destruction of brain tissue. To think, as to engage in any other human activity, one must care, one must be excited, one must be continually rewarded. There is no human competence which can be achieved in the absence of a sustaining interest"
TOMKINS (1962, 43)

Elizabeth A. Wilson

INTEREST

In 1946 Alan Turing was in correspondence with the British neurologist and cyberneticist W. Ross Ashby about the possibility of making mechanical models of the brain.[1] Turing had recently taken up a position at the National Physics Laboratory (NPL) where he was part of a team working on the construction of the Automatic Calculating Engine (ACE) – one of the first computing machines to be built in Britain. Turing didn't last long in this job. A passing comment to Ashby about the orientation of his interests at the NPL holds a clue to the brevity of Turing's tenure:

> In working on the ACE I am more interested in the possibility of producing models of the action of the brain than in practical applications to computing (Hodges 1983, 363).

Within a year it was evident that Turing's peculiar vision for the ACE was incompatible with the institutional constraints emerging at the NPL. While Turing understood himself to be engaged in "building a brain," the NPL was insisting on a more pragmatic, less fanciful regard for computational machines. Nonetheless, Turing's interests were not blunted by his lack of professional success. His formal report on the ACE, submitted after his departure from the NPL (1947), contained the following paragraph:

> One way of setting about our task of building a 'thinking machine' would be to take a man as a whole and try to replace all the parts of him by machinery. He would include television

---

1   The historical and textual data in this chapter are drawn from the papers and correspondence of Alan Mathison Turing (1912-1954) held at the Modern Archive Centre, Kings College, Cambridge (Turing Papers, hereafter cited by the archive coding "AMT") and from Andrew Hodges' exhaustive biography and website on Turing (Hodges 1983; and http://www.turing.org.uk). Some of the Kings College materials, and a full catalog of the collection, are available at www.turingarchive.org. A new catalog for the Kings College Turing material was completed in September 1999. The archive citations in this chapter reflect those new designations; there may be some variation between these new citations and those in Hodges (1983).

cameras, microphones, loudspeakers, wheels and 'handling servo-mechanisms' as well as some sort of 'electronic brain.' This would of course be a tremendous undertaking. The object if produced by present techniques would be of immense size, even if the 'brain' part were stationary and controlled the body from a distance. In order that the machine should have a chance of finding things out for itself it should be allowed to roam the countryside, and the danger to the ordinary citizen would be serious. Moreover even when the facilities mentioned above were provided, the creature would still have no contact with food, sex, sport, and many other things of interest to the human being. Thus although this method is probably the 'sure' way of producing a thinking machine it seems to be altogether too slow and impracticable (Turing Papers AMT C/11:16-17).

Meltzer and Michie (1969) report that Turing's document caused a "furore" at the NPL; it was claimed by some that "Turing is going to infest the countryside … with a robot which will live on twigs and scrap iron." In 1948, after a sabbatical back at Kings College Cambridge, Turing took up a new post at the University of Manchester to work on their automatic digital machine; he held this position until his untimely death in 1954.[2]

CURIOSITY

Turing's place in the history of 20th century computational science has been cemented by two eponymous contributions: the Turing machine and the Turing test. These two philosophical and computational milestones bookend a short but diverse career: the publication in 1937 of a precocious and important paper on the *Entscheidungsproblem* that outlined the logical parameters for a universal calculating machine (a Turing machine); the years spent at Bletchley Park during the second world war code-breaking and constructing a speech encryption device; the post-war appointments working with teams building some of the world's first electronic digital computers; and the publication in 1950 of the massively influential article "Computing machinery and intelligence," which formalized the conditions for evaluating intelligence in a computer (the Turing test).

While Alan Turing is known most simply as the inventor of the computer,[3] a closer examination of his interests indicates a more expansive intellectual life. As his comment to Ashby and his

---

2 Beside his work on computers, Turing's death by suicide (at the age of 41) is the other thing for which he is most commonly remembered. Turing died from asphyxia due to cyanide poisoning. Half an apple, with several bites taken from it, was found by the bed where he died. The coroner thought that the apple had been used to take away the taste of cyanide (Turing Papers AMT K/6); Hodges has suggested that the apple may have been dipped in cyanide (no tests were done on the apple). While Turing had appeared to be in good health and a stable frame of mind, the two years prior to the suicide had been difficult. In particular, he had been convicted of "gross indecency" (i.e., homosexuality) in unfortunate circumstances arising from a burglary of his house, and he had undergone 12 months of "organo-therapy" (chemical castration) in lieu of a prison sentence. While he maintained a cheerful front through these difficulties, he became depressed. In a letter to a friend, Turing wrote: "I am afraid that the following syllogism may be used by some in the future:

Turing believes that machines think
Turing lies with men
Therefore machines do not think.
Yours in distress, Alan"

(AMT D/14a; as read by Norman Routledge in the documentary *The Strange Life and Death of Dr. Turing* [1992]. See also Hodges 1997, 54).

Turing's fear of a hostile confusion of his sexual and intellectual ambitions was realized at the inquest into his death. The *Manchester Guardian* reported that the Coroner was "forced to the conclusion that [the death] was a deliberate act for, with a man of that type, one would never know what his mental processes were going to do next" (*Manchester Guardian*, June 11, 1954; Turing Papers AMT A/1). Whether Turing's type was homosexual or academic is unclear; and perhaps it is this very confusion (and the insinuation of an identity precariously formed) that makes both tendencies credible sources of suicidal intent.

3   Of course, the question of "who invented the computer" requires careful historical analysis and cannot be answered with a single name and a specific date. Hodges summarizes Turing's contribution as follows: "Alan Turing had not invented a *thing*, but had brought together a powerful collection of ideas … Alan Turing's invention had to take its place in an historical context, in which he was neither the first to think

report to the NPL suggest, he was perhaps less focused on computers than we might presume and more captivated by the inter-relation of certain mathematical, emotional, social, and engineering puzzles. Computational logic, the building of mechanical devices, and fantastic anticipation were always intimately allied for Turing. One aspect of the code-breaking work at Bletchley Park that he relished was that the fusion of conceptualization, implementation, and speculation was institutionally encouraged and rewarded – Turing spent as much time with wires and valves and reverie as he did with mathematical calculation.[4] During the post-war years at the NPL and the University of Manchester, when he was employed solely as a computational theoretician, he was consistently frustrated by the institutional distance placed between his theoretical labor and hands-on engineering. However, even in these emotionally and intellectually restrictive occupations there was a flowering of his curiosity; in particular, his post-war research in computing was amplified by an active interest in both neurology and morphogenesis (the emergence of pattern in developing biological organisms). While many of these ideas found shape in standard academic publication, many others took form in less formal ways: he was an occasional member of the London-based Ratio Club that discussed cybernetic research; he had met and become friendly with Claude Shannon while on a research trip to the Bell Laboratories during the war; he attended lectures at the University of Manchester given by Jean Piaget in 1952; he entered into public debate about the future of computers (Turing 1951, 1952); he corresponded with neurologists and biologists; he conducted all manner of chemistry experiments in a makeshift laboratory in his house; and he entered into a Jungian analysis that focused his interest on the interpretation of his dreams. It will be the argument of this chapter that any assessment of Turing's computational achievements and historical legacy needs to be located within the broader terrain of these enthusiastic and eccentric concerns.

It will also be the argument of this chapter that these concerns bear on – and are fashioned by – psychological questions, particularly questions about affect. Sometime in 1922, when Turing was 10 years old, he was given a copy of *Natural Wonders Every Child Should Know* – a child's introduction to science (Brewster 1912). The book contains 51 short and curiously diverse chapters with titles such as "Things That Don't Have to Be Learned," "What Plants Know," and "Why the Blood is Salt." According to Sara Turing (Alan's mother) this book made a lasting impression on Turing: "a book well worn and greatly valued by him and of which even in his last years he spoke highly" (Turing 1959). One of the chapters in *Natural Wonders* deals with the structure and function of the brain; it is called "Where We Do Our Thinking." The questions of where we do our thinking, how we do how thinking, and whether thinking, intelligence and feeling are commensurate became central, motivating aspects of Turing's life and work.[5]

---

about constructing universal machines, nor the only one to arrive in 1945 at an electronic version of the universal machine of *Computable Numbers*" (Hodges 1983, 295). A thorough analysis of this particular period of computational invention would also need to examine cybernetic research in the US – including the canonical work of Wiener, von Neumann, McCulloch and Pitts, and Shannon and the development of ENIAC (Electronic Numerical Integrator and Calculator) at the University of Pennsylvania from 1943. For introductions to cybernetic research in the US in this period see Hayles 1999, and Heims 1991; see also the publications from the Macy Conferences on cybernetics from 1943 to 1954 (e.g. von Foerster 1949-1955), which reveal the eclectic collection of researchers involved in these early discussions (mathematicians, physicists, neurologists, psychoanalysts, psychiatrists, statisticians, psychologists, philosophers, engineers, zoologists, physiologists, anthropologists, chemists, linguists).

4   One of Turing's obituaries notes: "He combined in a rare and remarkable way great powers of abstract reasoning and analysis with a very concrete imagination, and a keen desire to make with his own hands things that would 'work'" (Annual Report of the Councils, Kings College Cambridge, November 1954:5-6; Turing Papers AMT K/5).

5   Hodges reports two episodes, at the beginning and the end of Turing's academic life, that disclose Turing's sustained interest in psychological questions. A few years after the death of his dearest school friend, Christopher Morcom, and when Turing was newly ensconced at Cambridge, he wrote to Morcom's mother expressing his thoughts on the relation of spirit and body: "consider that the body

Turing's contribution to a philosophy of machine intelligence is widely documented. Nonetheless, it has become commonplace to censure Turing and the computational theories of mind he incited for their inert conceptualizations of where and how we do our thinking.[6] It is sometimes claimed that such theories too naïvely reproduce the inclinations of a 10-year-old science geek: unduly logical, fantastically embodied, emotionally attenuated, asocial, puerile. I would like to locate Turing's contributions to computational science within a more empathic context, wherein the geeky ambition for thinking machines can also be understood as imaginatively motivated, affectively animated and childishly astute.[7] While Turing's formulation of intelligence as symbolic manipulation separable from embodiment, emotion, and sociality is philosophically and empirically moribund, any critical assessment of his work and influence that becomes preoccupied with such an assessment is constricting in at least two ways: it diminishes the richly inventive, richly heterogeneous context within which his work on machine intelligence was conceived and it narrows the significance of that inventiveness and heterogeneity for the contemporary scene.

IMAGINATION

Turing's canonical 1950 paper approaches the question of computing machinery and intelligence from three different directions: the postulation of a test for assessing intelligence in a machine (the Turing test); a critique of a priori arguments against machine thinking; and a discussion of the importance of learning as a basis for machine intelligence.

Turing opens the paper by stating that he will not directly address the question "can machines think?"[8] Instead, he substitutes this question with a kind of game, arguing that an imitation task could be the basis for testing intelligence in a computer. If a judge is unable to tell the difference between the responses of a computer and the responses of a man to a set of questions – that is, if the computer can successfully imitate the responses of a man – then the computer can be said to be intelligent. Although, as Turing imagines it, such a test is more complicated

by reason of being a living body can 'attract' and hold on to a 'spirit,' whilst the body is alive and awake the two are firmly connected. When the body is asleep I cannot guess what happens but when the body dies the 'mechanism' of the body, holding the spirit is gone and the spirit finds a new body sooner or later perhaps immediately" (Turing Papers AMT C/29; Hodges 1983, 64). At the beginning of his affair with Arnold Murray, the lover who was the catalyst for Turing's arrest, Turing inquired excitedly of Murray "Can you *think* what I *feel*? Can you *feel* what I *think*?" (Hodges 1983, 452).

6 Traditional computational theories of mind come under attack from a number of different quarters. There are those who argue that the metaphor of the mind as a computer is inadequate philosophically and/or neurologically. (See Dreyfus 1992, Edelman 1992, Penrose 1989, and Searle 1980 for canonical arguments of this kind.) Some researchers working in robotics have also made a strong empirical case against the efficacy of symbolic theories of mind for the construction of intelligent machines (here Rodney Brooks' research on embodied, behaviorally based intelligent robots has been exemplary: Brooks 1999; and http://www.ai.mit.edu). In the realm of cultural studies, it has been argued, similarly, that computational theories of mind do not adequately contextualize psychological functioning: mind is taken to be isolated from corporeal and social constraints. For example, Hayles notes that "here, at the inaugural moment of the computer age, the erasure of embodiment is performed so that 'intelligence' becomes the property of the formal manipulation of symbols rather than enaction in the human life-world. The Turing test was to set the agenda for artificial intelligence for the next three decades. In the push to achieve machines that can think, researchers performed again and again the erasure of embodiment at the heart of the Turing test. All that mattered was the formal generation and manipulation of informational patterns" (Hayles 1999, xi).

7 The designation "childish" was applied to Turing not only as an adult, but tautologically even when he was a child. Much of Turing's childhood correspondence with his parents is preoccupied – somewhat depressingly–with his mother's concern about the untidy and child-like character of his handwriting (AMT K/1). Turing's childishness is accentuated in his mother's biography (Turing, 1959) in ways that are sometimes affectionate (110) and sometimes – when filtered through the responses of his professional colleagues – infantilising (105). A letter to Mrs. Turing from Professor Jefferson after Turing's death shares this latter tone, "He [Turing] was so unversed in worldly ways, so childlike, it sometimes seemed to me, so unconventional so non-conformist to the general <unclear>. His genius flared because he had never quite grown up" (Letter from Professor G. Jefferson, Manchester University to Mrs. Turing, 18 October, 1954: AMT A/16).

8 Turing says he avoids the questions "Can machines think?" as it would incite a long and probably fruitless effort to provide a definition of the terms "machine" and "thinking." While Turing does eventually narrow the meaning of "machine" to digital computer, "thinking" remains loosely specified and is taken to be synonymous with intelligence, or what these days might be called "cognition."

than simply being a comparison between a human and a machine. First of all, this comparison is established through a comparison of a man and a woman:[9]

> [The test] is played with three people, a man (A), a woman (B), and an interrogator (C) who may be of either sex. The interrogator stays in a room apart from the other two. The object of the game for the investigator is to determine which of the two is a man and which is a woman [by asking questions] … The ideal arrangement is to have a teleprinter communicating between the two rooms … Now we ask the question, 'What will happen when a machine takes the part of (A) in this game?' Will the interrogator decide wrongly as often when the game is played like this as he does when the game is played between a man and a woman? These questions replace our original, 'Can machines think?' (Turing 1950, 433-434).

The parameters for the test are not configured more definitely than this. Turing's goal is not to enact such a test, but rather to encourage serious consideration of the possibility of machine intelligence. That is, in response to the frequent, habitual and often vociferous rejections of the idea of machine intelligence that surrounded him, Turing asks his audience to consider the puzzle of thinking machines more carefully. Is there a way in which this fanciful notion could be subject to sustained philosophical inquiry? In a 1952 BBC Radio discussion, Turing's reply to an inquiry about the types of question that could be put to a computer makes the general aim of his test clear:

> BRAITHWAITE : Would the questions have to be sums, or could I ask it what it had for breakfast?

> TURING : Oh yes, anything. And the questions don't really have to be questions, any more than questions in a law court are really questions. You know the sort of thing. 'I put it to you that you are only pretending to be a man' would be quite in order. Likewise the machine would be permitted all sorts of tricks so as to appear more man-like, such as waiting a bit before giving an answer, or making spelling mistakes, but it can't make smudges on paper, any more than one can send smudges by telegraph. We had better suppose that each judge has to judge quite a number of times, and that sometimes they really are dealing with a man and not a machine. That will prevent them from saying 'it must be a machine' every time without proper consideration.

> Well, that's my test. Of course I am not saying at present either that machines really could pass the test, or that they couldn't. My suggestion is just that this is the question we should discuss. It's not the same as 'Do machines think?' but it seems near enough for our present purpose, and raises much the same difficulties (Turing 1952, 4).

First and foremost, Turing's test is a philosophical or conceptual exercise. It is less the means for testing any particular machine than it is the means for delineating the hypothetical space

---

9   See Curtain 1997, Halberstam 1991, Lassegue 1996, and Wilson 1996 for discussions of the implications of gender in the Turing test. Most philosophically orthodox commentators consider the role of a gender comparison to be immaterial to the main objective of the Turing test: "Turing's gender-guessing analogy detracts from his own argument" (Hodges 1997, 38; see also Dennett 1998). Two small, but important, pieces of historical data need to be considered in evaluating the Turing test, and they may help to make sense of the reference to women in the test. First, in 1950 the word "computer" refers to both a human and a machine. That is, up until this point large-scale computations (such as the calculation of ballistic ranges) were done long-hand by individuals with basic mathematics training. These individuals were called computers. One of the primary incentives for developing mechanical computers was to speed up the work and decrease the errors of these human computers. Turing's insistence that there is no radical separation of the capacities of a human and a machine is no doubt partly a consequence of this linguistic condensation. Second, most of the people employed as computers were women (Light 1999).

that these new inventions might occupy: "we are not asking whether all digital computers would do well in the game nor whether the computers at present available would do well, but whether there are *imaginable computers* which would do as well" (Turing 1950, 436, emphasis added).[10] The test is a way of exploring the imaginable limits of computational machines; and for Turing there is no justification for imagining that these limits must exclude the capacity for intelligent thought.

Indeed Turing's computational imaginings were only occasionally constrained by the prosaic opinions of his peers, or by the limits of foreseeable technological advancement. More often, his conception of intelligent computing machinery was motivated and amplified by seemingly inexhaustible excitement. Discussing the Manchester machine in *The Times* (London) on June 11, 1949, Turing said:

> This is only a foretaste of what is to come, and only the shadow of what is going to be. We have to have some experience with the machine before we know its capabilities. It may take years before we settle down to the new possibilities, but I do not see why it should not enter any one of the fields normally covered by the human intellect, and eventually compete on equal terms. I do not think you can even draw the line about sonnets [Professor Jefferson's Lister Oration delivered two days earlier had argued against mechanical intelligence on the basis that a machine would not have the emotional disposition to compose a sonnet] though the comparison is perhaps a little bit unfair because a sonnet written by a machine will be better appreciated by another machine (Turing Papers AMT A/1).[11]

Turing's own thinking about intelligent, thinking machines was a condensation of high-level expertise, child-like enthusiasm, and fantastic contemplation. While it has been common enough for commentaries on Turing to canvass the first of these talents, it has been less common to find an appreciation of the latter two talents outside of conventional psychobiography.[12] Eve Kosofsky Sedgwick and Adam Frank (1995), in a discussion of the work of Silvan Tomkins (a psychologist of affect who worked with cybernetic and computational theories in the early 1960s), point to the instructive effects of attending to the specific historical and affective conditions in which early cybernetic research developed. They suggest that Tomkins' research emerged out of what could be called 'the cybernetic fold' – a historical period they date provisionally from the late 1940s to the mid-1960s. The cybernetic fold encompasses

> the moment when scientists' understanding of the brain and other life processes is marked by the concept, the possibility, the *imminence*, of powerful computers, but the actual computational muscle of the new computers isn't available (Sedgwick and Frank 1995, 12; italics in original).

---

10 Since 1991 Hugh Loebner has been running a competition offering prize money for the first program to pass the Turing test (http://www.loebner.net/Prizef/loebner-prize.html). To date no program has passed the test, and the general consensus is that no program has even come close. See Platt 1995 and the postscript to Dennett 1998 for commentary from those involved in the early years of the Loebner competition.

11 See Jefferson (1949) for the Lister Oration.

12 There are a number of biographical details about Turing that appear with regularity in cultural commentaries on his work: the gift of *Natural Wonders* when he was a child; Turing's passionate attachment to an older boy (Christopher Morcom) at school; Turing's homosexuality; his eccentric behavior (for example, during the war he would ride a bicycle to Bletchley Park wearing a gas mask – to help alleviate his allergies); his arrest and suicide. Such anecdotes are often yoked to Turing's cognitive or intellectual development, but the nature of their influence is often not clearly spelt out. For example, the death of Morcom is usually treated as emotionally shocking but intellectually animating for Turing. In most analyses of these life-events it is argued that affect and imagination *catalyze* intellectual development; yet affect and imagination usually remain outside the intellectual particularities thus invoked. Homosexuality, in particular, stands in an awkward relation to Turing's intellectual accomplishments – it is constantly placed adjacent to his work on thinking machines but its influence on the specifics of Turing's philosophies has yet to be carefully delineated. To revisit Turing's earlier syllogism to his friend Norman, the following deduction has yet to be tested: Turing believes that machines think/Turing lies with men/Therefore machines do think.

This cybernetic moment – rather than being a period of idle narcissism or oblivious projection ("the machine is like me") – is a time of immensely productive cogitation and imagination. Sedgwick and Frank (1995, 12) continue:

> The prospect of virtually unlimited computational power gave a new appeal to concepts such as feedback, which had been instrumentally available in mechanical design for over a century but which, if understood as a continuing feature of many systems, including the biological, would have introduced a quite unassimilable level of complexity to descriptive or predictive calculations. Between the time when it was unthinkable to essay such calculations and [the] time when it became commonplace to perform them, there intervened a period when they were available to be richly imagined.

For Sedgwick and Frank, Silvan Tomkins' theories of affect "bear the mark of this moment of technological imagination" (12), and they argue that this historical-affective aspect of his work needs to find new consideration in a critical (post-structuralist) landscape that has come to be dominated by cynical, if not thoroughly paranoid, accounts of cybernetic and cognitive research. Likewise, commentary on computational landmarks like the Turing test could be energized by an understanding of the exuberant imaginings that developed in the post-war cybernetic fold (infesting the countryside with a robot that will live on twigs and scrap iron! sonnet-appreciating machines!) – imaginings utterly transformed by the force of technological and computational invention since the 1960s. More specifically, such commentary could do worse than attempt to delineate the positive and expansive ambitions of Turing's test. I don't anticipate that Turing's work will somehow be unburdened of its many axiomatic difficulties by such a strategy,[13] but I do hope that such an approach would return to Turing's texts, to Turing's historical period and to our critical repertoire a sense of the inventiveness of computational fabrication.

SURPRISE

In the middle of his 1950 paper Turing takes time to directly refute some common objections to the notion that machines might think. He lists nine complaints: the theological objection ("thinking is a function of man's immortal soul"); the "heads in the sand" objection ("the consequences of machines thinking would be too dreadful; let us hope and believe that they cannot do so"); the mathematical objection ("there are limitations to the powers of discrete-state machines"); the argument from consciousness ("not until a machine can write a sonnet or compose a concerto because of thoughts and emotions felt, and not by the chance fall of symbols, could we agree that machine equals brain"); arguments from various disabilities (a computer cannot be "kind, resourceful, beautiful, friendly"); Countess Lovelace's objection (a computer "has no pretensions to originate anything"); the argument from the continuity of the nervous system ("the nervous system is certainly not a discrete-state machine"); the argument from the informality of behavior ("it is not possible to produce a set of rules purporting to describe what

---

13 For example, one of the most serious limitations of the Turing test is the presumption that flesh is peripheral to the generation of intelligent thought. As Turing sees it, the *advantage* of his test is that it draws "a fairly sharp line between the physical and intellectual capacities of a man." He continues: "No engineer or chemist claims to be able to produce a material which is indistinguishable from the human skin. It is possible that at some time this might be done, but even supposing this invention available we should feel there was little point in trying to make a 'thinking machine' more human by dressing it up in such artificial flesh" (Turing 1950, 434). In his 1947 report to the NPL he had already claimed "we propose to try and see what can be done with a 'brain' which is more or less without a body" (Turing Papers, AMT C/11:17). In ways that are commensurate with this sentiment, Turing also explicitly rejects the importance of intersubjectivity and affect to the fabrication of intelligence. Too often in Turing, intelligence is autonomous, unemotional, and detached from its material instantiations.

a man should do in every conceivable circumstance"); and perhaps most oddly of all, the argument from extra-sensory perception ("the [computer's] random number generator will be subject to the psycho-kinetic powers of the interrogator").

Let me examine just one of these responses – for my goal is less an assessment of how well Turing dispatches each of his critics, than an account of what drives his defense of thinking machines. Countess Lovelace's objection – that a calculating engine cannot *originate* anything, that it can only do what it has been programmed to do – has been a persistent counter-argument to claims about machine intelligence, from Babbage's Analytic Engine to Deep Blue.[14] Indeed, one of the most frequent figurations of computation is that it is the opposite of, or a stranger to, originality or novelty or surprise. As such, it is argued, computational machines are intelligent in only the most narrow, the most culturally privileged, the most *unsurprising* sense of the word.

Turing disagrees: "Machines take me by surprise with great frequency" (Turing 1950, 450). In fact, he takes the capacity for unexpected behavior to be a necessary component in any intelligent machine. Max Newman's obituary for Turing touches on these intellectual convictions and contextualizes them affectively:

> In conversation he had a gift for comical but brilliantly apt analogies, which found its full scope in the discussions on 'brains v. machines' of the late 1940s. He delighted in confounding those who, as he thought, too easily assumed that the two things are separated by an impassable gulf, by challenging them to produce an examination paper that could be passed by a man, but not by a machine. The unexpected element in human behavior he proposed, half seriously, to imitate by a random element, or roulette wheel, in the machine. This, he said, would enable proud owners to say 'My machine' (instead of 'My little boy') 'said such a funny thing this morning' (Newman 1955, 255).

Not only is it important to register that Turing was interested in the unexpected (and childlike) character of machines (and here the familiar indictment of envious reproductive longings seems to inadequately grasp Turing's drift), it is also notable that he retained for himself the capacity *to be surprised*. His account of the origins of unexpected behavior in his own calculating machine emphasizes not the machine's psychic state, but his own:

> Machines take me by surprise with great frequency. This is largely because I do not do sufficient calculation to decide what to expect them to do, or rather because, although I do a calculation, I do it in a hurried, slipshod fashion, taking risks. Perhaps I say to myself, 'I suppose the voltage here ought to be the same as there: anyway let's assume it is.' Naturally I am often wrong, and the result is a surprise for me for by the time the experiment is done these assumptions have been forgotten. These admissions lay me open to lectures on the subject of my vicious ways, but do not throw doubt on my credibility when I testify to the surprise I experience (Turing 1950, 450-451).

---

14 Charles Babbage (1791-1871) was an English mathematician and inventor who designed (but only partially built) a mechanical computing device (the so-called Analytic Engine) in the 1830s. The Analytic Engine was an important precursor to 20th century digital computers. Countess Lovelace (1815-1852) was an English mathematician and associate of Babbage; she wrote the first "program" for the Analytic Engine. In 1996 the world chess champion Garry Kasparov played a specially designed and built chess computer: IBM's Deep Blue. Kasparov lost the first game but won the series. There was discussion (with reference to Turing) in the media at this time as to whether Deep Blue could be considered intelligent or whether it was simply a mindless number-crunching machine. Kasparov himself commented: "Although I think I did see some signs of intelligence, it's a weird kind, an inefficient, inflexible kind that makes me think I have a few years left" (*Time*, April 1, 1996, 55). In a rematch the following year, against overhauled programming and hardware ("Deeper Blue"), Kasparov lost the series (see *Time*, May 19 & 26, 1997).

What binds man to machine, then, may be less envy or narcissism than surprise. It may be surprise (and its conjoint relations to the other affects, especially interest and excitement) that preferentially cultivates an interface of the human and the machine.

Silvan Tomkins nominates surprise as one of the primary positive human affects.[15] A less intense form of the startle response, surprise has a particular effect on psychological activity – it interrupts and reorients attention:

> [Surprise] is ancillary to every other affect since it orients the individual to turn his attention away from one thing to another. Whether, having been interrupted, the individual will respond with interest, or fear, or joy, or distress, or disgust, or shame, or anger will depend on the nature of the interrupting stimulus and the interpretation given to it (Tomkins 1962, 498).

The nature of the machine's interrupting stimulus and the interpretation Turing gives to it are governed by his vicious and curious ways, such that his surprise is coassembled with, and amplified by, the positive affects of interest and joy. And in large part the argument in "Computing machines and intelligence" urges the reader to be likewise engaged with computational devices. This affective appeal (to be diverted and delighted) is perhaps the most potent use for the 1950 paper in the contemporary scene. Given the decreasing importance of the technical or empirical aspects of the paper, Turing's petition for a positive affective orientation to new computational research (surprise! interest! excitement!) has become one of the important lessons to be drawn from "Computing machinery and intelligence" – for the engineer as well as the cultural critic.

Despite his explicit disagreements with the philosophical and engineering ambitions of Turing's 1950 article and the mainstream AI research that it inspired, Rodney Brooks – the Director of the AI Lab at MIT – has a very similar orientation to the nature of computational invention. In the documentary *Fast, cheap and out of control* (1997), Brooks recounts how his childhood interest in building robots was substantiated through surprise:

> I sometimes ask myself why do I do this. And I trace it back to my childhood days. I used to try and build electronic things in a little tin shed in the back garden. I was always trying to build computers … I just had this tremendous feeling of satisfaction when I switched the things on, the lights flashed and the machine came to life.

> When I was at MIT, building these robots, there was an even more dramatic moment. One night the physical robot actually moved. I mean, it was one I was working on for days but it completely surprised me – it moved! It had that magical sort of thing. It worked. And the best part was that it completely surprised me.

What starts out as a familiar Freudian story of geeky sublimation (the boyish preoccupation with building autonomous creatures) actually turns on a more Tomkins-like affective experience: the importance of being emotionally surprised, attentionally diverted and therefore intellectually interested and engaged. Since the mid 1980s, Brooks has been engaged in robotics research that is often surprising: small, autonomous robots that can scramble like insects over rough terrain

---

15 In the early part of his four-volume work *Affect, imagery, consciousness*, Tomkins (1962, 1963a, 1991, 1992) names eight basic or innate affects: interest-excitement; enjoyment-joy; surprise-startle; distress-anguish; shame-humiliation; anger-rage; fear-terror; contempt-disgust (and later he adds a ninth, dissmell).

and perhaps then invade the solar system (Brooks and Flynn 1989); robots that can wander through the MIT office space and retrieve soda cans from cluttered work benches (Brooks 1990); and most recently a large humanoid robot that will integrate these lower-level capacities with human-like emotional and visual-motor systems (Brooks et al. 1998). While the ideas about what constitutes machine intelligence have changed dramatically – for Brooks (contra Turing) intelligence is always situated, embodied, and interactive – the affective compass of their research is strikingly similar; and this kinship seems worth exploring.

CHILDISHNESS

If the affects of surprise, interest, and enjoyment occupy a prominent place in both Brooks' and Turing's instantiation of machine intelligence, this is due in no small part to their shared belief in the importance of learning and infant behavior for the construction of intelligent machines.

The third part of Turing's 1950 paper – a discussion of education and learning in machines – is perhaps the least widely discussed part of the paper, despite the fact that these are central concerns elsewhere in Turing's discussions of machine thinking.[16] Drawing on some hasty neurological and computational calculations (the capacity of the human brain; the speed of computational machines; the output of an expert programmer), Turing concludes that trying to program the intellectual abilities of an adult human mind would just take too long. Attention could be turned more profitably to the construction of a less pretentious machine and the processes by which it might be trained: "Instead of trying to produce a programme to simulate the adult mind, why not rather try to produce one which simulates the child's? If this were then subjected to an appropriate course of education one would obtain the adult brain" (Turing 1950, 456).

As he closes the 1950 paper, Turing suggests that there are two domains from which the construction of intelligent machines could emerge: from the domain of chess and from the domain of the child.

> Many people think that a very abstract activity, like the playing of chess, would be best [as a basis for machine intelligence]. It can also be maintained that it is best to provide the machine with the best sense organs that money can buy, and then teach it to understand and speak English. This process could follow the normal teaching of a child. Things would be pointed out and named, etc. Again I do not know what the right answer is, but I think both approaches should be tried (Turing 1950, 460).

---

16 In his NPL report Turing notes: "If we are trying to produce an intelligent machine, and are following the human model as closely as we can we should begin with a machine with very little capacity to carry out elaborate operations or to react in a disciplined manner to orders (taking the form of interference). Then by applying appropriate interference, mimicking education, we should hope to modify the machine until it could be relied on to produce definite reactions to certain commands. This would be the beginning of the process" (Turing Papers AMT C/11:14).

And the 1952 BBC Radio discussion spends some time on the question of education and learning:

"TURING: It's quite true that when a child is being taught, his parents and teachers are repeatedly intervening to stop him doing this or encourage him to do that. But this will not be any less so when one is trying to teach a machine. I have made some experiments in teaching a machine to do some simple operations, and a very great deal of such intervention was needed before I could get any results at all. In other words the machine learnt so slowly that it needed a great deal of teaching.

"JEFFERSON: But who was learning, you or the machine?

"TURING: Well, I suppose we both were. One will have to find out how to make machines that will learn more quickly if there is to be any real success. One hopes too that there will be a sort of snowball effect. The more things the machine has learnt the easier it ought to be for it to learn others. In learning to do any particular thing it will probably also be learning to learn more efficiently. I am inclined to believe that some other things which one has planned to teach it are happening without any special teaching being required. This certainly happens with an intelligent human mind, and if it doesn't happen when one is teaching a machine there is something lacking in the machine" (Turing 1952, 7-8).

Most researchers since Turing have taken the first path: they have constructed artificial intelligence systems as disembodied, fully functioning simulations of adult minds (or, more commonly, as *parts* of adult minds). Chess-playing ability, in particular, has remained an exemplary sliver of machinic intelligence. For the majority of these AI researchers, intelligence is pre-programmed rather than learned, innate rather than acquired, a cause rather than an effect; and so the developmental aspects of both human and machine intelligence have been neglected in the AI literature. A return to the child-like (or developmentally dependent) aspects of intelligence reorganizes these longstanding computational presumptions. In particular, a renewed interest in learning has bought with it fresh regard for the infant's particular expertise: the expression and regulation of affect.[17]

Since the mid 1990s, a group of researchers in Brooks' AI lab have been interested in integrating emotional character into a humanoid robot. "Kismet" is a robotic head with facial features (lips, eyes, eyebrows, eyelids, ears) that move to express one of nine emotions: anger, surprise, fear, happiness, calm, interest, tiredness, disgust, sadness. Rather than being pre-programmed responses to stimuli, Kismet's emotional expressions are an effect of changes in the robot's drive state. When its drives are within a homeostatic range and its social drive is stimulated (by the perception of a face), Kismet appears calm and interested. A reduction in social/facial stimulation elicits a response from Kismet that a naive observer would interpret as "lonely"; overstimulation provokes expressions of disgust and then anger. Having established reliable and plausible expression of emotions in this robot, Kismet's designers are interested in using these expressions and the interactions they invoke with a human "caretaker" as scaffolding for learning:

> By consistently and repeatedly engaging in this process [the caretaker's reinforcement of positive affective expressions in the robot], the robot could eventually learn to associate a positive emotional state with the desired behavior. This effectively 'tags' that behavior as being worthy of pursuit in its own right (Breazeal [Ferrell] and Velásquez 1998).

Leaving aside the questions of how successfully the Kismet robot instantiates affect and what kind of learning it is that affect facilitates, I would simply like to note here the emergence of the child (a learning, developing creature) as a credible phenomenon in AI research. If Kismet's appearance (imagine a cutesy, mechanical Yoda) reinforces already existing prejudices about the geeky appetites of some AI researchers, the serious consideration of infant development research is more surprising:

> Our work focuses not on robot-robot interactions, but rather on the construction of robots that engage in meaningful social exchanges with humans. By doing so, it is possible to have

---

17 This interest in affect and computation is happening in conjunction with a similar renewal of interest in affect in neuroscience. Antonio Damasio's *Descartes' error: Emotion, reason, and the human brain* (1994) is frequently cited as a seminal text for both the neuroscientific and computational recuperations of the emotions (see for example, Picard 1997). My preference for discussing the emotions via Tomkins is to draw citational attention to much earlier, often influential and yet commonly disregarded work in theorizing affect. Tomkins' prescience regarding affect, development and artificial intelligence was evident in 1963:

"How then should one devise [a human automaton]? He must, first of all, be equipped to function with much less certainty than our present automata. He would require a relatively helpless infancy followed by a growing competence through his childhood and adolescence. In short, he would require time in which to learn to learn through making errors and correcting them. This much is clear and is one of the reasons for the limitations of our present automata. Their creators seem temperamentally unsuited to create and nurture mechanisms which begin in helplessness, confusion, and error. The automaton designer is an overprotective, overdemanding parent who is too pleased with precocity in his creations. As soon as he has been able to translate a human achievement into steel, tape and electricity, he is delighted with the performance of his brain child. Such precocity essentially guarantees a low ceiling to the learning ability of his automaton, despite the magnitude of information incorporated in its design and performance. A more patient designer would suffer through the painful steps which are required to nurture the learning capacities of the machine" (Tomkins 1963b, 11).

a socially sophisticated human assist the robot in acquiring more sophisticated communication skills. Specifically, the mode of social interaction is that of a caretaker-infant pair where a human acts as the caretaker for the robot. By treating the robot, Kismet, as an *altricial* system whose learning is assisted and guided by the human caretaker, this work explores robot learning in a similar environment to that of a developing infant (Breazeal [Ferrell] and Velásquez 1998).

In a research domain known for its hard-headed, unsentimental approach to human psychology and identity, the advent of the child – and the attendant emphasis on dynamic, affect-focused interaction – as a central concern in the construction of intelligent systems is somewhat eccentric. Of course, Kismet's creators are interested in affect as it assists learning; and the ability to learn is in turn expected to promote a more robust system of intelligence (that is, the affects are mobilized not so much for their own intrinsic interest, but in order to facilitate higher-order cognitive processing and complex behavior). Nonetheless, the robot child brings more to this scene than simply the capacity for learning and well-balanced intelligence. Like the human child, the robot child amplifies the caretaker's capacity to be surprised, interested, and excited. And it is this – the use of affect to render intelligence both intelligible and captivating – that connects the MIT research to Turing, despite the clear philosophical and engineering differences that separate them. Even if Turing and the MIT researchers do not fully appreciate the child's native talents,[18] nonetheless they each demonstrate in their own way the powerfully amplifying effects of affect on intelligence.

Whether machines like Kismet will one day pass the Turing test is perhaps beside the point. However, it is worth noting that these robots have moved the focus of that test from cognitive appraisal ("are these the responses of a man?") to facial and emotional appraisal ("is this creature really happy?"); and perhaps this turn to affect can more accurately be considered to be a return to what Turing always held dear to his computational inventions. Regardless of how competent the contemporary instantiations of affect turn out to be, their use of infant structure and its affective corollaries indicates an inventiveness in the field of computational and robotics research that would distract and delight Turing no less than did his fantastic electronic brains in 1950.

---

18 Turing denudes the child psychologically: "Presumably the child-brain is something like a note-book as one buys it from the stationers. Rather little mechanism, and lots of blank sheets" (Turing 1950, 456). In more subtle ways, Breazeal (Ferrell) and Velásquez (1998) judge the knowingness of the infant by the standards of adult cognition: "The infant does not know the significance his expressive acts have for his mother, nor how to use them to evoke specific responses from her." This renders the infant-caretaker dynamic primarily cognitive (rather than affective): "The infant's basic needs, emotions, and emotive expressions are among the few things his mother thinks they share in common. Consequently, she imparts a consistent meaning to the infant's expressive gestures and expressions, interpreting them as meaningful responses to her mothering and as indications of the infant's internal state."

REFERENCES

Brewster, E. T. 1912. *Natural wonders every child should know*. New York: Grosset & Dunlap.

Breazeal (Ferrell), C. and Velásquez, J. 1998. Toward teaching a robot "infant" using emotive communication acts. In *Proceedings of the 1998 Simulation of Adaptive Behavior workshop on socially situated intelligence*. Zurich, Switzerland: At http://www.ai.mit.edu/projects/sociable/publications.html (accessed Aug. 1, 2000).

Brooks, R., Breazeal, C., Marjanovic, M., Scassellati, B., & Williamson, M. (1998). *The Cog project: Building a humanoid robot*. In C. Nehaniv (Ed.), 1998. *Computation for metaphors, analogy and agents* (Vol. 1562, Springer Lecture Notes in Artificial intelligence). Heidelberg: Springer-Verlag (http://www.ai.mit.edu/projects/cog/publications.html [accessed Aug. 1, 2000]).

Brooks, R. and A. Flynn. 1989. Fast, cheap and out of control: A robot invasion of the solar system. *Journal of the British Interplanetary Society* 42:478-485.

Curtain, T. 1997. The 'sinister fruitiness' of machines: *Neuromancer*, Internet sexuality, and the Turing Test. In *Novel gazing: Queer readings in fiction*, ed. E. K. Sedgwick. 128-148. Durham: Duke University Press.

Damasio, A. 1994. *Descartes' error: Emotion, reason and the human brain*. New York: Avon.

Dennett, D. 1998. Can machines think? In *Brainchildren: Essays on designing minds*. 3-29. Harmondsworth: Penguin.

Dreyfus, H. L. 1992. *What computers still can't do: A critique of artificial reason*. Cambridge, MA: MIT Press.

Edelman, G. 1992. *Bright air, brilliant fire: On the matter of the mind*. Harmondsworth: Penguin.

Fast, cheap and out of control. 1997. Documentary film. Dir. Errol Morris. Fourth Floor Productions, Inc. in association with American Playhouse.

Foerster, H. von (ed.). 1949-1955. *Cybernetics: Circular causal and feedback mechanisms in biological and social systems*. Proceedings of five of the "Macy Conferences" on cybernetics 1943-1954. 5 vols. New York: Josiah Macy Jr. Foundation.

Halberstam, J. 1991. Automating gender: Postmodern feminism in the age of the intelligent machine. *Feminist Studies*, 17(3):439-460.

Hayles, N. K. 1999. *How we became posthuman: Virtual bodies in cybernetics, literature and informatics*. Chicago: University of Chicago Press.

Heims, S. J. 1991. *The cybernetics group*. Cambridge, MA: MIT Press.

Hodges, A. 1983. *Alan Turing: The enigma*. New York: Simon and Schuster.

————— 1997. *Turing: A natural philosopher*. London: Phoenix.

Jefferson, G. 1949. The mind of mechanical man. Lister Oration, delivered at the Royal College of Surgeons of England on June 9, 1949; reprinted in the *British Medical Journal* 1 (June 25):1105 [AMT B/44].

Lassegue, J. 1996. What kind of Turing test did Turing have in mind? *Tekhnema* 3:37-58.

Light, J. S. 1999. When women were computers. *Technology and Culture* 40(3):455-483.

Meltzer, B. and D. Michie. 1969. Preface. In *Machine intelligence 5*, ed. B. Meltzer and D. Michie. Edinburgh: University of Edinburgh Press.

Newman, M. H. A. 1955. Alan Mathison Turing, 1912-1954. *Biographical Memoirs of Fellows of the Royal Society* 1:252-263 [Turing Papers AMT A/36].

Penrose, R. 1989. *The emperor's new mind: Concerning computers, minds, and the laws of physics*. Oxford: Oxford University Press.

Picard, R. W. 1997. *Affective computing*. Cambridge, MA: MIT Press.

Platt, C. 1995. What's it mean to be human anyway? *Wired* 3.04 (April).

Routledge, N. 1960. Review of S. Turing's *Alan M. Turing*. *Cambridge Review* 11 June: 632-635 [AMT A/32].

Searle, J. 1980. Minds, brains, and programs. *Behavioral and Brain Sciences* 3:417-424.

Sedgwick, E. K. and Frank, A. 1995. Shame in the cybernetic fold: Reading Silvan Tomkins. In *Shame and its sisters: A Silvan Tomkins reader*. 1-28. Durham NC: Duke University Press.

The strange life and death of Dr. Turing. 1992. Documentary film. Dir. Christopher Sykes. Christopher Sykes Productions, BBC TV and WGBH Boston.

Tomkins, S. 1962. *Affect, imagery, consciousness*. Vol. 1, *The positive affects*. New York: Springer.

————— 1963a. *Affect, imagery, consciousness*. Vol. 2, *The negative affects*. New York: Springer.

————— 1963b. Simulation of personality: The interrelationships between affect, memory, thinking, perception, and action. In *Computer simulation of personality: Frontier of psychological theory*, ed. S. Tomkins and S. Messicck. 3-57. New York: John Wiley.

————— 1991. *Affect, imagery, consciousness*. Vol. 3, *The negative affects – anger and fear*. New York: Springer.

————— 1992. *Affect, imagery, consciousness*. Vol. 4, *Cognition*. New York: Springer.

Turing, A. M. 1937. On computable numbers, with an application to the Entscheidungsproblem. *Proceedings of the London Mathematical Society* 42:230-265.

————— 1950. Computing machinery and intelligence. *Mind* 59 (236):433-460.

————— 1951. Can digital computers think? Lecture delivered on BBC Radio Third Programme. 15 May [AMT B/5].

————— 1952. Can automatic calculating machines be said to think? BBC Radio discussion between A. M. Turing, M. H. A. Newman, Sir Geoffrey Jefferson, and R. B. Braithwaite. 14 and 23 January [Turing papers AMT B/6].

————— Papers and correspondence held at the Modern Archive Centre, Kings College, Cambridge. Some of the materials and a full catalog of the collection are available at www.turingarchive.org. Archive citations (AMT) in this chapter reflect designations of a catalog completed in September 1999; there may be some variation between these new citations and those in Hodges (1983).

Turing, S. 1959. *Alan M. Turing*. Cambridge: Heffer.

Wilson, E. A. 1996. Loving the computer: Cognition, embodiment and the influencing machine. *Theory and Psychology* 6(4):577-599.

ACKNOWLEDGMENTS

My thanks to Peta Allen Shera for her enthusiastic and comprehensive research assistance on Turing, and to Rosalind Moad at the Modern Archives Centre, Kings College, Cambridge, for access to and assistance with the Alan Turing collection. I am also indebted to the Silvan Tomkins Research Group (www.silvantomkins.org) for their intelligence on affect.

MARRYING THE PREMODERN TO THE POSTMODERN : COMPUTERS AND ORGANISMS AFTER WWII

Historically ... there could be no mechanical explanation of life functions until men had constructed automata: the very word suggests both the miraculous quality of the object and its appearance of being a self-contained mechanism ...
CANGUILHEM (1994, 293)

Evelyn Fox Keller

Despairing of the possibility of a fully physico-chemical (or mechanist) account of vital phenomena, Claude Bernard felt obliged to conclude in 1878 that "some invisible guide seems to direct [the living phenomenon] along the path it follows, leading it to the place it occupies." But within less than 100 years, biology found its homunculus, and it was, after all, a molecule. Bernard's statement is quoted in 1970 by the molecular biologist François Jacob, who is able to claim:

> There is not a word that needs be changed in these lines today: they contain nothing which modern biology cannot endorse. However, when heredity is described as a coded programme in a sequence of chemical radicals, the paradox disappears (Jacob [1970] 1976, 4).

Jacob is no mathematician, but when turning to history he adopts the most rudimentary of mathematical principles: given two points, he draws a straight line between them, giving us a marvelously linear narrative that starts with Bernard, develops through genetics and early molecular biology, and leads – inexorably, as it were – to the resolution of 19th century biology's paramount dilemma. Here, Bernard's "invisible guide," the agent responsible for the apparent purposiveness of biological organization, is recast as the genetic "program" encoded in a string of nucleotide bases. In the DNA's sequence, one finds the long sought source of living order, the self within a self. No telos here, only the appearance of telos (or teleonomy, as it is now renamed); no purpose, only, as Monod (1971) put it, "chance and necessity."

Yet history lends itself to linear narrative only retrospectively – or perspectively, if you will. Shift the vantage point, and other narratives come into view. In this essay I want to focus on a rather different (and far less linear) biological history than Jacob's now canonical version; a narrative that is rendered distinctive only in retrospect, evinced first by the advent of the computer in the mid 20th century (and with it, of new meanings of mechanism), and later, lent particular significance by late 20th century convergences between computer and biological science. This other narrative begins not with Bernard's "invisible guide," but with the notions of *organization* and *self-organization* that had been formulated almost a century earlier to counter both mechanistic and design accounts of life, and which, as such, were built into the very definition of *biology*. I begin, therefore, with a brief overview of the declining fate of these notions over the period 1790-1940 (as it were, their prehistory), before turning to their relegitimation by the advocates and architects of the new machine: the cyberneticists like Wiener.

THE PROBLEM OF ORGANIZATION: 1790 TO 1940

What is an organism? What is the special property, or feature, that distinguishes a living system from a collection of inanimate matter? This was the question that first defined biology as a separate and distinctive science at the beginning of the 19th century. And by its phrasing (implicit in the root meaning of the word *organism*), it specified at least the form of what would count as an answer. For what led to the common grouping of plants and animals in the first place – that is, what makes the two genres of "organized beings" (as they had earlier been referred to) *organisms* – was a new focus on just that feature: their conspicuous property of being organized, and of being organized in a particular way. As Jacob observes, by the end of the 18th century, it was

> [b]y its organization [that] the living could be distinguished from the non-living. Organization assembled the parts of the organism into a whole, enabled it to cope with the demands of life and imposed forms throughout the living world (Jacob [1970] 1976, 74).

Only by that special arrangement and interaction of parts that brings the wellsprings of form and behavior of an organism *inside* itself, could one distinguish an organism from its Greek root, *organon*, or tool. A tool, of necessity, requires a tool-user, whereas an organism is a system of organs (or tools) that behaves as if it had a mind of its own – that governs itself.

Indeed, the two words, *organism* and *organization,* acquired their contemporary usage more or less contemporaneously. Immanuel Kant, in 1790, gave one of the first modern definitions of an organism – not as a definition *per se,* but rather as a principle or "maxim" which, he wrote, "serves to define what is meant as an organism" – namely *"an organized natural product is one in which every part is reciprocally both end and means.* In such a product nothing is in vain, without an end, or to be ascribed to a blind mechanism of nature" (Kant 1993, 558, sec. 66; italics in original).

Organisms, wrote Kant, are the beings that "first afford objective reality to the conception of an *end* that is an end of *nature* and not a practical end. They supply natural science with the basis for a teleology … that would otherwise be absolutely unjustifiable to introduce into that science – seeing that we are quite unable to perceive *a priori* the possibility of such a kind of causality" (558, sec. 66).

Elaborating on this kind of causality, he wrote:

> In such a natural product as this every part is thought as *owing* its presence to the *agency* of all the remaining parts, and also as existing *for the sake of the others* and of the whole, that is as an instrument, or organ … [T]he part must be an organ *producing* the other parts – each, consequently, reciprocally producing the others … Only under these conditions and upon these terms can such a product be an *organized* and *self-organized being,* and, as such, be called a *physical end* (Kant 1993, 557, sec. 65; italics in original).

Indeed, it is here that the term *self-organized* first makes its appearance in relation to living beings. It is invoked – and underscored – to denote Kant's explicit opposition to argument by

53

design. No external force, no divine architect, is responsible for the organization of nature, only the internal dynamics of the being itself.

The beginnings of biology thus prescribed not only the subject and primary question of the new science, but also the form of answer to be sought. To say what an organism is would be to describe and delineate the particular character of the organization that defined its inner purposiveness, that gave it a mind of its own, that enabled it to organize itself. What is an organism? It is a bounded body capable not only of self-regulation, self-steering, but also, and perhaps most importantly, of self-formation and self-generation. An organism is a body that, by virtue of its peculiar and particular organization, is made into an autonomous and self-generating "self." The obvious task for biology was to understand the character of this special kind of organization or self-organization. At the close of the 18th century and the dawn of the 19th, it was evident – to Kant, as to his contemporaries – that neither blind chance nor mere mechanism, and certainly no machine then available, could suffice. "Strictly speaking," Kant wrote, "the organization of nature has nothing analogous to any causality known to us" (1993, 557, sec. 65). Necessarily, the science of such a mechanism would have to be a new kind of science, one which, given the technology of his age, Kant not surprisingly assumed to be irreducible to physics and chemistry.

Tim Lenoir has persuasively argued that, for the first half of the 19th century, such Kantian notions of "teleomechanism" (Lenoir's term) provided German biologists with a "fertile source for the advancement of biological science on a number of different fronts" (Lenoir 1982, 2). But by the second half of that century, the authority of the teleomechanists (for example, Blumenbach, Kielmeyer, Reil, Miller, von Baer, Bergmann, and Leuckart) gave way to that of a younger generation of biologists who espoused the sufficiency of more conventional mechanism (most notably, Ludwig, Helmholtz, and DuBois-Reymond), and who read the invocation of any form of teleology as inherently theological, and hence, as beyond the realm of science. By the end of the century, Kant's founding notion of "self-organization" had all but disappeared from biological research, except, that is, in embryology.

Lenoir attributes the decline of this tradition in part to generational conflict, and in part to the increasing specialization and competition prompted by changes in the structure of professional advancement in German universities. But the arrival of a new kind of machine, the steam engine, and with it, a new kind of physics, thermodynamics, was at least equally important. By the second half of the century, the first law of thermodynamics succeeded in providing a stunningly persuasive account of (at least some) vital functions – most notably metabolism and respiration. The production of heat, once thought to be the essence of animal life, was effectively shown to be no different in animals than in machines (see, for example, Brain and Wise 1994). As Helmholtz wrote in 1854:

> The animal body therefore does not differ from the steam-engine as regards the manner in which it obtains heat and force, but [only] in the manner in which the force gained is to be made use of (Helmholtz 1995, 37).

If animal heat, once the essence not only of ardor but of life itself, is itself mechanical, if the sustenance of animal life accords with the first law of thermodynamics, then any notion of life's special qualities might seem to be rendered superfluous.

But organisms do more than convert fuel into sustenance; even lowly organisms are not merely workers but lovers and creators. Above all, they must form themselves anew in every generation. From the simplest beginnings, they must construct themselves in the forms appropriate to their kind, and once formed, craft the parts needed to repeat the cycle. In short, they must procreate. And not even the steam engine could do that. Moreover, while metabolic processes obeyed the first law of thermodynamics well enough, it remained impossible to reconcile the second law – the law which prescribes that entropy reduces everything, ultimately, to disorder – with the manifest increase in order evidenced in embryogenesis. Here, in the emergence of a fully formed individual from a fertilized egg, remained as conspicuous a demonstration as ever of the seeming purposiveness of living matter, leading even so assiduous a student of the role of chemical and physical processes in physiology as Claude Bernard to conclude in 1878:

> What characterizes the living machine is not the nature of its physicochemical properties, complex though they may be, but the creation of that machine, which develops before our eyes under conditions peculiar to itself and in accordance with a definite idea, which expresses the nature of the living thing and the essence of life itself (Canguilhem 1994, 297).

But in embryology itself, a different tradition was invoked to deal with the dilemma of apparent purposiveness – more epigenetic rather than preformationist; indebted more to Kant, say, than to Haller. Indeed, at the close of the 19th century, embryology was one of the few areas of biology in which notions of teleomechanism still survived.

One might even say that the need to give precise meaning to older notions of organization, *epigenesis* and *regulation* constituted one of the dominant impulses of *Entwicklungsmechanik*. Hans Driesch and Wilhelm Roux championed different strategies for doing this, with Driesch focusing on the problem of regulation, and Roux on that of determination. In particular, Driesch sought to resurrect Aristotle's concept of an *entelechy* as a way both of making sense of his own experimental findings and of characterizing the essential purposiveness of the life force. For Driesch, the principle of entelechy captured the difference between the "whole-making causality" (Ganzheitkausalität) evident in embryological development and "merely mechanical causality": *entelechy* was the essence of the self-organizing capacity of biological organisms (see, for example, Driesch 1914). But Driesch's influence was short-lived. His views were contested within embryology itself (especially, by Roux), but more importantly, by the early decades of the 20th century, his concerns and even the entire tradition of experimental embryology came under direct challenge by the rising discipline of genetics.

*

The story of the disarticulation of genetics from embryology has been frequently told and does not need repeating here. In the immediate aftermath of his conversion, T. H. Morgan, the American geneticist who established the chromosome theory of heredity, seemed content to grant embryology a separate but equal status, but by the early 1930s, he and his colleagues had the confidence to take on the older discipline, recasting its preoccupations with organization and regulation as problems of "gene action" (see, for example, Keller 1995). By 1932, Morgan was ready to dismiss the entire tradition of *Entwicklungsmechanik*, faulting it for running "after false gods that landed it finally in a maze of metaphysical subtleties" (Morgan 1932, 285). Especially, he faulted Driesch's notion of entelechy, and with it, other, more recent attempts to define a neo-mechanist organicism (for example, Needham, Woodger, Watson):

> The entelechy … was postulated as a principle, guiding the development toward a directed end – something beyond and independent of the chemical and physical properties of the materials of the egg; something that without affecting the energy changes directed or regulated such

changes, much as human intelligence might control the running or construction of a machine. The acceptance of such a principle would seem to make it hardly worthwhile to use the experimental method to study development …[1] In fact, the more recent doctrine of the 'organism as a whole' is not very different from the doctrine of entelechy … (Morgan 1934, 6-7).

The growth of experimental genetics – its accumulation of solid experimental results – offered a healthy contrast. Even without tangible successes in embryology, an extraordinary confidence held sway in the community of American geneticists – bolstered in part from their success in correlating many phenotypic traits with particulate genes, in part, from the availability of a large agricultural market for these successes, in part from the heady expectation that their work would lead to the breeding of new, and better, races of men, and in part from the efficacy of their discourse of gene action in focusing attention away from the questions they could not answer. The question of the spatial and temporal organization so conspicuously manifested in embryogenesis was one of these.

Less than two decades later, Morgan and his school could claim a stunning vindication in the identification of DNA as the genetic material, and in Watson and Crick's 1953 identification of a mechanism for self-replication in the double helical structure of that molecule. In the 1930s, geneticists had to contend with constant challenges from embryologists over the inadequacy of genetics for dealing with their concerns, but by the 1950s, few embryologists were left to argue, and one might even say that the entire question of spatial and temporal organization had effectively disappeared from view.

WORLD WAR II AND THE COMPUTER

The years between Morgan and Watson and Crick were the years in which the revolution of molecular biology took shape; they were also the years in which a revolution of far broader proportions erupted, one that left in its wake a world irrevocably transformed. After WWII, no aspect of human existence could be recognized in pre-war terms, not even in science. Historians of science have devoted much attention to the ways in which WWII transformed the status, the structure, even the content of the physical sciences, but they have paid rather less attention to its impact on the biological sciences. Yet even here, shock waves were felt. The influx of physicists into molecular biology has of course been well noted. But equally importantly, the disciplinary status of embryology, already substantially weakened by its long contest with genetics, was reduced to an all-time low – in part because of its strong association with European (and especially with German) biology. Driesch's "vitalism" and his postulate of an entelechy were seen as standing for the forces of darkness, irrational, and anti-scientific. To a new generation of molecular geneticists, even Spemann's notion of an "organizer," in which Spemann hoped to find the causal locus of development, carried the taint of an anti-scientific metaphysics. Indeed, so discredited had this subject become in the aftermath of WWII that, later, when conditions seemed propitious for its revival, it was deemed tactical to rename it "developmental biology." But far and away the most consequential impact of WWII on scientific life – even for the life sciences – came about as a result of the development of the computer.

This newest of machines – a machine that was to transform the very meaning of mechanism – had itself been developed to cope with the vast increase in complexity of life outside the bio-

---

1 Driesch seems to have concurred. In 1912, he abandoned experimental research and joined the philosophy faculty at Heidelberg.

logical laboratory that had come with the war. The *computer*, as we now understand the term, came into being precisely because those people (mostly women) who had by tradition been responsible for data processing and computation – the original meaning of the word computer – could no longer handle the masses of data required to coordinate wartime military operations. *Cybernetics* originated in the self-steering and goal-seeking gunnery designed to meet the need for more effective control; and information theory arose out of the need to maximize the efficiency of communication.

"Out of the wickedness of war," as Warren Weaver put it, arose not only a new machine, but a new vision of science: a science, contra Descartes, based on principles of feedback and circular causality, and aimed at the analysis and mechanical implementation of exactly the kind of purposive "organized complexity" so vividly exemplified by biological organisms. Emphasizing its indebtedness to the history of servo-mechanisms, Wiener called this new science *cybernetics*. He envisioned a science of command, control, and communication that would permit the harnessing of biological resources for human use – just as an older science had succeeded in harnessing inanimate resources.

I have argued that these young Turks (or "cyberscientists") were inhabiting a different world from molecular biologists (Keller 1995, chap. 3). To the latter, life still seemed simple. Insulated equally from problems of military coordination and from the "organized complexity" of multicellular organisms, they retained the confidence that life could be managed by "gene action," by a unidirectional chain of command issuing from the instructions of a central office. Not surprisingly, the efforts of these two disciplines pointed in different directions.

Though happy to borrow the metaphors of information and program from the new computer science (see Kay 1996; 2000), molecular biologists remained committed to the strategy that had so spectacularly proven itself in physics (for many, their home discipline): that is, of streamlining their new subject and reducing it to the simplest possible units. This new breed of life scientists sought further elucidation of the essence of life by narrowing their focus to organisms so rudimentary and so simple as to be maximally immune from the mystifying and recalcitrant chaos of complex organisms. Their strategy led to the study of life in test tubes and Petri dishes populated by bacterial cultures of *E. coli* and their viral parasites the bacteriophage – forms of life that seemed simple enough to preserve the linearity of simple codes and telegraph-like messages.

Such arch reductionism had ample support in the 1950s, and not only from the unprecedented status of physics: it had the cultural support of a massive, almost visceral, reaction against the horrific associations with the revival of holism that had flourished in Germany in the 1930s and 40s, with what Anne Harrington (1995) calls the German "hunger for wholeness." Molecular biologists sought to build a new biology – in clear and often loudly voiced opposition to an older, organismic biology. Even more fully than Morgan and his colleagues had succeeded in doing, they would irrevocably rid descriptions of the organism of residual traces of vitalism and functionalism; especially, they would expunge from biological language those conspicuously teleological notions of purpose, organization, and harmony.

But elsewhere, in response to the exigencies of war (see, for example, Galison 1994), cyberscientists were avidly appropriating these older preoccupations for their own uses, leaning heavily on the very images, language, and even conceptual models of organicism that were now being so vigorously discredited in biology. Like Kant, they wished to develop new paradigms of circular (or reciprocal) causality, but unlike (and even contra) Kant, they did so with explicitly practical ends in mind. Their mission was to build a new kind of machine that could put the principles of Life to work in the world of the non-living. The ending of the war did little to dampen this mission: by then, it had taken on a life of its own.

BODY AND MIND : SELF-ORGANIZATION REDUX

Even more conspicuously than in Wiener's visionary musings of the 1940s, this appropriation of an older organicism for a new mission is evident in the concerted efforts of the late 1950s and 1960s to understand, and to build, "self-organizing systems." Such systems would marry the ultra-modern with the premodern, entirely bypassing the successes of molecular biology. Between 1959 and 1963 a flurry of conferences and symposia were convened under this mandate, mostly under the aegis of the Information Systems Branch of the Office of Naval Research (ONR) (see, for example, Yovits and Cameron 1960), and drawing from a wide array of scientific disciplines, both in the U.S. and in England.[2] A major impetus behind these conferences was the Navy's hope of building a new kind of machine, more adept and more versatile – indeed, more autonomous – than the digital computer developed in the war years.[3] If one wanted to build a machine that could handle truly complex problems – that could learn from experience – the trick, they argued, would be to study the natural systems (living organisms) that did so with such obvious success.[4] For the ONR, Frank Rosenblatt at Cornell's Astronautical Laboratory provided the direct inspiration. Rosenblatt, who had in turn been inspired by Donald Hebb's claim that an ensemble of neurons could learn if the connection between them were strengthened by excitation, attempted to build a machine that could learn in just that way. The Perceptron – a simple neuron-like learning device – is now seen as the first connectionist machine: it should be able, Rosenblatt argued (1958), to "perceive, recognize, and identify its surroundings without human training or control." Inspired by the work of Princeton mathematician John von Neumann in the late 1940s on self-reproducing automata, he went on to suggest that: "In principle, it would be possible to build Perceptrons that could reproduce themselves on an assembly line and which would be 'conscious' of their existence."

Kant's term "self-organization" had been resurrected in the 1930s: in 1933 by Ludwig von Bertalanffy, the founder of systems theory, to characterize the central feature of organismic development; and by Gestalt psychologists to describe the way humans process experience. At one and the same time, a single word denoted the key feature of minds and of organisms. And it was to these authors that the new cyberneticians referred. But a crucial difference is now evident: the hope of building a "self-organizing" machine required a new order of precision for notions of organization and purpose.

In a review of the first two conferences, both held in 1959, Rapaport and Horvath wrote:

> The notion of the organism ... is taken to be central in most holistic expositions of method ... [But] How does placing the organism at the center of one's conceptual scheme help one get away from 'limited mechanistic concepts'? ... What ... can a philosopher mean when he

---

2 Especially from a group formed in England in the mid-50s under the name "ARTORGA" for the study of artificial organisms.

3 C. A. Muses (1962), the organizer of one of the early conferences on self-organizing systems, suggested a slightly different aim, one he described as "man's dominating aim," namely "the replication of himself by himself by technological means" (114). Indeed, he introduced the published volume with the suggestion that "Man Build Thyself?" might be a better title (v).

4 The U. S. Army's interest in bionics had a similar basis. In 1960 Harvey Saley of the Air Force addressed a Bionics Symposium as follows: "The Air Force, along with other military services, has recently shown an increasing interest in biology as a source of principles applicable to engineering. The reason clearly is that our technology is faced with problems of increasing complexity. In living things, problems of organized complexity have been solved with a success that invites our wonder and admiration" (Bowker 1993, 118).

suggests that the method of biology may have more to offer than physics to the science of the future, the science of organized complexity? Can he mean that the modern scientist should place more reliance on those aspects of biology which are carry-overs from a previous age? Such a position would be difficult to defend. Yet a defense of the 'holist's' position can be made, provided we can spell out just how the older schemes ... can fit into the new modes of analysis ...

The question before us is whether we can ... extend systematic rigorous theoretical methods to 'organized complexity,' with which the holists are presumably concerned. There are at least two such classes of concepts discernible, and both are reminiscent of the ways of thinking of the biologist ... namely, teleology and taxonomy ... (Rapaport and Horvath 1959, 88-89).

Of course, as the authors go on to explain, "When we say teleology, we mean teleology in its modern garb, made respectable by cybernetics," defining cybernetics as "the theory of complex interlocking "chains of causation," from which goal-seeking and self-controlling forms of behavior emerge (Rapaport and Horvath 1959, 90).

It is striking, however, that post-war molecular biologists had little interest in and even less patience for such an agenda. Their own discipline was represented neither in the earlier Macy conferences, nor in the conferences on self-organization. They may have been happy to appropriate and adapt to their own uses some of the lexicon of information theory, but they sharply distanced themselves from the focus on complexity, organization, and purpose. Their interests, and their agenda, lay elsewhere. But the ears of many of those who were still interested in the problems of embryogenesis did perk up, recognizing here a new kind of scientific ally – an ally who was willing, for whatever reasons, to restore the organism-as-a-whole to conceptual primacy, and even eager to acknowledge its apparent "mind of its own." And what's more, an ally promising to bring clarity to the meaning of organization.

As J. H. Woodger – the philosopher of biology responsible for introducing von Bertalanffy to English-speaking audiences – had written in 1929:

If the concept of organization is of such importance as it appears to be it is something of a scandal that biologists have not yet begun to take it seriously but should have to confess that we have no conception of it ... Biologists in their haste to become physicists have been neglecting their business and trying to treat the organism not as an organism but as an aggregate ... (291).

In 1950 much the same could still be said – and it was: Edmund W. Sinnott, for example, wrote in *Cell and Psyche: The Biology of Purpose*:

The word *organism* is one of the happiest in biology, for it emphasizes what is now generally regarded as the most characteristic trait of a living thing, its *organization* ... But this central stronghold, we must ruefully admit, has thus far almost entirely resisted our best efforts to break down its walls.

Organization is evident in diverse processes, at many levels, and in varying degrees of activity. It is especially conspicuous in the orderly growth which every organism undergoes and which produces the specific forms so characteristic of life (Sinnott 1950, 22).

Like embryogenesis, or like mind. Indeed, Sinnott asserts "that biological organization ... and psychical activity ... *are fundamentally the same thing*" (48).

Now, however, a new breed of scientists appeared who were eager to help out. A direct response to the challenge can be found in the work of a young German refugee by the name of Gerd Sommerhoff who credits Uexhull for his inspiration. In the same year (1950) Sommerhoff published a work called *Analytical Biology*,[5] which begins by endorsing the primacy accorded by holists and organicists to organization and purpose as the key features of life, but then adds that "none of these schools manage to tell us in precise scientific terms what exactly is meant by 'organization.'" Sommerhoff undertakes to fill this gap, to create an "analytical biology" – that is, "to form a set of deductively employable concepts which will enable theoretical biology to deal with the really fundamental characteristics of observed life, viz. its apparent purposiveness" (13-14).

In the mid 1950s, Sommerhoff joined the ARTORGA group in England; and his book became one of the most frequently cited sources for participants of the "self-organization" conferences in the late 50s. Years later, W. Ashby described Sommerhoff's formulation as "of major importance, fit to rank with the major discoveries of science" for he had shown that "[t]he problem of the peculiar purposiveness of the living organism, and especially of Man's brain" could be represented with "rigour, accuracy, and objectivity" (from the introduction written for a planned re-edition that never appeared, but exists in Sommerhoff's files). To be sure, it was the properties of mind that were of primary importance to these would-be architects of self-organizing systems, but, like Sinnott, they viewed the design principles of biological organization and of mind as one and the same thing. And even C. H. Waddington seemed to concur: "The behaviours of an automatic pilot, of a target-tracking gunsight, or of an embryo, all exhibit the characteristics of an activity guided by a purpose" (1973, 20). The solution of a problem, at once so difficult and so general, clearly required ignoring minor disciplinary differences and pooling the resources of psychologists, embryologists, neurophysiologists, physical and mathematical scientists, and engineers.

Like the wartime architects of "command, control, and communication" systems, a chronic slippage is evident in these discussions between control and self-control. What distinguished them, and retrospectively earned them the appellate of "second-order cybernetics," was, paradoxically, their greater emphasis on autonomy, and at the same time, their explicit acknowledgement of the meaninglessness of autonomy. In the very first conference on self-organizing systems (convened in May, 1959), Heinz von Foerster,[6] a post-war émigré from Vienna who was soon to become the leading name associated with such systems, opened with the provocative claim: "There are no such things as self-organizing systems" (Yovits and Cameron 1960, 31). Systems achieved their apparent self-organization by virtue of their interactions with other systems, with an environment. In the same vein, Robert Rosen (1959) noted the "logical paradox implicit in the notion of a self-reproducing automaton," and two years later, Ross Ashby reiterated the point: "No system reproduces itself," he wrote. What we call "self-reproduction" is a process resulting from the interaction between two systems, or between the part and the whole (in von Foerster and Zopf 1962, 9).

---

5   Originally intended as his doctoral dissertation for the Biology Department at Oxford, but rejected (according to Sommerhoff) on the grounds that it was "too mathematical" for biology. The work interested others, however, and, instead of a degree, earned him a research fellowship in neuroscience (personal communication, April 12, 1996).

6   Trained in physics and engineering, von Foerster spent the war years employed in "top secret" research on electromagnetic wave transmission in Berlin. But his interests were far broader: after the war, he introduced himself to Warren McCulloch with a short volume he had written on quantum mechanics and memory. McCulloch was impressed; he sponsored his emigration in 1949, recruited him to edit the Macy conference proceedings, and helped him obtain a faculty position in Electrical Engineering at the University of Illinois.

Another player in these early conferences, Gordon Pask, reminded his colleagues that the crucial point in the development of an organism ("a self-organizing system from any point of view" [Pask 1961, 103-104]) is not its components (for example, nucleus and cytoplasm), but the interaction between and among them:

[W]hen we speak of an organism, rather than the chemicals it is made from, we do not mean something *described by* a control system. An organism *is a control system* with its own survival as its objective (Pask 1961, 72).

For von Foerster, Ashby, and Pask, the goal was to understand the logic of the coupling (or "conversation") between sub-systems that would generate the capacity of such a system to have its own objectives – that is, to be a natural system. Sometime in 1959 or 1960, von Foerster founded the Biological Computer Laboratory (BCL) at the University of Illinois at Urbana,[7] and for the next 15 years until his retirement in 1975,[8] the BCL (with support from the ONR, National Institutes of Health (NIH), and the Air Force) served as hearth and home for those mathematicians, logicians, and biologists interested in the problems of self-organizing systems.

Thirty years later Pask (1992, 24-25) summarized the differences between the old and new cybernetics as a shift in emphasis, from information to coupling; from the reproduction of "order-from-order" (Schroedinger 1944) to the generation of "order-from-noise" (von Foerster 1960); from transmission of data to conversation; from stability to "organizational closure"; from external to participant observation – in short, from "CCC" to an approach that could be assimilated to Maturana and Varela's concept of *autopoiesis* (see, for example, 1979).[9] The term "second-order" is von Foerster's own, introduced in hindsight to denote the explicit preoccupation of the new cybernetics with the nature of self-reflexive systems.

What came of their efforts? As it happens, not very much, at least not in the short run. Throughout the 1960s a number of developmental biologists attempted to make use of this new alliance, but by the end of that decade, their effort collapsed, as indeed (at least in the U.S.) did cybernetics itself. The 1960s were the years in which molecular biology established its hegemony in biology, and its successes lent its own agenda an authority that seemed impossible to contend with. Even its appropriation of cybernetic language worked effectively to eclipse the inherently global questions of developmental biology – especially, the question of what it is that makes an organism self-organizing. The cybernetic vision also collapsed in computer science, where it gave way to competing efforts from what has been called "high-church computationalism." The starting assumption of the latter was that, just like digital computers, minds are nothing more than physical symbol systems, and intelligent behavior could therefore be generated from a formal representation of the world.[10] A fierce funding war raged between the two schools in the early 1960s, championed respectively by the ONR and the Air Force; by the mid 60s, with the award of more than $10 million from the Advanced Research Projects Agency (ARPA) to Marvin Minsky

---

7 The exact date is unclear: von Foerster's papers suggest that the official date of its origin, January 1,1958, was decided retroactively – probably based on the starting date for his first major contract from the ONR for the Realization of Biological Computers. Once named, however, a wide assortment of books and papers written by visitors to his lab from 1957 on could be claimed as products of the BCL.

8 Von Foerster retired in 1975.

9 Maturana was a frequent visitor at the BCL during the 1960s, and after 1968, so was Varela; indeed Varela thanks von Foerster for his "role in the gestation and early days of the notion of autopoiesis" (1996, 407). Both Maturana and Varela have remained close personal friends of von Foerster ever since.

10 As Newell and Simon asserted in 1960 (quoted in Dreyfus and Dreyfus 1998, 16): "A physical symbol system has the necessary and sufficient means for general intelligent action."

(through Project MAC, created in 1963) for the development of artificial intelligence at MIT, it was effectively over.[11] Funding for the Perceptron and its requisite hardware (especially, for parallel processors) dried up (Yovits says "it was starved to death" [personal communication, April 14, 1996]), and the ONR-supported conferences on self-organizing systems ceased. In 1969, Marvin Minsky and Seymour Papert published a devastating attack on the entire project, explaining their intent in clear terms:

> Both of the present authors (first independently and later together) became involved with a somewhat therapeutic compulsion: to dispel what we feared to be the first shadows of a 'holistic' or 'Gestalt' misconception that would threaten to haunt the fields of engineering and artificial intelligence as it had earlier haunted biology and psychology (Minsky and Papert 1969, 19).

Shortly afterwards (July 11, 1971) Frank Rosenblatt, its guiding inspiration, died in a boating accident that some say was a suicide. But within the AI community, early connectionism had already met its demise (see, for example, Dreyfus and Dreyfus 1988, 24).[12] In the absence of funding, and of the kinds of concrete successes in embodying biological principles in non-biological systems that might have generated new alliances and new sources of funding, the union promised between organisms and computers failed to materialize. Neural net models, which had been the heart of Rosenblatt's effort, had to wait another 15 years to be revived; interactionist models of development have had to wait even longer. Evidently, the soil of American science in the 1960s was not quite right for this largely European transplant to take root. Even so, as Hubert and Stuart Dreyfus write (1988, 24), "blaming the rout of the connectionists on an antiholistic bias is too simple." In a rather expansive sweep, they fault the influence of an atomistic tradition of Western philosophy, from Plato to Kant; von Foerster (1995), by contrast, blames both the funding policies of American science (the short funding cycle, the emphasis on targeted research) and the excessively narrow training of American students. (Historians have yet to have their say on the matter.)

Today, however, the winds have shifted once again. With the explosive development of computer technology over the last 15-20 years, the roots of computer culture have spread wide and deep. Now, with the computational hardware needed to implement it,[13] neural net modeling has become a flourishing business, and thanks to the avidity of the media industry, complexity is becoming everyone's favorite buzz word. Fed the term "self-organization," my search engine turns up more than a dozen books and conference proceedings over the past decade. One of these is from a conference held in Bielefeld in 1988 on "Selforganization: Portrait of a scientific revolution." Heinz von Foerster emerges here as a retrospective hero of this revolution, credited not only as "one of the founders of modern self-organization theory" and the founder of the BCL (Krohn et al, 1990, 3), but also as the inspiration for Maturana and Varela's theory of autopoiesis. Other names – for example, Manfred Eigen, Hermann Haken, Ilya Prigogine, Erich Jantsch – also figure prominently.

---

11 Minsky recalls that their "original funding stream was created by J. C. R. Licklider, who went to ARPA in order to realize his dream of a nationwide network of computer research and application. He arranged the MIT collaboration with Professor R. M. Fano at MIT, who became the first director of MAC ... I should add that I had been a student and worshipper of Licklider from my sophomore days at Harvard in 1947" (personal communication, March 26, 1997).

12 Von Foerster, however, was able to keep the BCL going for another five years, largely by forging new alliances (for example, with the NIH and the Department of Education).

13 Thanks in good part to private corporations like Thinking Machines, Inc.

The scope of this latest "revolution" is dizzying. The 1990s, we were repeatedly told, was the era of "post-reductionism." The notion of life as an emergent phenomenon has become a truism. As physicist P. C. W. Davies writes:

> Because living organisms are nonlinear (for a start!) any attempts to explain their qualities as wholes by the use of analysis is doomed to failure … This is not to denigrate the very important work of molecular biologists who carefully unravelled the chemical processes that take place in living systems, only to recognize that understanding such processes can not *alone* explain the collective, organisational properties of the organism (Davies 1989, 110-111).

In fact, the vision of molecular biology has itself changed radically since Jacob's early description of an "invisible guide" encoded in the sequence of nucleotide bases. For those now engaged in the molecular study of development, "[t]he challenge is to link the genes and their products into functional pathways, circuits, and networks" (Loomis and Sternberg 1995, 649). Miklos and Rubin suggest that sequence information "will not, by itself, be sufficient to determine biological function. What is needed is to unravel the multi-layered networks that control gene expression and differentiation" (1996, 521). And even the aims of molecular biology have begun to converge with those of the first generation of biological computer scientists. As Miklos and Rubin conclude:

> In the Post-Sequence Era, we may eventually be able to move beyond what evolutionary processes have actually produced and ask what can be produced … That is, we may be able not only to discern how organisms were built and how they evolved, but, more importantly, estimate the potential for the kinds of organisms that can still be built (Miklos and Rubin 1996, 527).

Moreover, just as the early cyberneticians envisioned, in this new age of complexity, life is not limited to biological organisms. Davies continues his commentary by noting:

> The most cogent argument against vitalism is the existence of self-organizing processes in nonbiological science … Mathematically we can now see how nonlinearity in far-from-equilibrium systems can induce matter [here quoting Charles Bennett 1986] to "transcend the clod-like nature it would manifest at equilibrium, and behave instead in dramatic and unforeseen ways, molding itself for example into thunderstorms, people and umbrellas" (Davies 1989, 111).

Indeed, for some, life is not even limited to physico-chemical materiality. Artificial life – "life as it could be," as its supporters put it – can exist as pure simulation.

CONCLUSION

Two hundred years ago, biologists invoked the notion of organization to distinguish the living from the non-living, replacing an earlier tripartite division of the world (animal, vegetable, and mineral) with a bipartite division (animate and inanimate); today that same notion is employed to unify the animate and inanimate, arranging its objects not along a great chain-of being, but along a new succession, ordered solely by complexity. Born of the exigencies of war and its need for better technologies of control, this marriage of premodern categories with ultramodern technology has given birth to a postmodern ontology. In the old chain of being, we knew clearly enough who, or what, defined the end point, but this new succession has no end point, no limit to the possibilities of being. It is an organismic ontology, to be sure, but one which needs neither the vital force of generation nor an "invisible guide" to direct its beings along their journey. What now is an organism? No longer a bounded, organic body (for some, not necessarily even a material body). Instead, it is a nonlinear, far-from-equilibrium system that can mindlessly (or virtually) transcend the clod-like nature of matter and emerge as a self-organizing, self-reproducing, and self-generating being. It might be green or gray, carbon- or silicon-based, real or virtual.

Also one other difference: first-order cybernetics, born of concrete military needs, gave rise to command-control systems; second-order cybernetics, bred from the new alliances of the cold war, fostered a new preoccupation with self-organizing systems. For both, systems of control (with all the ambiguities of the term in place) were the primary preoccupation. But the revival of "self-organizing systems" in the 1980s was celebrated with promises of transcendence, even of self-transcendence. In Erich Jantsch's vision of this New Age, we are told:

> Life, and especially human life, now appears as a process of self-realization ... the inner, coordinative aspect of which becomes expressed in the crescendo of an ever more fully orchestrated consciousness ... Each of us would then, in Aldous Huxley's [1954] terms, be Mind-at-large and share in the evolution of this all-embracing mind and thus also in the divine principle, in meaning (Jantsch 1980, 397-398).

Today, in the post-cold war era of the third millennium, a new breed of self-organizing systematists offer us "life at the edge of chaos," celebrating not command and control, but life "out of control."[14] Meanwhile, less interested in meaning and more interested in products, an entire industry waits poised for the new technologies of design and construction given birth by their dreams.

If we are in fact in the midst of a revolution, the question is, what kind of revolution? In what domain is it taking place? How much of "life on the edge of chaos" is a biological, a technological, a cultural, or even a political vision? And, finally, where does vision leave off and reality begin? Or does it not any longer – for self-organizing, self-steering systems like ourselves and our progeny – even make sense to ask?

---

14 "Out of control" is the title of a recent popular account notable for its celebratory (and perhaps obfuscatory) moral. As the author concludes: "To be a god, at least to be a creative one, one must relinquish control and embrace uncertainty ... To birth the new, the unexpected, the truly novel – that is, to be genuinely surprised – one must surrender the seat of power to the mob below" (Kelly 1994, 257).

REFERENCES

Ashby, W. n.d. Unpublished recommendation. Holdings of G. Sommerhoff.

Bennett, C. 1986. On the nature and origin of complexity in discrete, homogeneous, locally-interacting systems. Paper. *Foundations of Physics* 16, (6) (June): 585-592.

Bertalanffy, Ludwig von. 1933. *Modern theories of development*, trans. J. H. Woodger. London: Humphrey Milford.

Bowker, G. 1993. How to be universal: Some cybernetic strategies, 1943-1970. *Social Studies of Science* 23:107-27.

Brain, R. and N. Wise. 1994. Muscles and engines: Indicator diagrams in Helmholtz's physiology. In *Universalgenie Helmholtz: Ruckblick nach 100 Jahren*, ed. L. Kruger. 124-145. Berlin: Akademie Verlag.

Canguilhem, G. 1994. *A vital rationalist*. Cambridge, MA: MIT Press.

Davies, P. C. W. 1989. The physics of complex organisation. In *Theoretical biology: Epigenetic and evolutionary order from complex systems*, ed. Brian Goodwin and Peter Saunders. 101-111. Edinburgh: Edinburgh University Press.

Dreyfus, H. L. and S. E. Dreyfus. 1998. Making a mind versus modeling the brain: Artificial intelligence back at a branchpoint. *Daedalus* 117 (1, winter): 15-44.

Driesch, H. 1914. *The problem of individuality*. London: Macmillan and Co.

Foerster, H. von. 1960. In *Self-organizing systems*, ed. M. Yovits and S. Cameron. New York: Pergamon Press.

————— 1995. Interview by Stefano Franchi, Güven Güzeldere, and Eric Minch. *SEHR* (special issue of the *Stanford Humanities Review*) 4(2). Available at http://www.stanford.edu/group/SHR/4-2/text/interviewvonf.html (accessed Feb. 28, 2001).

Foerster, H. von and G. W. Zopf (eds.) 1962. *Principles of self-organization*. New York: Pergamon Press.

Galison, P. 1994. The ontology of the enemy: Norbert Wiener and the cybernetic vision. *Critical Inquiry* 21 (autumn): 228-266.

Harrington, A. 1995. *Hunger for wholeness*. Princeton: Princeton University Press.

Helmholtz, H. von. 1995. *Science and culture*. Chicago: University of Chicago Press.

Hebb, D. O. 1949. *The organization of behavior*. New York: Wiley.

Jacob, F. [1970] 1976. *The logic of life*. New York: Vintage.

Jantsch, E. 1980. *The self-organizing universe*. Oxford: Pergamon Press.

Kant, Immanuel. 1993. *Critique of teleological judgement*. Vol. 39 of *Great Books*. 550-613. Chicago: Encyclopedia Britannica.

Kay, L. 1996. Who wrote the book of life? In *Exper. in den Biologishe-Medizinishcen Wessenschaften*, ed. M. Hagner and H.-J. Rheinberger. Berlin: Akademie Verlag.

————— 2000. *Who wrote the book of life? A history of the genetic code*. Stanford: Stanford University Press.

Keller, E. F. 1995. *Refiguring life: Metaphors of 20th century biology*. New York: Columbia University Press.

Kelly, K. 1994. *Out of control: The new biology of machines, social systems, and the economic world*. New York: Addison Wesley.

Krohn, W., G. Kuppers, and H. Nowotny (eds.). 1990. *Selforganization: Portrait of a scientific revolution*. Dordrecht: Kluwer Academic Publishers.

Lenoir, T. 1982. *The strategy of life*. Chicago: University of Chicago Press.

Loomis, W. F. and P. Sternberg. 1995. Genetic networks. *Science* 269 (August 4): 649.

Maturana, H. and F. Varela. 1979. *Autopoiesis and cognition*. Dordrecht: Reidel.

Miklos, G. L. G. and G. M. Rubin. 1996. The role of the genome project in determining gene function. *Cell* 86:521-529.

Minsky, M. and S. Papert. 1969. *Perceptrons*. Cambridge, MA: MIT Press.

Monod, J. 1971. *Chance and necessity*. New York: Alfred A. Knopf. Inc.

Morgan, T. H. 1932. The rise of genetics: II. *Science* 76:285-288.

————— 1934. *Embryology and genetics*. New York: Columbia University Press.

Muller, H. J. 1929. The gene as the basis of life. *Proceedings of the International Congress of Plant Sciences* (Ithaca, 1926) 1:897-921.

Muses, C. A. (ed.). 1962. *Aspects of the theory of artificial intelligence*. New York: Plenum.

Pask, G. 1961. *An Approach to cybernetics*. New York: Harper and Brothers.

————— 1992. Different kinds of cybernetics. In *New perspectives on cybernetics*, ed. G. van de Vijver. Dordrecht: Kluwer Academic Publishers.

Rapaport, A. and W. J. Horvath. 1959. Thoughts on organization theory and a review of two conferences. *General Systems* 4:87-93.

Rosen, R. 1959. On a logical paradox implicit in the notion of a self-reproducing automaton. *Bulletin of Mathematical Biology* 21:387-393.

Rosenblatt, F. 1958. In Electronic 'brain' teaches itself. *New York Times*, July 13. Section 4:9.

————— 1962. *Principles of neurodynamics. Perceptrons and the theory of brain mechanisms*. Washington D.C.: Spartan Books.

Schroedinger. E. 1944. *What is life?* Cambridge: Cambridge University Press.

Sinnott, E. W. 1950. *Cell and psyche: The biology of purpose*. Chapel Hill: University of North Carolina Press.

Sommerhoff, G. 1950. *Analytical biology*. Oxford: Oxford University Press.

Varela, F. 1996. The early days of autopoiesis: Heinz and Chile. *Systems Research* 13(3):407-416.

Waddington, C. H. 1973. *O.R. in World War 2: Operational research against the U-boat*. London: Elek.

Woodger, J. H. 1929. *Biological principles*, New York: Harcourt, Brace.

Yovits, M. and S. Cameron (eds). 1960. *Self-organizing systems*. New York: Pergamon Press.

CASSANDRA AMONG THE CYBORGS, OR, THE SILICON TERMINATION NOTICE

Samuel J. Umland & Karl Wessel

In the spring of 1974, Philip K. Dick – science fiction's most brilliant if most haunted imagination – experienced a series of disorienting visions that were by turns ecstatic and terrifying. Unfortunately they were also unintelligible to him, as if he had received a message intended for someone else, composed in a language he didn't understand. He spent the last seven years of his life writing ceaselessly, often deep into the night, producing two million words of prose that he called his "Exegesis" in a heroic but inevitably failed attempt to make sense of what had happened to him.[1] His essay, "Man, Android and Machine," is a single snapshot, taken in 1976, of the point he had arrived at by then in his quixotic metaphysical journey. Why does Dick's singular attempt to explain his unaccountable visions seem so uncannily prescient of postmodern cyberculture?

> We hold now no pure categories of the living versus the non-living ... one day we will have millions of hybrid entities which have a foot in both worlds at once (Dick [1976] 1988, 202).[2]

Although this passage seemed to refer to a safely distant future when it was written a quarter of a century ago, perhaps we should no longer feel safe. In a recent book, *Fuzzy Future* (1999), the neural net theorist and futurist Bart Kosko has suggested the route to personal immortality must follow a silicon road that winds its way from man to superman through the country of the cyborg, a contention that evokes the memory of the semi-human title character of Philip Dick's most riveting novel, *The Three Stigmata of Palmer Eldritch* (1965), who claimed, "GOD PROMISES ETERNAL LIFE. WE DELIVER IT" (Dick 1965, 180).

---

1 A small portion of Dick's "Exegesis" has been published. See Sutin (ed.) 1991.

2 Dick's "Man, android and machine" was first published in Peter Nicholls, ed., *Science fiction at large* (London: Gollancz, 1976), and subsequently in the posthumous collection by Dick, *The dark-haired girl* (1988); it is this latter text from which all quotations are taken. We accept Paul Williams' dating of the composition of the essay as early 1975, completed in late March of that year. See Williams' introduction to *The dark-haired girl* for more information about the context of the essay's composition, especially xix.

Section One

Dr. Kosko has noted a troubling catch, however.

Suppose you can load your brain into a chip. Where do you go? Do you die first? Do you stay in the brain or in the chip or both? The mind seems to split or bifurcate in the same way that some cells split and grow as new cells. How can you be sure it is still you in the chip? Suppose we make a hundred copies of the chip brain. Which one are you? (Kosko 1999, 250).

Which indeed? The proliferation of identical clones of the *self* – of the self, note, and not of the *body* – was anticipated by Dick within the dark, ontological labyrinth of the aforementioned novel, and earlier than that in several of his short stories. But perhaps the original credit should go to Dostoevsky, who'd already presented us with multiplying hordes of Golyadkins in his hallucinatory short novel, *The Double* (1846).[3]

Kosko's clones admit of three unpalatable interpretations:

1. The self is reborn as a metaphysical hydra which co-consciously occupies many different places at once, a conclusion that both contradicts the laws of physics and renders the notion of unitary subjectivity unintelligible.

2. A kind of macro-level quantum mechanical dice throw supervenes in which only one of the clones inherits the old, organic psyche, without cause or explanation. So much for the principle of sufficient reason.

3. None of the clones becomes conscious and can therefore only provide the means to oper-ate zombie-bodies that mechanically go through the motions of life without actually living. The shade of B. F. Skinner appears, protesting that this is a distinction without a difference, but maybe dead is dead after all.

Dr. Kosko's solution to his own riddle employs the ancient Greek paradox of the sorites. How does a grain of sand become a sandpile? Through the slow and continuous accretion of its sub-stance, with no wrenching dislocation of its identity. Suppose that we replace the brain with a whole system of chips, one small circuit at a time. Our Immortal-To-Be – a fellow named Barney Mayerson, say, a salesman of household appliances – notices nothing untoward happening to him during the course of this experiment, except for a gradual increase in the information processing speed of his mind. Now there simply *is* no point at which one can say that Barney was himself before that point but became something else afterward. Why should we? After all, through ordinary metabolism his original, organic brain has already completely replaced its material substance three or four times during the course of his lifetime without any cost to his identity.

Alas, there still seems to be a hole in the argument.

Suppose that we rebuild Barney's brain following Dr. Kosko's prescription, but with triple circuit-for-circuit redundancy such that none of the three silicon systems are not connected directly to each each another, but only to Barney's remaining – if slowly disappearing – organic circuitry. To avoid stacking the deck unfairly, suppose that all three developing systems share equal time in operating Barney during the period of his reconstruction, though with no pair of

---

3   The relationship of Dick's early short story "Upon the Dull Earth" (1954) to later works such as *The Three stigmata of Palmer Eldritch* and to Dostoevsky's *The double* is explored in Umland 1995 (81-91). Samuel J. Umland, "To Flee From Dionysus: *Enthousiasmos* from 'Upon the Dull Earth' to VALIS," in Samuel J. Umland, ed., *Philip K. Dick: Contemporary critical interpretations* (Westport: Greenwood Press, 1995), pp. 81-99. Note that in *The double* Golyadkin's hallucinations are *autoscopic*, of Self rather than Other. The relatively rare autoscopic phenomenon consists of an encounter with a visual hallucination of oneself. This image may have derived from an epileptic hallucination; Dostoevsky is believed to have suffered from temporal lobe epilepsy (TLE).

them ever "on" at the same time. Finally, once the last organic bit of Barney is excised and replaced, just one of the three systems will be awake while the other two sleep. Now we'll separate the three systems and awaken them all.

Where's Barney?

> It is one and the same thing to be living or dead, awake or asleep … (Heraclitus, Fragment 113, in Wheelwright 1959, 90).

Apparently we must encounter the same nagging ontological doubts as before. We've just two choices, either to outlaw such philosophical thought experiments, permitting the naïve technophiles among us to sleep more soundly, or else admit to ourselves that our old intuitions about the nature of consciousness and identity have completely failed us, that we've entered a strange new universe whose operating manual is written in a language we cannot read.

Of course, all this has happened before. It happened with our realization that the laws of physics as we know them must break down within the singularity of a black hole. The Lyotardian upshot is that we have no unitary theory, no mathematical metanarrative, that explains all physical phenomena.

The problem of consciousness is the black hole of psychology.

> … [B]y experiencing time as we do … we get a totally wrong idea of the sequence of events, of causality, of what is past and what is future … (Dick [1976] 1988, 207-208).

Just what is consciousness, anyway? Of what use to us is it? If a zombie's simulation of conscious decision-making were as effective in getting its genes passed on to the future as our own subjectively experienced behavior, what evolutionary advantage would we possess over the zombie by having such experiences?

The mystery is only deepened by certain odd scientific facts first discovered by the neurophysiologist Benjamin Libet. Our commonsense understanding of our own willed actions is that, first, we decide to perform them, second, that this decision initiates brain processes that cause the body to do our bidding, and finally that the body performs as it is commanded. In a long series of experiments Dr. Libet showed – astonishingly – that this order of events is wrong. EEG analysis indicates a counterintuitive revelation, that our brains begin to prepare motor commands (evoke the so-called "readiness potential") up to one-half second before we become consciously aware of our intentions to act.[4] The psychologist Merlin Donald has put this conclusion starkly:

> In short, Libet claimed that the conscious self gets the news too late to affect what the body has already decided to do. Thus it cannot prepare intentions or make decisions. They are made unconsciously, in the brain, before awareness develops (Donald 2001, 44).

What's going on here? Is the notion of freedom of the will only a cherished illusion, one more failed intuition about the nature of human identity? But what is our purpose in telling ourselves these stories if they're not true?

> I believe these ontological dysfunctions in time do occur, but that our brains automatically generate false memory-systems to obscure them, at once … The veil or dokos is there to deceive us for a good reason … (Dick [1976] 1988, 209).

---

4 Libet's argument can be traced through several papers written over twenty years. See Libet, 1993.

In fact, Dr. Libet came to believe something very like what Philip Dick says above. It is unlikely that Dick was familiar with Libet's research; rather, he may have come to his conclusion on the basis of dissociative episodes he'd experienced, or from the desynchronization of his inner clocks due to drug experiments or epileptic seizures.

> By studying what goes wrong with a machine – for example, when two mutually exclusive tropisms function simultaneously in one of Grey Walter's synthetic turtles, producing fascinatingly intricate behavior in the befuddled turtles – one learns, perhaps, a new, more fruitful insight … (Dick [1972] 1988, 124).

The Cartesian ego presents a seamless face to the world, but this is another illusion. Most of us suffer to one degree or another from the conflict of internal forces, but such conflicts seldom emerge in the form of a full-blown dissociation of consciousness, volition, or identity. When they do, as in the bizarre neurological phenomenon known as the alien hand syndrome (AHS), the unreal nature of the stories we habitually tell ourselves about who and what we are becomes all too disturbingly apparent.

AHS is generally caused by a lesion within the corpus callosum – the huge tract of nerve fibers that connects the two cerebral hemispheres – resulting in their extensive disconnection. People who suffer this kind of trauma sometimes experience the Strangelovian phenomenon of the alien hand, in which one hand – usually the left – seems to develop a mind of its own. (A famous moment in Stanley Kubrick's 1964 film because it is so comically ludicrous, the phenomenon is actually real.) The alien hand may unexpectedly throw dangerous objects, systematically undo work being done by the opposite hand, tear up money or paychecks, grope strangers of the opposite sex, even attempt to choke its owner to death, all without its victim having any voluntary control over it. Sometimes the "good" and "bad" hands engage in violent struggles for domination, a sight that must seem unreal to anyone privileged to observe it.[5]

What are we to make of this carnival of neural horrors? The alien hand appears to have some kind of purposeful agenda, but its opposite number doesn't know what the agenda is. Is the alien cortex consciously aware of what it is doing? Since it doesn't speak – language skills usually being lateralized to the other (left) side of the brain – the question seems unanswerable.

But just what is the relationship between consciousness and the storytelling instinct, anyway?

> … [T]here is a possibility that we have two entirely separate brains, rather than one brain divided into two bilaterally equal hemispheres, that, in fact, whereas we have one body we have two minds … (Dick [1976] 1988, 215).

Lesions to the corpus callosum aren't always accidental. Sometimes the nerve tract is surgically severed in an effort to treat otherwise intractable cases of interhemispheric epilepsy. The study of these split-brain patients, which began with Nobel Prize winner Roger Sperry in the 1950s, has continued since then under his student, Michael Gazzaniga.

In this context, Dr. Gazzaniga's most remarkable discovery was of a modular system he calls the left-brain interpreter. The case of a patient called P.S. shows how this was achieved. First, P.S.'s visual field was divided in half by a barrier so that information fed to each eye wouldn't be available to the other, hence could not be processed by the corresponding half-cortex. A series of

---

5   Several of these cases are described in Feinberg 2001, 93-99.

card pictures was then shown to P.S. In one example, a picture of a claw was shown to the right eye – hence to the left hemisphere – since the optic tracts cross to the opposite-side cortices – while a picture of a snow scene was shown to the left. A second pair of cards was then shown to P.S. – that of a chicken and a shovel – this time to both sides of the brain. P.S. was next asked to say which of the cards best matched the card he had previously seen. He pointed to the chicken with his right hand and the shovel with his left, as his disconnection syndrome had led Gazzaniga to expect. The crux of the experiment now came when P.S. was asked to explain *why* he had answered as he had. "Oh, that's easy," he replied. "The chicken goes with the claw and you need a shovel to clean out the chicken shed" (Gazzaniga 1985, 70-72).

P.S.'s half-brain – the only half capable of speech – had been compelled to explain the choice made by the hand controlled by its opposite number. Given that the left cortex had had no access to the snow scene because of the callosotomy, it then *confabulated* an explanation for the right side's choice of the shovel, a story that connected it to the left's choice of the chicken.

Gazzaniga's conclusion, based on this and hundreds of other similar trials with P.S. and other split-brain patients, was that *human beings possess a profound compulsion to explain in story-form why they behave as they do even when such explanations have no connection to consensual reality*, a compulsion that is localized to an interpretive module that operates in the left cerebral cortex. In better integrated brains this fact isn't immediately obvious; in Gazzaniga's callosotomized epileptics it became starkly apparent.

Recently, Pascal Boyer has employed similar arguments from the cognitive sciences to help explain the nature of religious ideation (Boyer 2001, passim but esp. 145-148). The brain's different cognitive modules don't always act in concert – the effect, in part, of its explosive biological evolution. This recurrent internal clash of inference systems is never more evident than in how we think about death: the module that processes data about the activities of *agents* simply doesn't recognize irreversible cessation of activity as an acceptable datum, while the module that treats of entities as biological objects automatically accepts such information. The resulting conflict is resolved in the form of a supernatural belief in the survival of some non-physical aspect of the self – the soul or spirit – after the death of the body.

Another aspect of the "death" problem is shown by the reflexive dread the living have of the dead, in particular of their mortal remains. This seems to make no sense, for nothing would appear to be more harmless than an inanimate lump of disentegrating flesh. Yet numberless cultural narratives exist to explain the meaning of the funerary practices and rituals designed to detoxify the recently dead. Once again, these stories bear no relationship to the underlying reality of the matter as we now understand it, and yet they serve, and have always served, a clearly adaptive function.

Corpse-borne pathogens – cholera, typhus and other bacteria – were always a threat to the survival of our ancestors. However, these same ancestors weren't aware until lately that such things as bacteria, or other microscopic entities, actually exist. Unable to supply them with an adequate scientific explanation of why corpses should be handled with care, natural selection did the next best thing, made us all instinctively averse to handling dead bodies. Our compulsively active interpretive modules then confabulated stories about the malevolent spirits that swirl about the death scene, along with elaborate systems of ritual required to rid ourselves of them, thereby helping us make sense of our otherwise inexplicable emotions.

It is the tendency of the so-called primitive mind to animate its environment. Modern depth psychology has requested us for years to withdraw these anthropomorphic projections from what is actually inanimate reality, to introject – that is, bring back into our own heads – the living quality which we, in ignorance, cast out onto the inert things surrounding us. Such introjection is said to be the mark of true maturity in the individual … But one wonders: has he not also, in the process, reified – that is, made into a thing – other people? Stones and rocks and trees may now be inanimate for him, but what about his friends? (Dick [1972] 1988, 123).

Without doubt the most remarkable – because most consequential – revelation to emerge from Boyer's argument (due to his colleague, Justin Barrett) is that human beings are instinctively inclined to see purposes, designs, or intentions operating in the natural world when in fact none are present. For Boyer and Barrett, the naturalness of agent-like entities such as ghosts, demons, and gods is due to "the fact that our agency detection systems are biased toward over-detection" (Boyer 2001, 145). But why should these projections of mentality seem so characteristically human?

One answer comes out of decision theory, and is in fact much older than humankind. Suppose that a zebra comes to drink at a watering hole. A breeze rustles through the nearby grass, causing him to bolt in fear that it may conceal a hungry lion. When he returns half an hour later, the same thing happens. Perhaps similar false alarms play themselves out a dozen times during the course of a hot afternoon on the savannah. On the thirteenth occasion the zebra, grown tired and thirsty, ignores the warning. Waiting in ambush, a lion strikes and slays him.

The point of this evolutionary vignette is that the genetic cost of false positives – imagining lethal threats when none exist – is generally far less than the cost of false negatives, that of ignoring lethal threats that are real. Moreover, the most dangerous threats come from living agents: predators. The interpretive imaginations of highly encephalized prey animals – hominids, humans – work feverishly in this kind of environment, conceiving such notions as that of the surrogate sacrifice to a hidden deity, the Deus Absconditus. As René Girard ([1972] 1984) reminds us, the sacred is always born in violence.

When the utility of an event becomes decoupled from its probability, what is rational to believe may no longer be true, and what is true no longer rational to believe. The stories we confabulate may serve our biological purposes even when they're objectively false or illogical. Unfortunately, the pragmatic rationality of such confabulations is often compromised by the fact that the value they possessed within the natural context in which we first evolved may no longer exist in a world constructed of technological artifice and simulacrum. It no longer makes sense to fear spirits that haunt the graveyard; another kind of spirit – embalming fluid – has taken care of the problem. Lions no longer threaten us; AIDS does, ozone holes do. No reflective person can watch television news without being disconcerted by its obsession with celebrity gossip at the expense of fundamentally important issues such as global warming or nuclear terrorism. The point is that such macro issues are evolutionary johnny-come-latelies that have no purchase upon our instincts, whereas gossip must have played a critical role in establishing the status (hence breeding potential) of individuals living within the small hunter-gatherer groups in which we spent our first five million years.

No doubt Joseph Stalin spoke the truth when he cynically observed that the death of one person is a tragedy, that of a million only a statistic. Cro-Magnon man, alas, never had to deal with large numbers. This vertiginous dislocation between postmodern realities and our Paleolithic perceptions of them is an obvious invitation to disaster. What's worse, when we do attend to these new, unfamiliar realities the usual response is to fall into some kind of collective paranoid delusion because of our evolutionary inexperience in dealing with problems of extreme complexity. But how did this impenetrably mysterious state of being ever arise in the first place?

Perhaps something of basic importance has been overlooked.

Already in the 1940s Karl Popper observed – contrary to the totalizing theories then (as now) extant – that history has no discernible meaning or pattern, because the future is radically contingent.[6] His argument has never been answered because it is effectively unanswerable. What Popper said was this: the human future will be as it has always been, dominated by technological changes. However, we can't know what effects future technologies will bring, because if we knew them now we'd already possess those technologies now, a transparent *reductio ad absurdum*.

Who are the agents of technological change? Can any empirically relevant facts about this group of people, perhaps detailing their psychological and sociological traits, help illuminate the nature of our era? In particular can they help explain the seeming futility, the sheer helplessness of our storytelling instincts in the teeth of a refractory postmodernity?

A strange question. Perhaps it requires an equally strange answer.

A human being without the proper empathy or feeling is the same as an android built so as to lack it ... (Dick [1976] 1988, 201-202).

The British psychologist Simon Baron-Cohen and his colleagues have made a series of recent discoveries that are remarkable in their import. In one study of a thousand families it was found that the fathers and grandfathers of children suffering from autism or Asperger's syndrome were more than twice as likely to have worked as engineers when compared to control groups. In another study at Cambridge University, Baron-Cohen's group learned that students in the physics, engineering, and mathematics departments suffered six times the rate of autism in their immediate families as did students enrolled in humanities programs such as English or French literature, but did not show such elevated rates for other psychiatric conditions such as schizophrenia or manic-depression (Baron-Cohen et al, 2000, and Baron-Cohen 2000, 77). More recently, when Baron-Cohen ran an AQ (Autism-Spectrum Quotient) test on several hundred experimental subjects, mathematicians scored highest for autistic traits among all occupational groups, followed closely by engineers, computer scientists, and physicists (Else 2001, 42). Close readings of the biographies of many leading physicists and mathematicians reveal hosts of traits diagnostic of Asperger's syndrome, high-end autism or what has been called the "shadow syndrome" of autism. These individuals include Paul Dirac, the great quantum physicist; Paul Erdös, the most prolific mathematician of the 20th century; and Alan Turing, whose formal model of computation, the Turing machine, is the cornerstone of modern computer science (Pais et al 1998, Schechter 1998, Hodges 1992, Sebag-Montefiore 2000).[7] Since the single most glaring aspect of autism and its variants is the limited ability to understand other people's thoughts or emotions – a deficit in the capacity for empathy, identified by theorists in the field as the absence of an effective "theory of mind module" (TOMM) – it is not accidental that technology schools such as the Massachusetts Institute of Technology have begun offering their students courses in elementary social skills, so-called "charm schools" (Ratey and Johnson 1997, 219).[8]

---

6   We are referring, of course, to Karl Popper's *The poverty of historicism* (1977).

7   See Wiener 1953, Schechter 1998, Hodges 1992. Sebag-Montefiore (2000) claims: "One of [Turing's] closest associates suggested that if examined today, he might have been diagnosed to be suffering from a mild form of autism. Perhaps it was Asperger syndrome, otherwise known as high high-grade autism" (96).

8   The expression "shadow of autism" is due to these authors.

The incapacity to represent within one's own consciousness the intentional states of other minds in a normal way may have consequences that go far beyond the social ineptitude of eccentric physics professors or computer science "nerds." Recalling the Boyer and Bennett thesis discussed earlier, it would seem that, unlike the majority of human beings, such individuals lack agency detection systems that are biased toward over-detection.

This important insight inevitably raises questions about the psychological basis for conflicting positions taken by people with respect to teleological (or design) arguments for the existence of god, for example. In a recent survey that asked Americans the reason for their own theistic beliefs, the largest number in fact responded with some version of the design argument (Shermer 2000, 84). It is known that well upward of 90 percent of members of the American general public believe in god. Revealingly, a recent survey of natural scientists and mathematicians showed the contrary result, a far higher rate of incidence of agnosticism and atheism than in the public at large, with the highest figure of all – again exceeding 90% percent – found within the most gifted and celebrated subset of this group, members of the National Academy of Sciences (Larson and Witham 1999, 88). It would seem that some people are animists by inclination, others mechanists: those who see minds at work *within* things are likely to find little common ground for discussion with those who see minds *as* things – the source of a perennial rift within our culture that has resisted endless efforts at amelioration. The rancorous and historically critical debate over the onto-theological status of embryonic stem cells is only the most recent example of this vexing problem.

Such a surprising revelation barely scratches the surface of a very complex phenomenon. Overgeneralizations and arguments from weak analogies always pose a danger in such situations, but so does overcaution and the conservative desire to maintain a quotidian view of reality. Perhaps the problem is best initially approached through a point-by-point comparison of key aspects of autistic cognition with the epistemic features of scientific practice.

Agency and teleology. Scientific practice abjures any appeal to explanations in terms of purposes or Aristotelian final causes, rather than to efficient causes and probabilities. This axiomatic premise is contested by intelligent design (ID) theorists and other religiously motivated opponents of the Darwinian theory of natural selection, for example. Irrespective of the epistemic merits of the scientific position, it is clearly consistent with the deflation of notions of agency or animistic projection one would expect from people who live in the shadow of autism.

Joint attention and the principle of consensibility. All autists are known to suffer from a deficit in their capacity to establish joint-attention – that is, to attend to the same perceptual stimuli simultaneously with other persons. At the limit this is obviously a debilitating problem because it must make learning even the simplest social skills impossible. Among those who live in the higher reaches of the autistic spectrum it is easy to see how this problem could raise to the level of consciousness the need to *formalize* the idea of attending jointly to the same object at the same time or under the same conditions. But this is simply the principle of consensible observation (Ziman 1978, 42) or, at another level, of reproducible results. In its epistemic fundament, scientific reality is exactly that set of intersubjective agreements which occur in the absence of influence.[9]

---

9   For example, Kepler and Ursinus each drew pictures of what they had seen through a telescope while looking at the moons of Jupiter, while the other was physically absent, and afterward compared notes. See Caspar 1993, 197.

By contrast, most people seldom question whether their perceptions are veridical, since in most instances the social consensus works automatically to reassure them. An astonishing neurological discovery may help explain this discrepancy. Researchers at the University of Parma (Motluk 2001, 22) have recently investigated a class of neurons that have the remarkable quality of firing whenever their owner performs a particular task, but also whenever he sees someone *else* performing it. These "mirror neurons" may play a crucial role in helping us understand the intentions of other people by fusing the connection between volition and social perception, while on the other hand their absence could offer a natural explanation for the failure of the theory of mind module to develop normally in autism. The autist-scientist must then follow a strictly external route, through the sensoria, to the goal of social agreement.

If Alan Turing did in fact live in the shadow of autism, it would place a poignant new twist on the meaning of the famous Turing test for artificial intelligence, since the external simulation of other people's behavior is the best that autists can themselves hope to achieve, given that they must forever lack any direct insight into their neighbor's thoughts.

Language and literalism. Autists learn language late, if at all. Even Asperger types, who do much better, tend to speak in hyperliteral monotones. Ordinary language is densely figurative, polysemous, and contextual; the referential boundaries of our terms are often fuzzy, based on family resemblances or weakly transitive chains of analogy, or on radial divergences from prototypes along any one of a multiplicity of pathways (see Lakoff 1987). Every one of these features strongly depends upon our ability to interact and communicate with other human beings in complex and subtle ways, implying that we have some sort of grasp of their internal states.

Autists find all of this a torment, a sort of interminable chaos of deconstruction. The reaction is to insist that black is black and white is white. Similarly, scientific language aspires to be starkly literal, its referential boundaries established through formally defined necessary and sufficient conditions. At the limit this means mathematicization, one point being to eliminate contextual ambiguities and thereby make hypotheses empirically testable. Another objective of such literalism is to reduce the appeal to rhetorical influence rather than to independent, consensible agreements. Still another is to reduce the influence of tacit, unconscious knowledge as opposed to the explicit, conscious variety. The autist-scientist demands that all the chips be put on the table, in plain view. Literalism, of course, is the death of literature: even at this shallow level of analysis it seems clearer why contemporary cyberculture, whose entire infrastructure consists of a set of scientific and technological artifacts and algorithms, is so refractory of narrative explanations. But there are much deeper problems.

Truth and deception. Human intelligence is Machiavellian. Anthropologists have long argued that the art of lying skillfully is critical for advancing one's status within human dominance hierarchies, though the same lesson can be learned by attending to any modern political campaign or showcase courtroom trial. The adroit use of language obviously enchances this ability. The deceits of language, when combined with the entailed need to produce sophisticated decryption mechanisms and cheater-detectors (Cosmides and Tooby 1992), may in fact have generated the positive feedback loop that drove the rapid evolution of the human brain.

Autists are unable to lie, even when their verbal skills are reasonably well-developed. In order to lie one must be able to represent to oneself the likely reaction – hence the internal mental state – of the person at whom the lie is directed, for the manipulation of his beliefs is the whole point of the deception.

Scientists think of themselves as professional truth-seekers. Hence their baffled anger when outsiders suggest that their motives are as potentially duplicitous as anyone else's or when some atypical insider actually falsifies data to further his own ends. At the same time it cannot be denied that the technological, hence commercial success of modern science has recently brought more of these legal and political types into the field because of the corrupting influence of money.

Specialization and the obsession with detail. Autists are notorious collectors – whether of train schedules, bus transfers, or obscure birthdates. Scientists are compilers of data about beetle species, stellar parallaxes, and prime numbers. None of this material has any intrinsic social value or meaning, though it can sometimes be put to utilitarian use – usually after a temporal lag – through technological development. The psychology of this collection process is both eccentric and obsessive. The focus of attention is dramatically narrowed, with a corresponding increase in intensity. This is the essence of specialization, one of the defining characteristics of our era, decried by contrarians from Blake to Ortega y Gassett.

Decontextualization and universal knowledge. Autistic knowledge is profoundly decontextualized, a fact implied by its asocial nature and origin. A recent series of experiments conducted at Ohio State University (B. B. 2000, 72), for example, demonstrated that autists are far superior to others at remembering words occurring on random lists. However, when words on one list evoke recollections of those that occur on a second list because of contextually linked meanings, the situation is reversed, with non-autists outperforming autists.

Autists frequently excel at tasks such as the assembly of jigsaw puzzles; unlike non-autists they are neither helped nor hindered by the nature of the picture on the puzzle, whether of the *Mona Lisa* or Jackson Pollock's *Lavender Mist*. Rather, they employ some universally applicable curve-matching algorithm that doesn't depend on any holistic or gestalt characteristic of ordinary perception.

Scientific knowledge presumes to be universal or non-contextual in just this way, once again the source of protest from intellectuals who themselves operate within a radically different cognitive paradigm. In an intriguing anthropological study of computer scientists working in the field of artificial intelligence, for example, Diane E. Forsythe has made the following observation:

> The knowledge engineers seem to think of knowledge as universal, that is, they see it as consisting of 'rules' that will apply across contexts … Given this assumption about the universal nature of knowledge, information that is clearly situational in nature or is only locally applicable may not strike them as knowledge at all (Forsythe 2001, 53).[10]

Invariance and regularity. This feature is intimately related to the previous one. Autists are often put into a panic by any alteration or derangement of their familiar environment. In fact, a simple diagnostic test for autism has recently been designed by researchers at York University in Toronto that is based on the atypical responses of autists to regularity and novelty (Rodier 2000, 56). Children – autists and subjects from a normal control group – are seated in front of a bank of three video screens. A flashing light appears on the middle screen. Both groups attend to it. This light switches off and the right screen begins to flash. Again both groups attend to it, as indicated by eye shifts.

---

10 Forsythe's book is a goldmine of revelations on our subject. "Speaking of the knowledge engineer on a joint project, one expert commented: 'Engineers by and large have dreadful communication skills. One can succeed in technical fields without these skills'" (45).

The sequence is repeated. The middle screen is turned on. This time, however, it remains on while the right screen also begins to flash. The control subjects shift their attention to the new stimulus. The autists remain firmly fixed upon the old one, which is to say upon that element of the situation that does not change. The problem is that autists are too easily overwhelmed by things that *do* change, because they can't make any sense of them. The relationship between this kind of "executive dysfunction" and inadequacy in the theory of mind is still a subject of debate. (See relevant essays in Baron-Cohen, Tager-Flusberg and Cohen 2000.)

One is almost tempted to see in this the reprise of an ancient discussion between the pre-Socratic philosophers Heraclitus and Parmenides. Autists, it would seem, are obligate Parmenideans, forever seeking the nut of the eternal within the shell of the temporal. In his last book, Karl Popper (1998) theorized that the philosophical origins of all scientific thought lay in Parmenidean consciousness. Invariants – natural laws – are after all only the mathematical expressions of timeless relationships that lie within the superficial flux of phenomena. Thus *all* right triangles in the Euclidean plane are subject to the theorem of Pythagoras, irrespective of their dimensions; *all* pairs of points in the orbit of Mars are connected to the sun by imaginary chords that sweep out equal areas in equal times (Kepler).

"We physicists know that time is unreal," said Einstein to the grieving widow of his friend Michele Besso; but Einstein was among the most autistic of the great minds of the bygone century.[11] Autism chants the mantra of the timeless, a consummation devoutly to be wished. From this seed, however, has sprung an arborizing rainforest of historical ironies, among them the whole of cyberculture.

"There is a new theory about autism," Dr. Glaub said … "I wished to discuss it with you, because it seems to offer us a new avenue with your son, here."

"I doubt it," Steiner said.

Dr. Glaub did not seem to hear him, he continued, "It assumes a derangement in the sense of time in the autistic individual, so that the environment around him is so accelerated that he cannot cope with it …" (Dick 1964, 41).

Holism and reductionism. The way in which autists perceive the relationship between wholes and their parts is strikingly unusual. Many autistic children are fascinated with Lego blocks, as they were in earlier generations by Tinker Toys and Erector Sets. The analytical drive to reduce complex objects to their component parts, then to reassemble those wholes in modular, hierarchical stages, is manifestly autistic; it is also the mark of the professional engineer or computer programmer.

Most of the time most of us think holistically, in terms of complex gestalts and the inferences from them to special applications, but perhaps "think" is the wrong word here since so much of this process works unconsciously, within the vast parallel networks of the neocortex. Autists, by contrast, are obligate reductionists, favoring parts over wholes while being forced to reconstruct the latter from the former in conscious, explicit, rule-bound stages. There is nothing fluid or

---

11 Of the physicist's childhood, Folsing (1997) remarks that "Einstein's speech development was strikingly slow," in fact not occurring until he was at least three and possibly as old as five, a pattern also seen in Edward Teller and Richard Feynman. The great quantum physicist Paul Dirac was notorious for rarely speaking even as an adult, and then only in the most literal of terms. Folsing continues, describing young Einstein's "… laborious and self-critical acquisition of language, in contrast to most children's natural, unproblematical learning," further claiming that he was, "averse to play and social behavior appropriate to his age, and moreover with an occasional total lack of self-control" (11-12).

automatic in this process. Cognitive psychologists refer to this syndrome as "weak central coherence."[12] The situation is neatly illustrated by the performance of autists on block design tests, in which, according to Beatte Hermelin: "Hidden shapes in a drawing that are overlaid by a larger, structured design are more easily detected by those with autism that by those without" (Hermelin 2001, 47).

Traditional scientific practices are inherently reductive in the same way. In experimental design theory, for example, it is axiomatic that the first step in trying to establish the plausibility of a hypothesis is to alter the relevant variables one at a time and watch what happens. First, however, one must locate and isolate the relevant variables. Sometimes "parts" can be rather abstract – as in the case of the N components of a vector in N-dimensional space, or the nested subroutines of a complicated computer program – but the point remains substantially the same, which is a child's lesson in the assembly of Lego blocks. Traditional science is reductive, trading in well-defined, decomposable syntactic forms and structures; life is holistic, fluid, indecomposable, a fathomless texture of emotion and meaning. The cognitive style of traditional science is autistic in root and branch.

> Our world is a deluding projection by an artifact that does not even know it is an artifact, or what its purpose in projecting our world is (Dick [1978] 1995, 303).

> Time is a child moving counters in a game … (Heraclitus, Fragment 24).

A paradox: the holistic mind is scientifically sterile; the reductive mind, fertile. Everything is a matter of the combinatorics of building blocks, but perhaps this formulation is a bit opaque. The mystery is best understood in terms of an analogy between language and its autistic other, yet another paradox.

Language is hierarchical: we proceed from phonemes, morphemes, words, phrases, sentences to paragraphs, chapters and whole books. When Camus discusses the meaning of Kirilov or Ivan Karamazov in *The Rebel* he is thinking at a very high level, one would say in terms of the holistic sense of entire novels, or even of the intertextual relationships between them. If one were to rearrange physically the paragraphs and sentences of "The Grand Inquisitor," however, its sense would disappear; one would have nothing left, certainly not the literary genesis of existentialism. In this sense literary wholes are finitely bounded. One can only begin to appreciate the infinitude of literature by going lower, by reducing language to its constituent elements, to its very atoms, and then recombining them into other, unfamiliar new structures: Queneau's *Exercises in Style*, Kafka's *The Castle*, Roussel's *Locus Solis*.

Newton and Leibniz were both intrigued by the works of the 13th century Spanish priest Ramon Lull, whose system of the *Ars Magna* atomized and reconstituted all meanings through a combinatorial process (see Eco 1995, 271). In a sense, Lull was the distant progenitor of both modern particle physics and modern computer language theory. Much later, in his story, "The Library of Babel," Jorge Luis Borges (1998) would create the perfect metaphor for this Lullian consciousness, which shows us the nightmare of the autistic mind but also its infinite reach. At this point our metaphor begins to fail, because the relevant building blocks are no longer words and sentences but primitive geometric, arithmetic, and logical entities – the constituents of form but not of meaning. And so we have Albert Einstein's autistic testament:

---

12 Discussed in Hermelin 2001, 45-48. The term itself is due to Uta Frith.

The words of the language, as they are written or spoken, do not seem to play any role in my mechanism of thought ... Conventional words or other signs have to be sought for laboriously only in a secondary stage (Hadamard 1945, 142).

For *Time* magazine's Man of the Century, there was nothing inside the text.

Perhaps, unexpectedly, we've arrived at the crux of the matter. Postmodernity is infinite, unspeaking and unspeakable, an autistic child's vast Legoland. The old narratives can't find their footing on this strange new landscape. The problem is that the content of this landscape has become a matter of form rather than meaning because of its complete domination by the technological infrastructure, but more importantly because that infrastructure has become so complex, yet so occult in its complexity that our narrative instincts have become disoriented by it, able only to confabulate and delude. Without in any way rendering this mystery transparent, by explicating three generative principles we can at least begin to parse the differences between postmodern reality and everything that came before it.

The role of autistic reduction in enabling the growth of complexity has already been described; this is a relatively straightforward matter of combinatorial logic that shows how it is that many more types of things, and more complex things, can be built from simple building blocks than from complex ones. By way of comparison, if a writer were compelled to work only with pre-existent sentences (or worse, paragraphs), he'd be paralyzed. In failing to see the trees for the forest, holistic perception is in this sense uncreative. Of course this sense of the concept of reduction is potentially misleading because the states of most complex systems aren't themselves decomposable into independent parts, but are rather the emergent expressions of highly non-linear relationships and interactions. The question for the cognitive scientist who wishes to know how it is that mathematicians understand such systems then becomes: what sorts of entities interact to generate sequences of such states? In cellular automata, for example, these discrete entities are the explicit initial states and sets of transition rules that generate the future history of the system. In irreversible automata this "seed" cannot be recovered by time-reversal and is therefore not recognizable in the "tree" that sprouts from it. Rather, we begin with the seed and develop the history of the system from it; this is the synthetic, as opposed to the analytic side of autistic consciousness, the work of the autist-engineer rather than the autist-scientist.

The second principle is that of precision. At its lowest level the autistic fixation upon unvarying regularities translates to an obsession with precision, with the ability to fabricate objects and procedures – artifacts and algorithms – that are as measurably indistinguishable from each other as the human mind can make them.

The central idea here is that precision is anti-entropic, a means of resisting friction and other encroaching forms of disorder, including chaos, in whose domain small differences between causes lead to large and unpredictable differences between effects. If the parts of an automobile weren't precisely machine-tooled, its valves would stick, its pumps leak, its gears clash – in short, it would not operate. As systems become larger and more complex their component parts must become correspondingly more precisely engineered because the occasions for failure in such systems tend to increase exponentially with increasing complexity. The same rule applies to rules – that is, to algorithms or procedures – as to physical artifacts. Precision enables very complex, highly connected and coherent systems to exist. So does redundancy, the provision of backup systems and components to reduce the risk of system failures. Nonetheless, these identitarian biases of autistic-technical cognition have their dark side. For example, by eliminating sources of variation in the pursuit of optimality they simultaneously eliminate the ability to develop immunity to alien intrusions – whether by accident or design – into one's system, a feature that distinguishes nearly all complex technologies from nearly all living organisms. Autistic "narrowness" also plays a key role here because the functions of most technical systems are very narrowly defined; such systems are inherently brittle and susceptible to disintegrative collapses when their operating environments change.

Specialization also plays a crucial role in the evolution and development of complex systems. The narrowness and obsessiveness of autistic attention are well exploited in this context, since the division of labor made possible through the distributed intelligence of many different skill-types – of different savant domains – is obviously critical. Further, there is an economic advantage implicit in such divisions, first discovered by British economist David Ricardo and familiar to all economics students as the Principle of Comparative Advantage – the basis of international trade, hence of globalization. Of course natural selection discovered this principle of connection first, in the form of symbiosis. The cells of all higher organisms – including those of human beings – are in fact symbionts, chimerae created by the fusion of different organisms. (This is shown, for example, by the fact that mitochondrial DNA is genetically quite different than nuclear DNA.) Nature, however, works the bugs out of its systems in a manner not available to postmodern technologies, which are prone to make saltationist leaps that aren't responsive to any artificial analogs of natural selection.

There are great dangers lurking here, for though the world has become fathomlessly complex we continue to presume to understand it through dressed up versions of hoary myths that can only lead us down the primrose path. This response is understandable: history frightens us.[13] The mind reflexively recoils from the disturbing vision of radical historical contingency; from the inky opacity of the future. For instance, few people gave a thought to the Y2K problem – a seeming triviality based on the number of digits in a date – 30 years ago, when the problem was being hatched. A trillion dollars later we've managed to dodge that particular bullet. There have been other pieces of good fortune, still more unnerving yet seldom spoken of. At a recent conference of European climatologists, for example, one speaker described the results of a study done on the atmospheric effects of the release of refrigerant gases since the early 20th century, leading to the so-called ozone hole problem.[14] His conclusion was that if bromine had been substituted for chlorine in CFCs in the 1920s and 30s – a decision that turned on relatively minor scientific factors – the ozone holes we face today would have been far more extensive and long-lasting than they are, posing a major threat to life on earth. Since the technology needed to report the stratospheric effects of these substances wasn't available until several decades after they themselves were, we could easily have sealed our fate as a species by mid-century without ever having fathomed the meaning of what we'd done. No doubt we would have taken such a catastrophe to be a latter-day Biblical curse upon humanity: in other words, told ourselves a story about the tragedy that would not have made contact with its real explanation at a single point.

The mind recoils. The theory of complex systems, though unable to lift the veil of the future and show what lies ahead, may nonetheless give us some insight into the terata created by the old science and thereby serve as an instrument of warning.

---

13 This point has been eloquently made in by Mircea Eliade (1959), *Cosmos and history: The myth of the eternal return.*

14 *Los Angeles Times*, July 13, 2001.

One sense of the term "complexity" – that of self-organized criticality – has been investigated by the physicist Per Bak of the Santa Fe Institute. Bak's model, if it correctly applies to mass social interactions, may signal the epitaph of every totalizing theory and grand history that's been penned from the time of G. W. F. Hegel's *Philosophy of History* (1837) to Francis Fukuyama's *The End of History and the Last Man* (1992). On the basis of empirical data collected on stock market activity, military conflicts, and other historical events, it has slowly become evident why past efforts to perceive cyclical or progressive patterns, or for that matter any other kind of intelligible pattern in history, were foredoomed. Rather, it is likely that historical events tend to follow a ubiquitous set of power laws – that is, statistical correlations which show the frequency of occurrence of an event of a defined type to be an exponential function of its size under some defined measure. What this means, unfortunately, is that history more closely resembles a series of avalanches or earthquakes than it does a rising tide or set of waves rolling onto a beach (Iredale 2001, 29).[15] The disasters come in all sizes, and are unpredictable. Gavrilo Princip is always waiting for Franz Ferdinand's car to make a wrong turn on a side street, ready to initiate the chain reaction that leads to a world war.

Complexity theorists Brian Goodwin and Robert Sole have explained how these extinction threats are greatly exacerbated by the kind of hyperconnectivity that characterizes postmodern systems of communication, production, and exchange (Sole and Goodwin 2000, 301-302). Citing research in systems theory, cybernetics, and information theory begun in the 1970s, Sole and Goodwin suggest that complex societies may tend to evolve into a state of hypercoherency such that a failure in one part of the system can create a cascade of effects that topple subsystems downstream from it like so many falling dominos, finally leading to a general civilizational collapse. Ecologist and theoretical biologist Simon Levin underscores this threat when he claims that the trend toward globalization has led to a hyperconnection of world financial systems in which small economic ripples can lead to system-wide crashes, as well as to the reemergence of worldwide disease pandemics such as AIDS due to the ease and speed with which we can travel to nearly any place on earth (Levin 1999, 202). If there is one caveat to Karl Popper's thesis about future history, it is that technologically distinct types of complex systems may come to grief in the same way because they share the same connection topologies and network dynamics. For example, recent work by Albert Barabasi and others shows that power law networks occur ubiquitously in both natural systems (genomic regulatory networks) and artificial ones (the World Wide Web), and are especially vulnerable to deliberate attacks because of the critical importance of rare but highly connected "hub" nodes. This fact invites the possibility of a powerful new generation of anti-cancer drugs, but also of a devastating kind of cyber-terrorism.[16]

System failures can no longer be contained or quarantined: modular barriers and boundaries – the natural bulkheads that can keep the social ark afloat should it begin to take on water – have disappeared. Complexity theorist Duncan Watts adds a sinister twist to the argument by showing how human cooperation begins to deteriorate when the ratio of the number of human interactions to the number of interactors approaches unity in such systems (Watts 1999, 211). In Watts's view, in traditional societies recurrent relationships with a handful of familiar faces lead to mutually beneficial, reciprocal exchanges, as well as to the certain identification and punishment of defectors and malefactors. By contrast, one-time interactions with multitudes of strangers only

---

15 Iredale 2001, 29. The full treatment is given in Bak 1996.

16 Cohen, D. 2002. All the world's a net. *New Scientist* 174 (13 April): 24.

encourage cheating and exploitation – or worse, unidentifiable acts of physical, biological, and cybernetic terrorism (e.g., hacking, spreading computer viruses via the World Wide Web).

These creatures are among us … (Dick [1976] 1988, 201).

The hyperconnected domain of cyberculture, in short, is the natural dwelling place of the psychopath; its folk hero is Hannibal Lecter. Anarchy inevitably gives rise to tyranny, as Dostoevsky had already observed in the 19th century. We can even foresee the shape this tyranny may take in our own future should the inevitable system crash occurs: that of Foucault's universal, panoptical surveillance and security state (Foucault 1979). The harbingers are already here, with hidden cameras on every street corner in some American cities. Meanwhile, the US National Security Agency's advanced ECHELON system monitors millions of telephone calls surreptitiously, looking for suspicious keywords. All that is missing are AI systems sophisticated enough – in other words, that have scored well enough on the Turing test – to actually find the sense in all of this noise. Of course, when that happens, the stories these systems tell themselves are likely to be no more than paranoid confabulations based on their bias toward overdetecting agency. Perhaps one day they even will have their own cybernetic religions.

All of this complexity seems doomed by the autistic narrowness of its focus, by its vulnerability to the law of unintended consequences. Air conditioners don't only cool our houses and businesses; they also increase the likelihood that we'll contract skin cancers from overexposure to UV radiation because of the impact of CFCs on the ozone layer. The World Wide Web not only helps us obtain otherwise inaccessible information and conduct our affairs with once unimaginable speed and efficiency, but also threatens to transmit electronic analogs of the black plague, ruining businesses, costing billions, and disrupting the infrastructures of nations.

There simply is no organic "whole" to be found anywhere in all of this complexity. It is a pseudo-organism that possesses no immune system, no way to maintain homeostatic equilibrium. Why? Because it never evolved; it was created. As Stuart Kauffman has pointed out in a series of insightful books on the nature of complex systems, living organisms and their environments coevolve (Kauffman 1995, chap. 10, and Kauffman 2001, chap. 8, esp. 194-207). Modern organisms are *able* to evolve because ancient organisms altered their environments in such a way as to make them evolution-friendly. Of course there was no plan in any of this; countless evolutionary cul-de-sacs were explored, countless creatures died as this series of increasingly more benign environments was manufactured through the alchemy of natural selection and coevolution.

Unfortunately, it is likely that nothing like this is happening within the vast expanse of the postmodern cyberculture. Though Kauffman's rhetoric is relatively optimistic about the emergent properties of this system, his mathematics often speaks otherwise. Can one in fact speak of the adaptations occurring within a kaleidoscope? On the other hand, what disasters can we expect when such tightly coupled, highly optimized systems are even slightly perturbed, when the only direction to be found on the adaptive landscape is down? Individual and collective rationality all too often find themselves at war with each other in such environments. Nor is anything good likely to happen in the long run when the genetic bottleneck is so narrow, when the hyperconnection of all things to each other guarantees that the human future will have only one founding population. And then, if our luck finally fails, none.

The light became so bright that I could make out every detail around me. The slumbering woman, the little boy, the dozing cat – they seemed etched or painted, unable to move, pitilessly revealed by the light. And in addition something looked down at us as we lay as if on a purely two-dimensional surface; something which traveled and made use of three dimensions studied us creatures limited to two. There was no place to hide; the light, the pitiless gaze, were everywhere. We are being judged, I realized. The light has come on without warning to expose us, and now the judge examines each of us. What will his decision be? The sense of death, my own death, was profound; I felt as if I were inanimate, made of wood, a carved and painted toy ... we were all carved toys to the judge who gazed down at us, and he could lift any – and all – of us off our painted surface whenever he wished. (Dick [1976] 1985, 178-79).

Even when one warns against the dangers of storytelling, the warning somehow comes in the form of a story – in this case a latter-day version of the myth of Icarus. History is racing asymptotically toward an unknown limit referred to elsewhere in this volume as "The Spike." Inevitably, this realization evokes thoughts of apocalypse, stoking our fears of some final judgement, whether by deity or cold equation.

Harold Bloom says that all readings are misreadings. And so – quite consciously – we have misread Philip K. Dick, whose spirit nonetheless moves through the pages of this essay like a shuttle through a loom. Dick was neither a Marxist prophet, nor a postmodern polemicist as he is sometimes portrayed, but rather a gnostic mystic who very much believed that there is a story to be told about the world, though one that is both obscure and quite utterly strange. For in the interminability of Philip Dick's confabulations to explain the strange things that had happened to him in 1974, a fragment of which is titled "Man, Android and Machine," we can see a trope or figure of cyberreality, whose story, like his, eludes any final meaning.

Dedicated to the memory of Philip K. Dick

REFERENCES

Bak, P. 1996. *How nature works: The science of self-organized criticality*. New York: Springer-Verlag.

Barkow, J., L. Cosmides, and J. Tooby (eds.). 1992. *The adapted mind: Evolutionary psychology and the generation of culture*. New York: Oxford University Press.

Baron-Cohen, S. 1995. *Mindblindness: An essay on autism and theory of mind*. Cambridge, MA: MIT Press. 2000.
    Autism: Deficits in folk psychology exist alongside superiority in folk physics. In Baron-Cohen, Tager-Flusberg, and Cohen 2000.

Baron-Cohen, S., H. Tager-Flusberg, and D. J. Cohen (eds.). 2000. *Understanding other minds: Perspectives from developmental cognitive neuroscience*. New York: Oxford University Press.

B. B. 2000. When autism aids memory. *Science News* 158:72.

Borges. J. L. 1998. *Collected Fictions*. Trans. A. Hurley. New York: Viking.

Boyer, P. 2001. *Religion explained: The evolutionary origins of religious thought*. New York: Basic Books.

Caspar, M. 1993. *Kepler*. New York: Dover.

Cosmides, L. and J. Tooby. 1992. *Cognitive adaptations for social exchange*. In Barkow, Cosmides, and Tooby 1992.

Davison, R. 1997. *Camus: The challenge of Dostoevsky*. Devon U.K.: University of Exeter Press.

Dick, P. K. 1964. *Martian time-slip*. New York: Ballantine.

———— 1965. *The three stigmata of Palmer Eldritch*. Garden City, New York: Doubleday & Company.

———— 1972 [1988]. *The android and the human*. In Dick 1988.

———— 1976 [1988]. *Man, android and machine*. In Dick 1988.

———— 1978 [1995]. *Cosmology and cosmogeny*. In Sutin 1995.

———— 1985. *Radio free Albemuth*. New York: Arbor House.

———— 1988. *The dark-haired girl*. Willimatic, CT: Mark V. Ziesing.

Donald, M. 2001. *A mind so rare: The evolution of human consciousness*. New York: W. W. Norton.

Eco, U. 1995. *The search for the perfect language*. Cambridge, MA: Blackwell.

Eliade, M. 1959. *Cosmos and history: The myth of the eternal return*. New York: Harper and Row.

Else, L. 2000. Interview with Simon Baron-Cohen. *New Scientist*, 170 (14 April): 42:42.

Feinberg, T. 2001. *Altered egos: How the brain creates the self*. New York: Oxford University Press.

Folsing, A. 1997. *Albert Einstein*. New York: Viking.

Forsythe, D. 2001. *Studying those who study us: An anthropologist in the world of artificial intelligence*. Stanford: Stanford University Press.

Foucault, M. 1979. *Discipline and punish*. Trans. Alan Sheridan. New York: Random House.

Girard, R. [1972] 1984. *Violence and the sacred*. Trans. Patrick Gregory. Baltimore: Johns Hopkins University Press.

Hadamard, J. 1945. *The psychology of invention in the mathematical field*. Princeton: Princeton University Press.

Hermelin, B. 2001. *Bright splinters of the mind: A personal story of research with autistic savants*. Philadelphia: Jessica Kingsley.

Hodges, A. 1992. *Alan Turing: The enigma*. London: Vintage.

Gazzaniga, M. 1985. *The social brain: Discovering the networks of the mind*. New York: Basic Books.

Iredale, M. 2001. Sci-phi: Adventures in science and philosophy. *The Philosopher's Magazine* (summer): 42.

Kauffman, S. 1995. *At home in the universe: The search for the laws of self-organization and complexity*. New York: Oxford University Press.

———— 2001. *Investigations*. New York: Oxford University Press.

Kosko, B. 1999. *Fuzzy future: From society and science to heaven in a chip*. New York: Harmony Books.

Lakoff, G. 1987. *Women, fire and dangerous things: What categories reveal about the mind*. Chicago: University of Chicago Press.

Larson, E. and L. Witham. 1999. Scientists and religion in America. *Scientific American* 281(3):88.

Levin, S. 1999. *Fragile dominion: Complexity and the commons*. Reading, MA: Perseus Books.

Libet, B. 1993. *Neurophysiology of consciousness*. Boston: Birkhauser.

Lyotard, J.-F. 1984. *The postmodern condition: A report on knowledge*. Trans. G. Bennington and B. Massumi. Minneapolis: University of Minnesota Press.

Motluk, Alison. 2001. Read my mind. *New Scientist* 169 (27 January): 22.

Nicholls, Peter (ed.). 1976. *Science fiction at large*. London: Gollancz.

Pais, Abraham et al. 1988. *Paul Dirac: The man and his work*. New York: Cambridge University Press.

Popper, K. 1977. *The poverty of historicism*. New York: Harper and Row.

———— 1998. *The world of Parmenides: Essays on the pre-Socratic enlightenment*. New York: Routledge.

Ratey, J. and C. Johnson. 1997. *Shadow syndromes*. New York: Pantheon Books.

Rodier, P. 2000. The early origins of autism. *Scientific American*, 282(2):56.

Schechter, B. 1998. *My brain is open: The mathematical journeys of Paul Erdös*. New York: Simon and Schuster.

Sebag-Montefiore, H. 2000. *Enigma: The battle for the code*. New York: John Wiley.

Shermer, M. 2000. *How we believe: The search for god in an age of science*. New York: W. H. Freeman.

Sole, R. and B. Goodwin. 2000. *Signs of life: How complexity invades biology*. New York: Basic Books.

Sowell, T. 2001. *The Einstein syndrome: Bright children who talk late*. New York: Basic Books.

Sutin, Lawrence (ed.). 1991. *In pursuit of VALIS: Selections from the Exegesis*. Novato, CA: Underwood-Miller.

———— (ed.). 1995. *The shifting realities of Philip K. Dick: Selected literary and philosophical writings*. New York: Pantheon Books.

Tager-Flusberg, H. 1999. *Neurodevelopmental disorders*. Cambridge, MA: MIT Press.

Tainter, J. 1988. *The collapse of complex societies*. Cambridge U.K.: Cambridge University Press.

Umland, Samuel J. 1995. To flee from Dionysus: Enthousiasmos from 'Upon the Dull Earth' to VALIS. In *Philip K. Dick: Contemporary critical interpretations*, ed. Samuel J. Umland. Westport: Greenwood Press.

Watts, D. 1999. *Small worlds: The dynamics of networks between order and randomness*. Princeton: Princeton University Press.

Wheelwright, P. 1959. *Heraclitus*. Princeton: Princeton University Press.

Wiener, N. 1953. *Ex-prodigy*. Cambridge, MA: MIT Press.

Witelson, S., D. Kigar, D., and T. Harvey, T. 1999. The exceptional brain of Albert Einstein. *The Lancet* 353 :2149.

Ziman, J. 1978. *Reliable knowledge: An exploration of the grounds for belief in science*. Cambridge U.K.: Cambridge University Press.

CYBERQUAKE :
HARAWAY'S MANIFESTO

Zoë Sofoulis

The overused term "ground-breaking" always invokes for me images of pickaxes and bulldozers churning up undeveloped land. This is not quite the right word for Donna Haraway's "A Manifesto for Cyborgs" (1985), which behaved more like the seismic center of an earthquake that jolted many out of their categorical certainties as it shifted the terrain of debate about culture and identity in the late 20th century. A decade earlier Laura Mulvey's hugely influential essay "Visual Pleasure and Narrative Cinema" (1975) had had a comparable effect, opening up new directions for a generation of scholars of feminist film theory, art history, textual criticism, and cultural studies. But Haraway's Cyborg Manifesto set out even more multidisciplinary questions, connections, and directions for further research, and its rumbles in the field of cyberstudies, a field it helped to initiate, are still being felt at the beginning of the 21st century. Both Mulvey and Haraway achieved an author status still too rarely accorded feminist writers, a status partly measurable in the fact that they are usually named by lazy female students who are otherwise likely to refer simply to "feminists" or "feminist perspectives," and are cited even by male scholars, a population that, on the whole, successfully ignores vast forests' worth of feminist publishing.

This chapter is concerned with the earthquake effects of the Cyborg Manifesto in academia and cyberculture more generally. I cannot claim any overarching objectivity: I was a student of Haraway's during the 1980s, and have been familiar with the essay from its pre-history (Haraway 1991a, 244 n.1) to its breakthrough with the notion of the cyborg, following its career through various published, republished, and re-edited versions, not to mention spin-off interviews and elaborations. Rather, I practice what Haraway (1988) calls a "situated knowledge": partial, with inevitable blind spots, and very much part of the field it examines (in a very practical way as well, since my career has directly benefited from Haraway's renowned generosity in citing her students' works).

First I look at key ideas of the Manifesto itself, concentrating on the first section, then examine the academic and feminist contexts in which it was received, and finally indicate some of the ways it has been influential in cyberculture and the field of cybercultural studies. There is not enough space here to attempt a comprehensive and detailed survey of just how widely the ideas in Haraway's essay have been disseminated and how deeply they have influenced feminist and cybercultural scholarship. Instead, my more modest aim is to identify some of the theoretical and metaphorical features which made the Manifesto particularly valuable and relevant both to feminist scholars and a diversity of creative participants in the computerized world of the late 20th century.

HARAWAY'S IRONIC DREAM

The Cyborg Manifesto was first published in the journal *Socialist Review* in 1985 under the title "A Manifesto for Cyborgs: Science, Technology and Socialist-Feminism in the 1980s," although a closely related piece had appeared in German the year before (Haraway 1984). It grew a little over time until its definitive final version was published in Haraway's (1991a) edited collection *Simians, Cyborgs and Women* under the title "A Cyborg Manifesto: Science, Technology, and Socialist-Feminism in the Late Twentieth Century." It is to this edition that I will refer.

The Manifesto is divided into five sections. "Ironic Dream of a Common Language" is the most cited and least academic section, which outlines Haraway's myth of the cyborg in straightforward and at times vivid language, and considers the various types of boundary breakdowns which give rise to the hybrid and ambiguous figure of the cyborg. The second section, "Fractured Identities," situates the work in relation to issues within feminist theory, including the question of identities in multi-ethnic communities where essentialisms don't seem to work, at a time when the category "woman" has lost its "innocence" as a political, analytic, and epistemological starting point. Looking more closely at the context in which cyborgs emerge, the following section on "The Informatics of Domination" contrasts terms from modernity and "white capitalist patriarchy" with contemporary forms of technoscientific knowledge and power which understand (and, in effect, produce) genetics, people, and populations in terms of principles of information, coding, and communication. The fourth section, which was the original and most "socialist feminist" kernel of what became the Manifesto, is called "Women in the Integrated Circuit"; it examines the complexities of international gendered and ethnic divisions of labor in globalized economy. The cyborg myth of the first section is then further elaborated in a more utopian mode in the final part, entitled "Cyborgs: A Myth of Political Identity," which draws on a prior decade of exploring cyborg and other hybrid states in feminist science fiction and writings of U.S. women of color. This section is also cited quite often, though as I will later suggest, quite selectively.

The Manifesto's "ironic dream" of cyborg identity has captured the imagination of many outside the normal circuits of feminist and cultural studies academic readerships. But before examining the figure of the cyborg, I want to highlight the contexts in which Haraway situates it.

One of the central aims of the Cyborg Manifesto was to explore possible sources of empowerment for feminists in an era of postmodern technics, and to find something other than victim metaphors linking women with an idealized Nature from which technology was excluded. The essay's political starting point is explicitly socialist-feminist, and thus part of a tradition unafraid of the idea that political empowerment may be linked to "seizing the means of production." Acknowledging that her own educational opportunity to study science was afforded by the space race which motivated government sponsorship of science students in the 1950s and 1960s,

Haraway encourages feminists (especially of the first world, and especially ecofeminists and feminist goddess-worshippers) to stop pretending we can somehow occupy a position on technology separate from the institutional and communicative knowledges and practices that have produced us as certain kinds of historical subjects caught up in certain technological ensembles (Haraway 1991a, 173-176). Drawing on notions of power developed by philosopher Michel Foucault, especially the idea of "biopower," Haraway labels the postmodern technological configuration "the informatics of domination," governed by the military logic of the C³I (command, control, communication, intelligence/information), a configuration that not only allows for "top down" powers of control, command, negation, and repression, but also provides opportunities for perversion, resistance, productivity, and pleasure. A central explanatory feature of the Manifesto is the chart (Haraway 1991a, 162-163), which lists on the left side the terms concepts, practices, and preoccupations associated with "white capitalist patriarchy," and on the right side, the equivalent terms related to post–World War II biology and the "informatics of domination":[1]

Representation	Simulation
Bourgeois novel, realism	Science fiction, postmodernism
Organism	Biotic component
Depth, integrity	Surface, boundary
Heat	Noise
Biology as clinical practice	Biology as inscription
Physiology	Communications engineering
Small group	Subsystem
Perfection	Optimization
Eugenics	Population control
Decadence, *Magic Mountain*	Obsolescence, *Future Shock*
Hygiene	Stress management
Microbiology, tuberculosis	Immunology, AIDS
Organic division of labor	Ergonomics/cybernetics of labor
Functional specialization	Modular construction
Reproduction	Replication
Organic sex role specialization	Optimal genetic strategies
Biological determinism	Evolutionary inertia, constraints
Community ecology	Ecosystem
Racial chain of being	Neo-imperialism, United Nations humanism
Scientific management in home/	Global factory/electronic cottage
Family/market/factory	Women in the integrated circuit
Family wage	Comparable worth
Public/private	Cyborg citizenship
Nature/culture	Fields of difference
Cooperation	Communication enhancement
Freud	Lacan
Sex	Genetic engineering
Labor	Robotics
Mind	Artificial intelligence
Second World War	Star Wars
White capitalist patriarchy	Informatics of domination

1 Other versions of the chart appear in Haraway 1979 and Haraway 1989b.

Haraway identified the problem that feminist (and perhaps especially socialist-feminist) theorists were using a conceptual map and vocabulary based on terms on the left side of the chart, when we urgently needed to update to deal with the contemporary technoscientific realities on the right side of the chart. Instead of critiquing technology from a feminist position miraculously "outside" the (post)modern world, and nostalgically harking back to a prepatriarchal agricultural "golden age" of maternal fertility goddesses, feminists might admit our complicity with the current world systems – including of course the communications technologies that enabled feminism to become a global movement – and begin taking care and responsibility for the way we design and use technologies. Haraway argues that the goals of feminist political and technological empowerment might be better served by myths and metaphors more appropriate to the information age – not an essential or pure, natural organism, but a figure like the cyborg.

Haraway situates the cyborg within the context of postmodern technoscience, especially biology, in which comforting modernist dualisms of the left side of the chart – such as organism versus machine, reality versus representation, self versus other; subject versus object, culture versus nature – are broken down.

These breakdowns have implications for epistemology (theories of knowledge) as well as ontology (theories of being). The latter tend to be emphasized in the Manifesto, which explicitly seeks to forge a myth of political (non/plural/post)identity, while epistemological concerns are more central to Haraway's (1988) "Situated Knowledges" essay.[2] Unlike many contemporary enthusiasts of the "wired" culture of the information society, or some of her own fans, Haraway is not herself a "technological determinist" with faith that the technologies of the information "revolution" automatically produce liberatory effects. She shares with many feminist writers on technology a commitment to the "social constructionist" thesis that technologies are themselves expressive of prior social and institutional arrangements and decisions, and that their effects in turn will vary according to the social practices surrounding them and the political contexts in which they are deployed (see, for example, Cockburn 1992, Cockburn and Furst-Dilic 1994). Haraway makes no secret of her indebtedness to actor-network theory and particularly Bruno Latour, a long-term colleague of hers, especially for the idea that agency is not confined only to humans in sociotechnical systems:

> In a sociological account of science [like Latour's] all sorts of things are actors, only some of which are human, language-bearing actors, and … you have to include, as sociological actors, all kinds of heterogenous entities … Perhaps only those organized by language are *subjects*, but agents are more heterogenous. Not all the actors have language (Haraway 1991b, 3).

Not all the actors have language, but they nevertheless can be caught up in signification: for Haraway, Latour, and other actor-network theorists, the objects and bodies studied/produced by technoscience and biomedicine are "natural-technical objects" (165), or "material-semiotic actors," whose boundaries are not pre-defined but "materialize in social interaction" (Haraway 1991a, 200-201, 208).[3] The exact shape taken by an object of knowledge (a scientific fact or a

---

2   These epistemological questions are pursued more in feminist science studies (for example, Lykke and Braidotti 1996) and in feminist geography, such as various authors in Duncan 1996.

3   See Law and Mol 1995 (274-295) and Latour 1992 for examples of the way messages maybe encoded into artifacts and vice versa. Pickering (1995) discusses contingency and performativity in technological developments; while Nina Lykke (1996), Roberts (1999), and Sofoulis (2001) look at Haraway/Latour connections. Deleuze and Guattari also talk about similar concepts in notions of machinic assemblages, striated and smooth space, etc., on which see Lee and Brown 1994. For evaluations of and problems with actor-network theory (ANT) see Law and Hassard 1999; for some feminist problems with ANT see Cockburn 1992.

technological product) is the result of a specific and contingent set of interactions between its material character and the semiotic and/or technical operations to which it is subject, that is, how it is made to mean, and what is materially done to it or with it. This perspective stresses contingency and hybridity in the outcomes of networks, and concomitantly downplays the idea that every technology follows some necessary and pre-ordained trajectory or ideological program. Holding to such a perspective puts Haraway squarely against those who would interpret every technology developed or used within "white capitalist patriarchy" as inevitably playing out a white, capitalist, and/or patriarchal logic, and it opens the way for an imaginative leap to speculate about the political and epistemological possibilities for using technologies to develop alternative connections with each other and the lifeworld. Enter the cyborgs, "illegitimate offspring of militarism and patriarchal capitalism" who can be "exceedingly unfaithful to their origins" (151).

"Cyborg" is a term condensed from "cybernetic organism," and is typically defined as an entity comprising organic as well as machinic parts and information circuits. Introduced first by Clynes and Kline (1960), and developed further by Halacy (1965), the term is now loosely used to refer to many states of human-technology interface.[4] As Haraway sees it:

> Contemporary science fiction is full of cyborgs – creatures simultaneously animal and machine, who populate worlds ambiguously natural and crafted. Modern medicine is also full of cyborgs, of couplings between organism and machine, each conceived as coded devices, in an intimacy and with a power that was not generated in the history of sexuality (Haraway 1991a, 149-150).

Her argument is "for the cyborg as a fiction mapping our social and bodily reality and as an imaginative resource suggesting some very fruitful couplings" (150), but more boldly than this metaphorical or figurative role, Haraway proclaims the cyborg as an identity:

> By the late twentieth century, our time, a mythic time, we are all chimeras, theorized and fabricated hybrids of machine and organism; in short, we are cyborgs. The cyborg is our ontology; it gives us our politics (150).

Unlike the usual kinds of essentialism and oppressed identities of (especially U.S.) "identity politics," the cyborg's political and ontological commitments are not to wholes, essences, naturalness, or purity, but instead, to "partiality, irony, intimacy, and perversity"; it is "oppositional, utopian, and completely without innocence" (151). The metanarrative of redemption is not part of Haraway's cyborg myth, in which "[t]he cyborg incarnation is outside salvation history" (150) and "wary of holism" (though "needy for connection"), so that:

> Unlike the hopes of Frankenstein's monster, the cyborg does not expect its father to save it through the restoration of the garden; that is, through the fabrication of a heterosexual mate, through its completion in a finished whole, a city and cosmos. The cyborg does not dream of community on the model of the organic family, this time without the oedipal project. The cyborg would not recognize the Garden of Eden; it is not made of mud and cannot dream of returning to dust (151).

---

4   For an inventory of contemporary cyborg types see Gray, Mentor, and Figueroa-Sarriera 1995, or Gray and Mentor 1995.

In the Manifesto, Haraway examines the three main boundary breakdowns which she considers crucial for the emergence of such hybrid figures.[5] The first is the blurring of boundaries between human and animal. Late 20th century thought has nowhere near the same investment as late 19th or early 20th century western thought (or "white capitalist patriarchy") in defining the human as absolutely unique and distinct from other members of the animal kingdom, whether this be in terms of "language, tool use, social behavior, or mental events" (152). "The cyborg appears in myth precisely where the boundary between human and animal is transgressed" (152). In the vocabulary of actor-network theory, both animals and machines are examples of "nonhuman" actors with which humans may enter intimate relations in what Haraway describes as "disturbingly and pleasurably tight coupling" (152). Secondly, the distinction between the machine and the organism (animal or human) has been eroded, as machines become more "self-moving, self-designing, autonomous" (152), and as we (*pace* Latour 1993, 1994, and other actor-network theorists) come to understand ourselves not as purely human in a physical world of nonhuman entities, but as part of a lifeworld of *sociotechnical* hybrids. Finally, the distinction between physical and non-physical is broken down in electronic technology, and through miniaturization, where the silicon chip has become a surface for writing, and where:

> Our best machines are made of sunshine; they are all light and clean because they are nothing but signals, electromagnetic waves, a section of a spectrum, and these machines are eminently portable, mobile – a matter of immense human pain in Detroit and Singapore. People are nowhere near so fluid, being both material and opaque. Cyborgs are ether, quintessence (Haraway 1991a, 153).

This apparent cleanliness and immateriality is a dangerously deceptive phenomenon, making it hard for us to grasp the materiality of the politics involved in their production, distribution, and environmental aftermath (154).

As these formerly foundational distinctions have broken down, metaphors of information exchange and circulation, coding, language, translation, misreading, "noise," etc. have come to dominate postmodern biological theories, as well as theories of how scientific knowledge is itself produced. The "objects" of technoscience now appear more and more as leaky and unstable boundaries formed in the interaction of material and semiotic effects. Viruses, auto-immune responses (interpreted as problems of the self misrecognizing itself as "other"), and biomedical practices like organ transplant surgery, transgenic engineering (where bits of the DNA "code" of one organism may be spliced into that of other quite different organisms) generate models of selves in constant intimate interchange with others; living organisms whose every cell bears the mark of technoscientific intervention.[6]

Haraway argues that such a world calls for new forms of political thinking that do not re-entrench old dualisms (on the left side of Haraway's chart), but which respond to both the pleasures and the dangers of the "informatics of domination:"

---

5   The painting "Cyborg," which artist Lynn Randolph made after reading the Manifesto and which was used as the cover illustration for *Simians, cyborgs and women*; and also *Biopolitics of postmodern bodies* illustrates aspects of these breakdowns. A woman sits with her hands on a keyboard, with a semi-translucent big cat draped over her shoulders, its paws in parallel with hers. The woman's sternum is replaced by a panel of DIP switches from which emanate integrated circuit paths. The keyboard on which Randolph's cat-woman-computer hybrid has her hands seems to be resting amidst sand dunes and mountains; behind her are screens representing the structure of our galaxy, and a black hole's gravity well. The woman's hands and cat's paws seem to be luminous with energy. See also Haraway's discussion of this image in Haraway 1992.

6   These biomedical themes are further explored in Haraway's paper "The biopolitics of postmodern bodies" (in Haraway 1991a), and in her book *Modest_Witness* (1997).

From one perspective, a cyborg world is about the final imposition of a grid of control on the planet, about the final abstractions embodied in a Star Wars apocalypse waged in the name of defense, about the final appropriation of women's bodies in a masculinist order of war. From another perspective, a cyborg world might be about lived social and bodily realities in which people are not afraid of their joint kinship with animals and machines, not afraid of permanently partial identities and contradictory standpoints. The political struggle is to see from both perspectives at once because each reveals both dominations and possibilities unimaginable from the other vantage point (154).

Unlike the organic goddess/mother still favored in some domains of ecofeminism and feminist spirituality, as well as by some who are also interested in Haraway's cyborgism (such as Lykke 1996, Star 1996, Graham 1999 and forthcoming) the cyborg figure cannot be traced back to a pure origin or natural essence – which is not to say that it does not have a history (see Munster 1999, 127; Gonzalez 1995, 267-270; Halacy 1965; Huyssen 1982). It is an unnatural assemblage of heterogeneous parts, a "monstrous chimera" emergent from the very belly of the beast of global capitalism (including militarism). Haraway argues for the "friendliness" of the impure cyborg figure for feminist political thought, not only as a model of a potentially empowering connection between women and contemporary technologies, but also because its heterogeneity offers a model of political subjectivity alternative to that of an identity politics based on presumptions of natural unity and essential commonality between women. In the Manifesto, the cyborg myth is elaborated as a project whose aims share "the utopian of imagining a world without gender, which is perhaps a world without genesis"; the cyborg is defined as "a creature in a post-gender world." But in a later interview, Haraway concedes that her cyborg really is a girl (1991b, 20). Claudia Springer, among others, found that the cyborgs of popular film were not "post-gender" but had, if anything, exaggerated gender characteristics (Springer 1991):

The cyborg is a kind of disassembled and reassembled, postmodern collective and personal self. This is the self feminists must code (163).

Impure and not even "identical" to itself, the cyborg does not need to erase its differences from those to which it connects; a creature of parts, it can illustrate a widespread contemporary experience of having several partial and hybrid identities and axes of political and cultural affinity. As Chela Sandoval observes, a key feature of Haraway's theoretical apparatus is its references to US Third World Women's writings on partial and multiple identities and changing relations to power, and the approach Sandoval has variously named differential consciousness, oppositional consciousness, or the methodology of the oppressed (Sandoval 1991, 1995a, 1995b). It is Haraway's linking of "machines and a vision of first world politics on a transnational, global scale together with [this] apparatus of survival" which Sandoval suggests has the potential to help overcome the "theoretical apartheid" (1995a 419) between, on the one hand, "white male post-structuralism," "hegemonic feminism," and "postcolonial theory," and on the other "US Third World Feminism" (1995a 409) – not that this potential has yet been realized (1995a 415).

Haraway's coding of the postmodern self as cyborg was bold, evocative, and timely. The next sections will consider some of its effects in the field of feminist theory and beyond.

Something that becomes a "cult text" (Penley and Ross 1991, 1), read widely beyond feminist and academic circles, does so not just because it shakes things up (which it did), or because it contains pearls of universal wisdom (which it surely does), but also because in certain ways it fits in with the preoccupations of its times: there is a context that makes it relevant and meaningful; there are readers sharing questions with the text and ready to be moved by it; there are foundational concepts already shaky and ready to crumble. Exuberant, expansive, perhaps over-responsible, and certainly ambitiously synthetic, with its own suggestive flaws and fissures, the chimerical assemblage of elements that is Haraway's Manifesto was capable of bearing many readings by highly divergent audiences.

Haraway's ironic dream of a common language of technologically mediated hybridity and a politics of perverse affinity eventually became almost *de rigeur* reading for anyone within the growing field of "technocultural" studies and especially what some have called "cyberstudies" or "cybercultural studies," as well as cultural studies of technoscience (Penley and Ross 1991, Aronowitz et al 1996). Appealing to a wide range of readerships, the Manifesto's cyborg managed to insinuate itself into diverse discursive spaces, from feminist political theories of anthropology,[7] ethnic identities, and lesbian and queer sexualities (for example, Sandoval 1991, 1995a, 1995b; Morton 1999), to studies of science fiction literature and film (for example, Bukatman 1993, Springer 1996, McCaffery 1991, Csisery-Ronay 1991); she is invoked in papers on postmodern pedagogy (Bigum 1991, Green and Bigum 1993); in studies of cosmetic surgery, reproductive technologies, and female body building (Balsamo 1996), in relation to architecture (Vidler 1992), postmodern war (Gray 1995, Edwards 1995) and feminist theology (Graham 1999 and forthcoming).

The vocabulary of cyborgism was found relevant for discussing a range of cultural and technological phenomena, from mundane computer and videogame interactions, or human interactions mediated by computers and the Internet, to more extreme examples of anything involving physical and virtual intimacies between humans and machines, especially where the latter were seen to exert some type of agency. This included artworks involving bionic bodies such as Stelarc's third arm and more recent gestures towards the "posthuman" (Dery 1996, especially 153-167, 229-256), or interactions between human and nonhuman elements (such as in interactive artworks in digital media or installations fitted with responsive sensors). The cyborg could be invoked in relation to engineering and biomedical advances that produced new genetic hybrids (a favorite theme in cyberfiction and its precursors), or which coupled humans to bionic components, whether to prolong life beyond its "natural" term (as in Gibson's example of the "posthuman" person connected to all kinds of life-support systems), or to allow existence in inhospitable environments like the inside of a nuclear reactor, outer space, or the deep ocean (one of Haraway's favorite cyborg images was of a deep sea diver in a pupa-like casing), or to afford entry into virtual worlds (like the "jacked in" hackers of Gibson's and Cadigan's cyberfictions, directly interfacing their brains with cyberspace; see Balsamo 1993). Itself a science-fictional creature emergent at the interface of theory and imagination, the mythic figure of the cyborg was at home in the blurred boundaries between present reality and the heralded near-future presaged in the boosterist discourses around IT, the "technohype," and "cyberbabble."

---

7   See the articles on cyborg anthropology in Gray 1995.

The Manifesto was predictably of interest to feminists in the fields of labor studies as well as the already established area of feminist science and technology studies, with people interested in a range of matters from the history of women scientists, gender and computer education, non-traditional trades for women, and international divisions of labor by race and gender, to issues of gender, epistemology, and ontology in science. It helped clarify differences between techno- and eco-oriented feminists, and those who didn't subscribe to cosmic feminism could find in Haraway a voice that validated a range of other approaches to studying, interpreting, dreaming, and mythologizing about the woman-technoscience-world relation.

Haraway's manifesto also spoke to the (mainly academic) feminists who were getting bored with the old critiques of dualisms, dissatisfied with the essentialisms of identity politics and strictures of political correctness, pained by the impasses between white women and their many "others," intrigued by the "bad girls" of the radical sex movement, and in search of political and intellectual affinities that didn't depend on shared natural origins, victimhood, or oppression. The term "patriarchy" seemed pretty well past its use-by date for those theorists looking for new ways to frame feminist cultural critique of the globalized capitalist world. Just when the (mainly white male) postmodernists were proclaiming the death of the subject and the end of metanarratives (Jameson 1984), the Cyborg Manifesto – offering its own myths along with the various utopian and US Third World Women's writings it drew upon – celebrated a different vision of a new kind of fractured subject, for whom partiality, hybridity, and lack of a single smooth identity or wholeness did not imply death, but, on the contrary, invoked the possibility for connectedness and survival beyond innocence in an impure world.

Foucault was big and getting bigger in the US in the 1980s, and the Foucaultian dimensions of Haraway's work, including ideas of how the operations of power could involve resistance, pleasure, and perversity were legible to this broader academic readership. Perhaps because she labels the Manifesto "socialist feminist," thus putting the reader off the Foucaultian scent, Haraway is not lined up with "the usual suspects" of Foucaultian feminists. Arguably she deserves to be, since she takes Foucault on board, and critically develops his notions of biopower and biopolitics into her (post)Foucaultian notions of the "informatics of domination" and "technobiopower" (Haraway 1997, 11-12). Moreover, she does actually undertake a complex discourse analysis of a particular science – primatology (Haraway 1989a) – which is more than many a credentialed Foucaultian feminist!

The Manifesto's publication coincided with the peak of textual studies and the linguistic turn in cultural theory.[8] Its descriptions of cyborg as a kind of etched surface, and its notions of information and coding, fitted in well with some of the feminist psychoanalytic-textual analysis that understood gender in terms of language, codes, and signs. Hence the Manifesto's cyborg gained a life within feminist theories of gender and representation. For example, Judith Halberstam interprets this cyborg as something that "posits femininity as automation, a coded masquerade" (1991, 449). With reference to Turing's theories of computer intelligence, Halberstam argues that gender is also not innate or essential, but "a learned, imitative behavior that can be programmed" (443), so that both femininity and masculinity are "always mechanical and artificial" (454),

---

8   For example, in the 1981 History of Consciousness postgraduate Theory and Methods seminar all students had to read Josué V. Harari's *Textual Strategies: Perspectives in Post-Structuralist Criticism* (Ithaca, NY: Cornell University Press, 1979) which, in view of its woman-free line-up of post-structuralists, we renamed *Sexual Tragedies*.

forms of simulation or masquerade that become lived realities. The female cyborg figure offers resistance to "static conceptions of gender and technology":

> The intelligent and female cyborg thinks gender, processes power, and converts a binary system of logic into a more intricate network. As a metaphor, she challenges the correspondences such as maternity and femininity or female and emotion. As a metonym, she embodies the impossibility of distinguishing between gender and its representation (Halberstam 1991, 454).

Similar themes are explored by Sadie Plant (1994, 1995, 1998), who links women to machines, especially computers, not only in terms of metaphors of weaving, networks, and the matrix, but importantly in terms of their shared capacities for simulation (*pace* feminist philosopher Luce Irigaray's notion of mimesis): "it is always as machinery for the reproduction of the same that women and information technology first sell themselves" (Plant 1995, 59). Although she shares Haraway's interest in the libratory and subversive potentials of an alignment between women and the powerful communications and networking machines of the information age, Plant is also Irigarayan, and manages at once to be both more "machinic" and more "essentialist" than Haraway, linking the automatic qualities of intelligent machines to the experiences of species being and reproduction which have traditionally defined women. Haraway was looking for empowering metaphors that didn't rely on birthing and definitions of women as natural. In a related paper, Plant writes that while man may be shattered by the revelation of multiple and machinic intelligences after "two and a half thousand years of sole agency," woman is "already in touch with her own abstract machinery" for she:

> ... has never had a unified role. Mirror, screen, commodity, means of communication and reproduction, carrier and weaver, carer and whore, machine assemblage in service of the species, a general purpose system of simulation and self-stimulation. While man connected himself to the past, woman was always in touch with the virtual matter of her own functioning. This is the connection which has constituted her so-called missing part: the imperceptible necessity of every machine (Plant 1994, 6-7; see also Plant 1995, 58).

Thus the historical denial of agency and subjectivity to women, their relegation as machinic apparatuses obeying species imperatives, is seen to give them an "edge" with respect to the emergent future of cyborg life where both machines and women evade control by the men who have become peripheral to their functioning.

Whereas Plant's cyberfeminism celebrates the machinic agency of both woman and nature ("species" as mechanism), and Halberstam (1991) emphasizes the automated character of the feminine identity (gender and information as masquerades, codes), "environmental feminist" Stacey Alaimo (1994) rejects Haraway's cyborg metaphor as "inimical not only to an environmental feminism but also to any politics that opposes the military industrial complex"; she prefers Haraway's figure of nature as a witty agent like the Native American trickster figure, "the wily Coyote and an artifactual nature seem more effective agents for an environmental politics" (Alaimo 1994, 149). In Alaimo's post-Harawayan ecofeminism, women and nature are to be aligned again, not as passive victims, but as active agents in political affinity with each other (150) – though how exactly this happens is not spelled out.[9]

---

9   For a variety of other views and the cyborg/goddess issue see Lykke & Braidotti 1996; for a view on Haraway as an ecofeminist, albeit of a non-essentialist, non-goddess worshiping variety, see Sturgeon 1997.

Although Haraway's account of the cyborg and cyborg politics is amenable to a textualist reading, it also, as Alaimo emphasizes, disrupts and goes beyond textualism because of its insistence on a real and material dimension to the world that evades and often tricks language: for Haraway the world is emphatically *not* a *tabula rasa* which can be written on and reshaped any way we want it. Nonhuman entities set their own limits to their inscribability. Kathleen Woodward describes the Manifesto as "the most seminal [piece] in articulating the importance of biotechnology in contemporary cultural criticism" which achieved this by "smuggling, as it were, the subject of biotechnology into an essay that deals with communication technology" (Woodward 1999, 286). This seems to me an awkward oversimplification of the previously discussed perspectives on the centrality of communications metaphors in post-World War II biology, as well as the ANT notions of "material-semiotic objects." These perspectives understand entities to be shaped out of contingent interactions with heterogenous others, in what John Law describes as a "ruthless application of semiotics" (Law 1999, 3) to the material world, whose constituents, analogously to phonemes and morphemes of language, are not essentially given, but shaped in their distinctions from other elements. I attribute the "earthquake" effect of the Cyborg Manifesto (and its close relatives, Haraway 1988, 1989b, 1991b, 1992) as arising in large part from the way it introduces terms and ideas from social studies of science, especially actor-network theory, into debates in feminist theory and political struggles for identity and subjectivity (especially by US Third World Women; Sandoval 1991), and from there into a wide range of textual and cultural studies. Perhaps even more important than Haraway's inspirational and clarifying effects *within* her own science studies field are those enabling effects on feminist scholars more peripheral to that field. As Nina Lykke suggests (Lykke 1996, 20-22), putting constructionism onto an interdisciplinary feminist science studies agenda allows people from humanities fields, including semiotics, narratological, and discourse studies, to make a contribution from somewhere not completely outside science.

Despite the wildly inclusive scope of the Manifesto, readers who've looked to it or its author to provide some overarching theory of the lot are inevitably disappointed. Haraway is no great meta-theorist and she does not assume the omniscient god position on any field, not even her own work. In the early 1980s she could barely name what it was she was doing. I suggested it was a semiotics of technology, but we both knew that didn't quite capture all the dimensions she wanted to bring together when deciphering objects and artifacts: more than semiotics, she was working on the cultural logic of globalization as expressed in the shape of reality itself, condensed into the very being of the object. Becoming progressively more articulate about her method, Haraway was able to develop more fully the notion of figuration central to her method in *Modest_Witness*:

> I emphasize figuration to make explicit and inescapable the tropic quality of all material-semiotic processes, especially in technoscience. For example, think of a small set of objects into which lives and worlds are built – chip, gene, seed, fetus, database, bomb, race, brain, ecosystem. This mantra-like list is made up of imploded atoms or dense nodes that explode into entire worlds of practice. The chip, seed, or gene is simultaneously literal and figurative. We inhabit and are inhabited by such figures that map universes of knowledge, practice, and power. To read such maps with mixed and differential literacies and without the totality, appropriations, apocalyptic disasters, comedic resolutions, and salvation histories of secularized Christian realism is the task of the mutated modest witness (Haraway 1997, 11).

Later in the book she describes the objects on this list as:

> … stem cells of the technoscientific body. Each of these curious objects is a recent construct or material-semiotic "object of knowledge," forged by heterogenous practices in the furnaces of technoscience. To be a construct does NOT mean to be unreal or made up; quite the opposite. Out of each of these nodes or stem cells, sticky threads lead to every nook and cranny of the world. Which threads to follow is an analytical, imaginative, physical, and political choice. I am committed to showing how each of these stem cells is a knot of knowledge-making practices, industry and commerce, popular culture, social struggles, psychoanalytic formations, bodily histories, human and nonhuman actions, local and global flows, inherited narratives, new stories, syncretic technical/cultural processes, and more (129).[10]

Not all of Haraway's readers appreciate the significance of the dimension of her work I am highlighting here, namely those concepts from actor-network studies translated into cultural and political theory. That these could potentially help feminists go beyond a "merely" discursive or semiotic understanding of the object, and help rethink the subject-object relation in terms of agency and material-semiotic realities in a sociotechnical world, was not appreciated by Carol Stabile (e.g., Stabile 1994), who seemed to be looking for Haraway to provide a complete model for eco-techno-identity-politics. This author distinguishes between "technophobic" and "techno-maniac" approaches to technology: the former represented by the "reactionary essentialism" of ecofeminists such as Mary Daly and Susan Griffin, who act as "ventriloquists" for nature; and the latter by Donna Haraway, whom Stabile accuses of depoliticizing the field by being seduced by postmodern images of "fragmentary and destabilized identities" (the main source for which, as Sandoval [1995a] reminds us, comes from US Third World Women and utopian feminist writing). The "technophobes" can ignore the grim realities of poverty and dispossession in an urban environment in their celebrations of the links between women and nature, while the "techno-maniacs" can posit a cyborg subject whose high degrees of literacy, mobility, and choice are shared only by the most educated elites.[11]

Stabile attacks Haraway as a practitioner of discourse theory, which she associates with idealizing, aestheticizing, dehistoricizing, and depoliticizing moves within culture critique. Her argument implicitly relies on the old Marxist idealist/materialist distinction and fails to show any understanding of the way discourse theories (like Foucault's, from whom Haraway borrows) are precisely about how discourses (traditionally considered part of the nonmaterial "superstructure"), the structures and practices of speaking and acting in institutional contexts, have both historical determinants and real, material effects, and hence how telling new or different stories can potentially change sociotechnical realities. Moreover, Stabile studiously ignores those aspects of Haraway's work which don't fit her critique, as well as those in sympathy with her own socialist views – such as Haraway's attention to socialist-feminist studies of the gendered international division of labor, especially of women workers in the global electronic industry, her calls for the necessary continuance of "normal" kinds of political work (for example, affinity groups, collective actions, socially responsible practices, etc.), and her encouragement of fellow white western feminists to become more aware of the privileged and partial character of their/our own

---

10 Equivalent accounts of Haraway's method may be found in Haraway's *Primate Visions* (1989a, 14, 369), and her "Promises of Monsters" essay (1992; also in Wolmark 1999, esp. pp. 317-318).

11 Similar points are made in a more subtle and interesting way by Susan Leigh Star (1996), writing on themes of containment and being "homed" in the networked world.

standpoints, and to retell or transform existing stories into empowering political narrative appropriate for our own time and experience. Stabile's own perspective by contrast offers very little encouragement to political action or imagination and her conclusion is simply that better historical analysis of the same old stories is needed:

> Before we invent futurologies, we need to be able to tell stories about the fundamental and persistent narratives that continue to exclude, maim, and kill (Stabile 1994, 157).

Stabile accuses Haraway of installing a feminist avant-gardism (Stabile 1994, 145) and of promoting a cyborg subject that doesn't have to *do* anything in order to be political, since "[p]olitics are, so to speak, embedded in the cyborg body." This critique, I would counter, applies less to Haraway than to some of her over-enthusiastic and under-politicized readers, especially those caught in the "textualist turn" and hence liable to equate semiotic ambiguity with political subversiveness (bolstered in some cases by Derrida's deconstructionism or notions of the carnivalesque from Bahktin, Kristeva, or Stallybrass and White).

Whether or not we agree with Stabile, and diagnose the Manifesto as infected by the very textualism it claims to resist in the name of a more realist epistemology, there is no doubt that it tempts some readers to interpret the category transgressions of cyborg figures, including gender transgressions, or even enthusiastic interfacings with high-tech equipment, as somehow inevitably or inherently politically subversive. A closely related temptation, and a highly seductive one for budding cybertheorists (and I've read just one too many of their dissertations), is to lump the cyborg in with every other kind of semiotically undecidable creature, and to read it as equivalent or identical to every other supposedly teratological (monstrous), transgressive, abject, or some way "in-between" entity or sign. I agree with Anna Munster that through such readings "the 'cyborg' configuration has been reified to denote 'hybridity' as such rather than to seek out partialities and produce changing alliances with the technical" (Munster 1999, 127). Munster points out that:

> If there is something potentially exciting about hybridity, it is the sense in which it resists a capture into the mere grafting of two connecting points (the technological and the cultural, the natural and the artificial, women and technoculture) and encourages a sense of movement between them (127).

There is no doubt Haraway's writings give some support to the idea of grouping all kinds of monsters together as hybrids. One could cite the statement in the Cyborg Manifesto that in the late 20th century "we are all chimeras, theorized and fabricated hybrids of machine and organism; in short, we are cyborgs," or the postscript to the "Cyborgs at Large" interview where she describes as "monsters" the "boundary creatures – simians, cyborgs and women" which have destabilized "western evolutionary, technological, and biological narratives" (1991b, 21). My own reading of Haraway understands the emphasis to be not on hybridity as such, but on the specificity of hybrid forms that arise in particular situations. Frankenstein's creation, for example, could be described as a monstrous fabricated "chimera," but it is not a cyborg. It belongs on the left side of the chart and with the early 19th century, whereas the cyborg is the kind of chimera that becomes possible in the latter part of the 20th century, and in relation to powers and knowledges of the postmodern "informatics of domination." For Haraway, as for Latour, the mere fact of something's hybridity does not mean much in itself, for, as Latour argues (1993, 1994), there are always and everywhere sociotechnical hybrids; it's just that the project of modernity was to pretend there wasn't, in an effort to "purify" the messy interdependence of humans and nonhumans by making strict divisions between social and physical/natural sciences.[12] More salient questions are: what are the specific networks in which the hybrid is produced or enrolled? with which other actors? how extended are these networks? and to what extent does the prevailing worldview recognize, celebrate, or on the contrary deny, these hybridities? Frankensteinian science was close to field surgery and dealt with "found [biological] objects," dead organisms whose flesh is to be rejoined and reanimated. It aimed to find the secrets of life and death in order to create a new man, not change the nature of living being while turning it into a commodity, "life itself" (Haraway 1997, 13 and 276 n.8). In the late 20th century, the genetic engineering techniques and corporate capital that produce cyborgs and creatures like the OncoMouse™, whose being at the very cellular level bears the mark of commercial techno-scientific intervention, are part of complex and extended set of networks in the informatics of domination, involving many universities, research teams and agendas, technological developments, advertising campaigns, etc.

CYBERCULTURAL MANIFESTATIONS

In the opening of this essay, I stated that Haraway was one of the rare feminist writers to be accorded the status of author, quoted and named by both feminist and nonfeminist writers. Her ironic myth of the cyborg and related essays on epistemology and postmodern bodies have inspired many scholars, including a number of her students at the History of Consciousness, to get beyond the essentialist impasses of early 1980s feminist identity politics, and to explore new directions in cultural analysis and criticism in the late 20th century, with renewed attention to scientific, technological, biotechnological, and biomedical themes. A number of articles and edited collections that gather up writings on cyborg and cyberspace themes have come out over the last decade, including, of course, this collection.[13]

The mid-80s in the U.S. was the crest of the first big wave of diffusion of personal computers into academic, and increasingly, domestic life. We were starting to use email and online chat. My archived versions of Haraway's Manifesto and its precursors demonstrate through their fonts and print qualities the increasing sophistication of personal computing. People's personal experiences of cyberspace and their excitement at the potential in these new and networked machines made the subject matter of the Manifesto especially meaningful.

Adding to the impact of the Cyborg Manifesto was the growing popularity of what became known as "cyberpunk" science fiction. In the year before the Manifesto's publication, William Gibson scooped up all the major awards with his cyberpunk novel *Neuromancer* (Fitting 1991, 312 n.3), which explored various states of cyborg being, celebrated the obsessive hacker mentality (on which see Turkle 1984, 1996), and offered a definitive literary description of the virtual landscape he named "cyberspace," a translation into words of the imagery associated with high technologies and networks, especially computers and cities, in popular science, high-tech advertising

---

12 See also Lykke 1996, Sofoulis 2001 n.3.

13 Those with contributions by Haraway's students include Stone 1996 and Gray 1995, a gigantic compendium of key sources and writers in the in the field, including several History of Consciousness students. Two wide-ranging collections of cultural studies approaches are Dery 1993, and Brahm and Driscoll 1995, which includes Sandoval (1995b) on US Third World Feminism, and critical and philosophical essays on geography, cyborgs and aliens, rap, Shakespeare, etc. Jenny Wolmark (1999) gathers a number of "classic" essays exploring various aspects of gender, race, class, technological embodiment, and popular media, with helpful introductions to each section, while Lykke and Braidotti 1996 emphasizes science studies and related questions of knowledge, objectivity, and spirituality, with some critiques of and developments beyond Haraway.

and films like Ridley Scott's *Blade Runner*. Some readers enthusiastically linked and even confused Gibson's "five minutes into the future" cyberpunk visions with Haraway's technomyth of the cyborg. Just as Gibson was a little disconcerted by the leather-clad, technologically identified real-life cyberpunks who wanted him to sign their battered copies of *Neuromancer* (Gibson 1990), so was Haraway "galled" to encounter the technologically determinist ways her work had been interpreted by hardcore cyberpunks, hackers, and writers in *Mondo* and *Wired* who read the Manifesto "as a sort of technophilic love affair with techno-hype" and herself "as some blissed-out cyborg propagandist" (Haraway quoted in Olson 1996, 25). In 1984 (the year of the original "roadshow" where Haraway's "Ironic Dream of a Common Language" was performed) I also attended a science fiction conference in southern California. Gibson's awards had just been announced and the enthusiasm for *Neuromancer* was accompanied by a palpable sense of relief amongst the (80-90 percent male) academic SF fraternity, that finally a man had come up with an outstanding new novel, breaking the women writers' dominance of the awards since the mid-70s. "Find a man and you've found the origin," as my colleague Sarah Redshaw puts it: from then on Gibson's *Neuromancer* became the exemplar of a cyberfiction with only male-authored antecedents such as Philip K. Dick, J. G. Ballard or William Burroughs (for example, Fitting 1991). It became common to discuss Gibson's cyberpunk in relation to Haraway's Manifesto without mentioning the references in both texts to the previous decade of feminist science fiction cyborg figures, cyberspaces, and stories by people like Joanna Russ, Octavia Butler, Vonda McIntyre, Anne McCaffery, Marge Piercy, James Tiptree Jnr (Alice Sheldon), and so on – authors by whom Gibson had been influenced and from whom he borrowed.[14] Since Haraway's ideas about the cyborg's utopian possibilities had been developed through her engagement with the prototypes of cyberfiction, it would be difficult to trace any direct influence upon subsequent productions in the field, including cyberpunk-style fictions by female authors like Pat Cadigan, Katherine Kerr and Melissa Scott, although Piercy's (1991) novel *He, She, It* (retitled *Body of Glass* in Australia), which draws parallels between the Golem of Prague and a 21st century cyborg, gives explicit acknowledgment to the formative influence of the Cyborg Manifesto.

Already compatible through their shared history of female-authored speculative fictions about biotechnologies and virtual worlds, the Gibsonian and Harawayan visions, together with the real-world experiences of digital technologies, the Internet, and cyberspace expanding into education, workplaces, home, and the arts, all combined to produce a kind of mutant love child in the form of cyberfeminism, a term I loosely understand as a kind of feminism interested in exploring the theoretical and artistic potential of technologies and metaphors of the information age for women, and/or taking feminist activism into the virtual world and its real-world infra-structures. Cyberfeminists make the connection between the cyberpunk slogan "information wants to be free" and the feminist libratory dimensions of Haraway's cyborg myth. They tend to be enthusiasts of the new technologies, especially the web and its possibilities for networking (Kuni 1999), though most are critical of the idea that virtual worlds offered a seductive escape from embodiment, and many are explicitly interested in maintaining sight of the embodied, indeed "visceral" character of Haraway's cyborg and the leaky, penetrable bodies of Gibson's cyberpunk

14 On this point see Samuel R. Delaney interviewed by Mark Dery in *Flame wars* (Dery 1993, 743-763, see n.13) and Zoë Sofia (1993, 113-114); for a brief comparison of feminist SF and cyberpunk see Jenny Wolmark's "The Postmodern Romances of Feminist Science Fiction" in Wolmark 1999.

15 For example, in 1993 Virginia Barratt curated an exhibition *Tekno Viscera*, in which various contributors, including electronic artists, played on the theme of "putting guts into the machine" (Barratt 1993).

antiheroes.[15] The term "cyberfeminist" arose simultaneously in 1991 for scholar Sadie Plant in England, and the Australian feminist art group VNS Matrix. Like many Australian feminists, VNS Matrix became aware of Haraway's Manifesto when it was reprinted in the journal *Australian Feminist Studies* in 1987, and paid homage to it in their (1992) *Cyberfeminist Manifesto for the 21st Century*.[16] Produced using electronic image-making technologies, it featured a horned woman in a shell amidst a molecular matrix, with a text announcing "the clitoris is a direct line to the matrix" and proclaiming themselves as the "virus of the new world order ... saboteurs of big daddy main-frame ... terminators of the moral code ... mercenaries of slime ... we are the future [etc.]." The group went on to produce *All New Gen*, a series of lightboxes and the prototype of a computer game based on these themes.[17] During the mid 1990s, along with one of the VNS Matrix artists, Virginia Barratt, I conducted interviews with Australian women artists in digital media: we found that almost all our interviewees had read the Manifesto and been inspired in one way or another by it. Slightly later than us, and with a partially overlapping sample of interviewees, researcher Glenice Watson (2000) found that a number of Australian women pioneers of the Internet were committed to the practice of being feminist activists on and around the Internet, though not all identified as "cyberfeminists" per se.

While Australian interest in cyberfeminism had more or less peaked by around 1996, the term continued to attract attention in Europe and the international cyberarts scene. In 1997 and 1999 the first and second "Cyberfeminist International" events were organized by an intern-ational group of women (including former VNS Matrix member Julianne Pierce) who ironically call themselves The Old Boys Network.[18] Themes at the second event included "Split Bodies and Fluid Gender," computer hacking and whether feminists could "appropriate the practice for their purposes," and feminist responses to globalization (Volkart and Sollfrank 1999, 4-5), as well as discussions and papers on the problems of defining cyberfeminism, how it might be different from other feminist critiques of technology, its essential pluralism, its connections with praxis, etc. For writers in the *Next Cyberfeminist International* catalog (Sollfrank and Old Boys Network 1999), Haraway was a foundational author to be critiqued. However, one – Nat Muller – observed that Haraway's (1988) "Situated Knowledges" essay, which had been overlooked in favor of the Manifesto, offered some promising openings for cyberfeminists who wanted a more technologi-cally oriented approach than that offered in the discourse-heavy cybercultural studies of the mid to late 1990s. Likewise the early self-proclaimed cyberfeminist Sadie Plant was "the theorist we all love to slag off" (Muller 1999, 75), while cultural studies scholar Anne Balsamo was cited approvingly for her interests in the real conditions of women's working lives and technological engagements (Muller 1999, Ackers 1999). Rosi Braidotti's (1994) critical approach to cyborgism was also accepted. Various contributors expressed dissatisfaction with Haraway's and Plant's too-easy linkage under the "cyborg" banner of those cybergirls of the rich nations with Third World women producing the equipment (one of the standard critiques of the Manifesto):

I am very weary of making these celebratory gyno-social links as in: Ooowww look at us girlies

99

---

16 Versions of the VNS Matrix manifesto have been reprinted in Zurbrugg (1994, 427), and Sofia (1996, 61).

17 VNS Matrix and other Australian technological artists are discussed by Glenda Nalder (1993), and in pieces by Teffer, Bonnin , Kenneally, in the "Natural/Unnatural" special issue of *Photofile 42* (June 1994); by Jyanni Steffensen (1994), Bernadette Flynn (1994), Zoë Sofoulis (1994), and Zoë Sofia (1994, 1998). See also various contributors and interviewees in Zurbrugg 1994.

18 I am grateful to Irina Aristarkhova for sending me a copy of the *Next Cyberfeminist International* catalogue.

we're all digital divas whether we're slaving away in a chip factory or whether we're suffering from rep. strain injury or carpal tunnel syndrome. This makes me think of 70s sisterhood feminism ... before we start jumping around with terms like 'virtual sisterhood,' we should be sensitive to just how inclusive that sisterhood is (Muller 1999, 76).

Some writers rejected Plant's (1995, 1998) essentialist ideas about the posthuman and machinic character of women's subversiveness in the computer age, along similar lines to my critique, above (see Bassett 1999, Muller 1999), and perhaps best summed up by Volkart and Sollfrank:

Unlike approaches which assume that female resistance is already happening unconsciously in unknown, uncontrollable spaces, we insist on the idea of aware responsibility, reflection and of engaged motivation and intention (1999, 5).

Signaling disaffection with the earlier euphorics of VNS Matrix's *Cyberfeminist Manifesto*, Sollfrank wrote near the end of a report on her research on female hackers (she found few) that "My clitoris does not have a direct line to the matrix – unfortunately. Such rhetoric mystifies technology and misrepresents the daily life of the female computer worker" (Sollfrank 1999, 48). This interest in practice and the realities of cyberspace life was a theme in several other articles in the *Next Cyberfeminist International* catalog, which included besides the theoretical contributions I have highlighted others about more technical details of hacking and networking, discussions of various cyber-inspired artworks (or artworks interpretable in cyberfeminist terms), pieces about biotechnologies, the visible human project, and accessing cyberspace in the former Soviet Union; in short, a range of themes quite typical of those discussed in and around cyberarts, and in cybercultural studies (see note 13).

Coming in conjunction with cyberpunk fiction and the personal computer revolution, the Manifesto was well placed to give some focus to expressions of hope and fear about the emergent technoworlds. As I earlier suggested (after Lykke 1996), Haraway's work, especially the Manifesto, has been important in enabling people from outside the fields of science studies to feel empowered to talk, think, criticize, write, and make artworks about the new forms of being and experience, and new kinds of sociotechnical and biotechnical hybrids emergent into the 21st century. Even though some of the post-Harawayan cyberfeminists have found problems with the Manifesto and its utopian dreams, it is still the case that the work was enabling for many scholars, artists, critics, and activists, who can take from it concepts and vocabulary to help name some of the new experiences and possibilities – both scary and pleasurable – afforded by technologies of the late 20th century, and to put some of these into perspective in relation to historical developments in science and industry over the 19th and 20th centuries.

There was a certain infectious euphorics of impurity in the Manifesto's vision of a hybrid identity committed to "partiality, irony, intimacy, and perversity": the cyborg myth offered an appealing way out to those frustrated by the purisms of identity politics, as well as those white women who experienced forms of hybridity besides those lived by their non-white or non-Anglo sisters. The Manifesto's account of the "informatics of domination" outlined the broader power-knowledge context in which the breakdown of purisms became more legible than they were within the dualisms of "white capitalist patriarchy" and the kind of feminism generated in/against it. The former Catholic girl's celebration of blasphemy, irreverence, and iconoclasm fitted in well with the images of "bad girls," flirting on the edge of pleasure and danger, that were cultivated by sexual radicals like lesbianfeminist sadomasochists of the 1980s (and got a heterosexual re-run in the 1990s), and it continued to resonate with the pluralism, activism, and sense of fun in queer (and cyberqueer) political sensibilities emerging from the late 80s and 90s.

Haraway's poetic claim that the cyborg "gives us our ontology" captured the imagination of many who were beginning to experience prolonged interactions with computers, and starting to explore new identities and forms of social life and community made possible by the Internet. While the computer's "holding power" and the fuzzy boundaries experienced between self and machine were already being written about when the Manifesto first came out (for example, Turkle 1984, Gibson 1984), Haraway's contribution was to locate these experiences within a potentially utopian political landscape, in a thought experiment based on the idea that science fiction imaginings could be a form of feminist politics. Whereas a standard feminist line on technology had been to equate it with abstract masculinist rationality, militarism, and the rape of the Earth, Haraway insisted on the intimate physicality of our relations to nonhumans, and on what Claudia Springer (1991) would later call "the pleasure of the interface." Beyond the problematics of heterosexual desire, the political correctness (or not) of lesbian sexualities, and the hybrid identities of (Sandoval's 1991, 1995a) "US Third World Women," the Cyborg Manifesto acknowledged the pleasures and desires we hold in relation to the nonhuman entities that are part of our lifeworld, and the sociotechnical and material-semiotic hybrid entities and plural identities we might form with nonhumans. Haraway's political commitments to feminist, anti-nuclear and environmental politics, coupled with her bold determination to provide an alternative to what Stabile (1994) called "technophobic" feminists' essentializing equations between woman, reproduction, and nature, helped open up a more positive perspective on new technologies and their possibilities. Her socialist politics and familiarity with social studies of scientific practice and knowledge (especially in the ANT tradition) led Haraway to reject the idea of an inevitable trajectory of domination implicit in technologies, even those of military origin. If the cyborg could be unfaithful to its origins, then so could we. This perspective opened the way for women interested in new technologies like personal computers to explore their libratory, productive, and poetic possibilities in imagination, artwork, and cyberfeminist practice. In the characteristically 1980s spirit of the Cyborg Manifesto, complicity with "the system" was not an unmentionable crime nor a paralyzing political embarrassment, but understood as something inevitable, which did not necessarily prevent further effective political work for justice, equality, peace, and survival.

REFERENCES

Ackers, S. 1999. Cyberspace is empty – Who's afraid of avatars?. In Sollfrank and Old Boys Network 1999.

Alaimo, S. 1994. Cyborg and ecofeminist interventions: Challenges for an environmental feminism. *Feminist Studies* 20:1.

Aronowitz, S., B. Martinsons and M. Menser 1996. (eds). *Technoscience and cyberculture*. New York: Routledge.

Balsamo, A. 1993. Feminism for the incurably informed. In Dery 1993. (Also in Balsamo 1996.)

————— 1996. *Technologies of the gendered body*. Durham: Duke University Press.

Barratt, V. (ed.). 1993. *Tekno Viscera*. Exhibition catalog. Brisbane: Institute of Modern Art.

Bassett, C. 1999. A manifesto against manifestos?. In C. Sollfrank and Old Boys Network 1999.

Bigum, C. 1991. Schools for cyborgs: Educating aliens. In *Navigating in the nineties*. Proceedings of ACEC '91: Ninth Australian Computing in Education Conference, Bond University. Computer Education Group of Queensland.

Bigum, C. and B. Green 1993. Changing classrooms, computing, and curriculum: Critical perspectives and cautionary notes. *Australian Educational Computing* 8:1.

Brahm, G., Jr. and M. Driscoll (eds.). 1995. *Prosthetic territories: Politics and hypertechnologies*. Boulder: Westview Press.

Braidotti, R. 1994. *Nomadic subjects: Embodiment and sexual difference in contemporary feminist theory*. New York: Columbia University Press.

Bukatman, S. 1993. *Terminal identity: The virtual subject in postmodern science fiction*. Durham NC: Duke University Press.

Cadigan, Pat. 1991. *Synners*. New York: Bantam.

Clynes, M. E. and N. S. Kline. 1960. Cyborgs and space. *Astronautics* Sept. Reprinted in Gray 1995.

Cockburn, C. 1992. The circuit of technology: Gender, identity, and power. In *Consuming technologies: Media and information in domestic space*, ed. R. Silverstone and E. Hirsch. London and New York: Routledge.

Cockburn, C. and R. Furst Dilic (eds). 1994. *Bringing technology home: Gender and technology in a changing Europe*. Birmingham and New York: Open University Press.

Csisery-Ronay, Istvan, Jr. 1991. The SF of theory: Baudrillard and Haraway. *Science Fiction Studies* 18(3).

Dery, M. (ed.). 1993. *Flame wars: The discourse of cyberculture*. Special issue of *South Atlantic Quarterly* 92(4). Durham, NC: Duke University Press.

————— 1996. *Escape velocity: Cyberculture at the end of the century*. London: Hodder & Stoughton.

Duncan, N. (ed.). 1996. *BodySpace: Destabilizing geographies of gender and sexuality*. London: Routledge.

Edwards, P. 1995. *The closed world: Computers and the politics of discourse in postwar America*. Cambridge, MA: MIT Press.

Featherstone, M. and R. Burrows (eds). 1996. *Cyberspace, cyberbodies, cyberpunk: Cultures of technological embodiment*. London: Sage.

Fitting, P. 1991. The lessons of cyberpunk. In Penley and Ross 1991.

Flynn, B. 1994. Woman/machine relationships: Investigating the body within cyberculture. *Media Information Australia* 72 (May): 11-19.

Gibson, W. 1984. *Neuromancer*. London: HarperCollins.

————— 1986. *Count Zero*. London: HarperCollins.

————— 1988. *Mona Lisa overdrive*. London: HarperCollins.

————— 1990. Interviewed in *Cyberpunk* (film). Dir. P. von Brandenberg and M. Trench. Dist. Voyager Company.

Gonzalez, J. 1995. Envisioning cyborg bodies: Notes from current research. In Gray 1995. Reprinted in Wolmark 1999.

Graham, E. 1999. Cyborgs or goddesses? Becoming divine in a cyberfeminist age. *Information, Communication and Society* 2(4).

————— Forthcoming. *Representations of the post/human: Monsters, aliens, and others in popular culture*. Manchester: Manchester University Press.

Gray, C. 1993. *Postmodern war: Computers as weapons and metaphors*. London: Free Association Press.

————— (ed.). 1995. With S. Mentor and H. Figueroa-Sarriera. *The cyborg handbook*. New York: Routledge.

Gray, C. and S. J. Mentor. 1999. The cyborg body politic and the new world order. In Brahm and Driscoll 1995.

Gray, C., S. J. Mentor and H. J. Figueroa-Sarriera. 1995. Cyborgology: Constructing the knowledge of cybernetic organisms. In Gray 1995.

Halacy, D.S. 1965. *Cyborg: Evolution of the superman*. New York: Harper & Row.

Halberstam, J. 1991. Automating gender: Postmodern feminism in the age of the intelligent machine. *Feminist Studies* 17(3):439-460.

Haraway, D. 1979. The biological enterprise: Sex, mind and profit from human engineering to sociobiology. *Radical History Review* 20. Reprinted in Haraway 1991a.

————— 1984. Lieber Kyborg als Göttin: für eine sozialistisch-feministische Unterwanderung der Gentechnologie. In *Argument-Sonderband*, ed. B.-P. Lange and A. M. Stuby. 105, Berlin.

————— 1985. A manifesto for cyborgs: Science, technology and socialist feminism in the 1980s. *Socialist Review* 80:65-107. Reprinted in *Australian Feminist Studies* 4 (1987), in Nicholson 1990, and in Haraway 1991a.

————— 1988. Situated knowledges: The science question in feminism and the privilege of partial perspective. *Feminist Studies* 14(3):575-599. Reprinted in Haraway 1991a.

————— 1989a. *Primate visions: Gender, race and nature in the world of modern science*. New York: Routledge.

————— 1989b. The biopolitics of postmodern bodies: Determinations of self in immune system discourse. *differences: A Journal of Feminist Cultural Studies* 1(1). Reprinted in Haraway 1991a.

————— 1991a. *Simians, cyborgs, and women: The reinvention of nature*. New York: Routledge.

————— 1991b. Cyborgs at large: Interview with Donna Haraway, by Penley and Ross. In Penley and Ross 1991.

————— 1992. Promises of monsters: A regenerative politics for inappropriate/d others. In *Cultural Studies*, ed. L. Grossberg, C. Nelson and P. Treichler. New York: Routledge. Reprinted in Wolmark 1999.

————— 1997. *Modest _Witness@Second_Millenium. FemaleMan _Meets_OncoMouse*™. New York: Routledge.

Huyssen, A. 1982. The vamp and the machine: Technology and sexuality in Fritz Lang's *Metropolis*. *New German Critique* (fall/winter), 24-25.

Jameson, F. 1984. Post-modernism, or the cultural logic of late capitalism. *New Left Review* (July/August), 146.

Kuni, V. 1999. The art of performing cyberfeminism. In Sollfrank and Old Boys Network 1999.

Latour, B. 1992. Where are the missing masses? The sociology of a few mundane artifacts. In *Shaping technology/building society*, ed. W. E. Bijker and J. Law. Cambridge, MA: MIT Press.

———— 1993. *We have never been modern*. Trans. C. Porter. Originally published 1991. Cambridge, MA: Harvard University Press.

———— 1994. Pragmatogonies. *American Behavioral Scientist* 37(6).

Law, J. (ed.). 1991. *A sociology of monsters: Essays on power, technology and domination*. London: Routledge.

———— 1999. After ANT: Complexity, naming, and topology. In Law and Hassard 1999.

Law, J. and J. Hassard (eds) 1999. *Actor network theory and after*. Oxford: Basil Blackwell/Sociological Review.

Law, J and A. Mol. 1995. Notes on materiality and sociality. *Sociological Review* 43:274-295.

Lee, N. and S. Brown. 1994. Otherness and the actor-network. *American Behavioral Scientist* 37(6).

Lykke, N. 1996. Between monsters, goddesses and cyborgs: Feminist confrontations with science. In Lykke and Braidotti 1996.

Lykke, N. and R. Braidotti (eds). 1996. *Between monsters, goddesses and cyborgs: Feminist confrontations with science, medicine and cyberspace*. London: Zed Books.

McCaffery, L. (ed.). 1991. *Storming the reality studio: A casebook of cyberpunk and postmodern science fiction*. Durham, NC: Duke University Press.

Morton, D. 1999. Birth of the cyberqueer. In Wolmark 1999.

Muller, N. 1999. Suggestions for good cyberfem (house)-keeping: or how to party with the hyperlink. In Sollfrank and Old Boys Network 1999.

Mulvey, L. 1975. Visual pleasure and narrative cinema. *Screen* 16(3).

Munster, A. 1999. Is there postlife after postfeminism? Tropes of technics and life in cyberfeminism. Feminist Science Studies issue of *Australian Feminist Studies* 14(29):119-129.

Nalder, G. 1993. Under the VR spell? Subverting America's masculinist global hologram. *Eyeline* 21 (autumn): 20-22.

Nicholson, L. (ed.).1990. *Feminism/postmodernism*. New York: Routledge.

Olson, G. 1996. Writing, literacy and technology: Toward a cyborg writing. *Journal of Composition Theory* 16(1).

Piercy, M. 1991. *Body of glass*. Originally published as *He, she, it*. London: Penguin.

Penley, C. and A. Ross (eds) 1991. *Technoculture*. Minneapolis: University of Minnesota Press.

Pickering, A. 1995. *The mangle of practice: Time, agency, science*. Chicago: University of Chicago Press.

Plant, S. 1994. Cybernetic hookers. Paper presented at Adelaide Festival Artists' Week. Reprinted in *Australian Network for Art and Technology Newsletter* (April-May).

———— 1995. The future looms: Weaving women and cybernetics. In Featherstone and Burrows 1995. Reprinted in Wolmark 1999.

———— 1998. *Zeroes and ones*. London: Fourth Estate.

Roberts, C. 1999. Thinking biological materialities. Feminist Science Studies issue, *Australian Feminist Studies* 14(29).

Sandoval, C. 1991. US Third World Feminism: The theory and method of oppositional consciousness in the postmodern world. *Genders* 10.

———— 1995a. New sciences: Cyborg feminism and the methodology of the oppressed. In Gray 1995, reprinted in Wolmark 1999.

———— 1995b. Video production, liberation aesthetics, and US Third World Feminist criticism. In Brahm and Driscoll 1995.

Sofia, Z. 1993. *Whose second self? Gender and (ir)rationality in computer culture*. Geelong: Deakin University Press.

———— 1996. Contested zones: Futurity and technological art. *Leonardo* 29(1):59-66.

———— 1998. The mythic machine: Gendered irrationalities and computer culture. In *Education/ technology/ power: Educational computing as a social practice*, ed. H. Bromley and M. W. Apple. 29-51. Albany, NY: State University of New York Press.

Sofoulis, Z. 1994. Slime in the matrix: Post-phallic formations in women's art in new media. In *Jane Gallop Seminar Papers*, ed. Jill Julius Matthews. 83-106. Canberra: Humanities Research Centre, ANU.

———— 2003 (forthcoming). Cyborgs and other hybrids: Latour's nonmodernity. In *Encore le corps: Embodiment now and then*, ed. E. Probyn and A. Munster. London: Routledge.

Sollfrank, C. 1999. Women hackers. In Sollfrank and Old Boys Network 1999.

Sollfrank, C. and Old Boys Network (eds). 1999. *Next Cyberfeminist International*. Proceedings of conference in Rotterdam, March 1999. Hamburg: Old Boys Network and Hein & Co.

Springer, C. 1991. The pleasure of the interface. *Screen* 32(3).

———— 1996. *Electronic Eros: Bodies and desire in the postindustrial age*. Austin: University of Texas Press.

Stabile, C. 1994. *Feminism and the technological fix*. Manchester: Manchester University Press.

Star, S. L. 1996. From Hestia to home page: Feminism and the concept of home in cyberspace. In Lykke and Braidotti 1996.

Steffensen, J. 1994. Gamegirls: Working with new imaging technologies. *Mesh: Journal of the Modern Image Makers Association* 3 (autumn): 8-11.

Stone, A. R. (Sandy). 1996. *The war of desire and technology at the close of the mechanical age*. Paperback edition. Cambridge, MA: MIT Press.

Sturgeon, N. 1997. *Ecofeminist natures: Race, gender, feminist theory and political action*. New York: Routledge.

Turkle, S. 1984. *The second self: Computers and the human spirit*. New York: Simon & Schuster.

———— 1996. *Life on screen: Identity in the age of the Internet*. London: Weidenfeld & Nelson.

Vidler, A. 1992. *The architectural uncanny: Essays in the modern unhomely*. Cambridge, MA: MIT Press.

Volkart, Y. and C. Sollfrank. 1999. Editorial. In Sollfrank and Old Boys Network 1999.

Watson, G. 2000. *Just do IT: Australian women in cyberspace*. Ph.D. dissertation. School of Cultural Histories and Futures, University of Western Sydney.

Wolmark, J. (ed.). 1999. *Cybersexualities: A reader on feminist theory, cyborgs and cyberspace*. Edinburgh: Edinburgh University Press.

Woodward, K. 1999. From virtual cyborgs to biological time bombs: Technocriticism and the material body. In Wolmark 1999.

Zurbrugg, N. (ed.). 1994. *Electronic Arts in Australia*. Special issue of *Continuum: The Australian Journal of Media and Culture* 8 (1).

SECTION TWO

VIRTUALITY:
WEBWORLDS AND CYBERSPACES

> "I could be bounded in a nutshell and count myself a King of infinite space"
> HAMLET

Darren Tofts

In an age dominated by the concept of the virtual, what is reality? Simulation processes, both cultural and digital, have occasioned a new logic of appearances that dismantles the binary opposition anchoring our metaphysics in terms of a difference between reality and the seemliness of things. Virtuality is the name that has been given to this paradigm of experience, combining the discrete terms virtual and reality into a singular concept – a semantic fusion that enacts the dissolution of difference between the world and its representation. Virtuality implies that the seemly, the "almost real" no longer depends upon a reality to imitate, but is itself a state of being, an irreducible world orientation that brings with it the reassurance of intuitive, first-hand experience. If, as the essays in Section One suggest, cyborg subjectivity queers our sense of what constitutes life, virtuality prompts us to question the status of "reality" as the foundation and guarantee of what we now accept as being real.

The idea of reality as simulation, as model of a real, pre-dates the technological apparatus of VR, cyberspace, three dimensional interfaces and "immersive" experience – the ensemble of technologies and concepts, in other words, that we associate with computer networks, telecommunications and digital simulation. The immediate theoretical background to electronic virtuality can be traced to the 20th century critique of reproductive technologies and their impact on the defining principles of reality: presence, originality and authenticity. Institutionalized by Walter Benjamin in the 1930s, this critique exerted a profound influence on contemporary thought to do with the disappearance of reality and the emergence of a simulated virtuality. However, as the essays in this section suggest, the very logic of representation itself situates reality as merely one item in a series of appearances of the real, technological mediations that fuse reality with verisimilitude, actuality with virtuality. In other words, virtuality is a new name for an ongoing crisis *within* representation.

If reality is that which appears real, then electronic media, such as the computer network, have introduced a new dimension of the virtual into human life. The idea of an "outering" and distancing of the senses from the immediate context of the body, suggested in concepts such as telepresence – a distancing reminiscent of the disembodied model of mind discussed in Section One – suggests a dramatic change in human nature, an augmentation of the corporeal limitations of sensation. Have cultural phenomena such as cyberspace, with its promise of remote communication and perception at a distance, introduced new forms of social interaction as immediate – hence as real – as face-to-face communication? Virtuality, conceived in these terms, may be the portent of a real yet to come. That is, a real in excess of the real, to be delivered by powerful simulation technologies like the holodeck in *Star Trek: The Next Generation*, or the "jacked in" scenario of cyberpunk fiction – a real conceived on an infravisible, in-human scale of abstraction.

Returning to one of the defining moments in the history of representation, Gregory Ulmer enters Plato's cave in his "Reality Tables: Virtual Furniture," to explore what virtuality means in the context of an emerging paradigm of electracy. Less an essay, Ulmer presents an experiment in choragraphy, an abductive style of writing that draws on analogy and figurative juxtaposition to generate a theory, rather than assert an argument. Plato's cave is for Ulmer a relay, a point of departure for the invention of a theory of interface design appropriate to the social apparatus of the computer network. The displacement of speakers in electronically mediated dialogue is for Ulmer a recognition that the simulation of face-to-face immediacy is a metonym for embodiment, the interaction of whole bodies. Via a detour that seconds one of the more famous metonyms of the body into his model, Elvis Presley's pelvis, Ulmer develops, for the benefit of the human-interface design community, a theoretical "project of interpelvis (interbody) design for an online polylogue in cyberpidgin" – that is, a hybrid medium that fuses the residual traditions of speech (orality) and writing (literacy) to form a pidgin or expedient language that adapts the full social dynamics of FTF (face-to-face) to CMC (computer mediated communication).

Electrate dialogue, for Ulmer, resolves the differences between dispersal and synthesis, dilation and intimacy, since it allows for the simulation of proximity (virtuality) within embodied displacement (actuality). The notion of telepresence assumes an economy of relations between binary values: the local and the remote, here and there, fixed and distributed addresses. In "Porous Memory and the Cognitive Life of Things," John Sutton explores this economy in terms of the history of cognitive science and the development of a theory of the brain as a storage system for memory. Sutton draws on the exotic image of a 17th century sponge, reputedly capable of transmitting acoustic messages with precision when squeezed by a distant recipient. The idea of a porous, slippery storage mechanism for thought was attractive to the age of Descartes, for whom the brain was a fluid, dynamic cognitive tool of memory. But as Sutton notes, Descartes was aware of the fact that this fluidity was not particularly reliable and the flow of spirits through the brain's networks was fickle – writing three centuries before Norbert Wiener, Descartes knew only too well that information has a tendency to leak in transit. Seeking a responsive "solution" to this problem of memory storage, Sutton goes backward, rather than forward in time, to the ancient *ars memoria*, or arts of memory. In reviewing the memory palaces of St. Augustine and the Renaissance practitioners of memory architecture, Sutton draws a vital analogy between these ancient practices and contemporary information architecture as cognitive technologies. In addressing the relationship between memories that are stored or distributed, Sutton emphasizes the subtlety and interplay of these two apparent oppositions, noting that an integrated, networked approach to memory and information architecture is more in keeping with the "porous" nature of memory phenomena.

For Sutton, then, there is a rich historical legacy of thinking about memory and cognition that we need to draw on in our thinking about issues to do with cognitive technologies – as an extension of mind – and their interface with theories of locality and distribution. In "Becoming Immedia: The Involution of Digital Convergence," Donald Theall extends this idea of technological augmentation further by exploring the cognitive-sensory transformations precipitated by media convergence. Drawing on Pierre Teilhard de Chardin's concept of the "noosphere," Theall explores the ways in which mind, memory and consciousness have historically been modified by electromagnetic and, latterly, digital technologies. For Theall, the paleontologist-theologian's theory of a world shrunk into simultaneous presence anticipated the emergence of a new cybernetic age or "chaosmology," a further evolutionary stage in the becoming informatic. In the compression and connectedness of people's minds, we witness the emergence of the ultrahuman – a state of being that all the essays in this book are, in one way or another, attempting to describe and understand. Theall argues, however, that Teilhard, writing in the 1950s, didn't fully develop what he had intuited and that, in a curious anachronism, it was in the writings of James Joyce, in particular *Finnegans Wake* (1939), that we find a more complete exploration of his evolution of mind. And it is in *Finnegans Wake* that we also experience the first articulation of the cognitive/sensory writing of informatic subjectivity, a hypertextual, "dynamic ideography" that inscribes Teilhard's theory of extended communication (of which, too, Ulmer's choragraphic writing could be an example).

In Teilhard's noosphere Theall identifies a world beyond media, a multisensory "postelectric mimesis," or virtual realization. McKenzie Wark, in his discussion of Ray Bradbury's short story "The Veldt," identifies a similar realignment of the real, asking the question, "If virtual reality is mimesis, then mimesis of what?" Wark questions the mimetic context of discussions of the virtual and seeks instead to identify a different basis of relation between the real and the virtual. Wark sees the holographic nursery in the story as an allegory of the "too real," a phenomenological space of transition from the fantastic and improbable to the oppressive inevitability of the "more than real." For Wark the story is an instantiation of the becoming informatic, in which the programmed, holographic game-world of the nursery is not a space of representation but of emergence, a mechanism for the creation of new worlds and the alarming, unexpected metaphysics they bring with them. In the mythos of the story, the death of the real is a consequence of the irruption into the world of something new, a transubstantiation of the abstract into the corporeal. This dramatic shift from metaphor to reality, if it is representative of anything, concerns the end of one media paradigm (representation), and the becoming into the world of another (virtuality).

The idea of simulated reality being in excess of, or beyond the real, is also developed in Scott McQuire's discussion of William Gibson's *Neuromancer* and his iconic concept of cyberspace. McQuire sees cyberspace as a figure for a post-televisual, de-territorialized culture still coming to terms with the social and cultural changes associated with the rapid implementation of computers and information technology. The parallels between urban space and data space prompt a dramatic reconsideration of our experience of experience itself – how, for example, do we come to terms with the notion of "telepresence"? Focussing on Gibson's portrait of the matrix, McQuire argues that while cyberspace is a space beyond representation, its figuration as the image of an illuminated city at night sustains, rather than defuses, the abstract nature of simulated experience. Gibson's portrait of the "unthinkable complexity" of cyberspace invokes the experience of the sublime, of awesome, unbearable excess. In the realpolitik of an increasingly telematic culture, accelerated telecommunication spaces bring with them real anxieties about what has happened to the real, to space, place, and identity.

In pursuing the virtual, these writers face the same challenge that beguiled the great Jorge Luis Borges, who, in his extraordinary 1949 text "The Aleph," grappled with the idea of an "other" space that defied logic and baffled description. In a nondescript domestic cellar in Buenos Aires, all time and space is glimpsed in a small iridescent sphere no bigger than a coin – domesticity and "intimate immensity" indelibly juxtaposed, without seeming in any way unusual. The passage from Plato's cave to Borges' cellar is a long one, but the quest to understand and accept the real as a postulate of the virtual, the "as if real," is much the same. Borges' Aleph is a vision of the infinite, a way of seeing all things glimpsed in an instant. Such "allatonceness" is a reminder of what the real has become in an age dominated by electronic communications and media networks, a real apart from the real, a virtuality reducible to nothing outside itself and as compelling as the voice on the other end of the line: "Eventually the telephone lost its terrors" (Borges 1998, 280).

REFERENCES

Borges, J. L. 1998. *Collected fictions*, Trans. A. Hurley. London: Penguin.

REALITY TABLES :
VIRTUAL FURNITURE

Table: An arrangement of numbers, words, or items of any kind, in a definite and compact form, so as to exhibit some set of facts or relations in a distinctive and comprehensive way, for convenience of study, reference, or calculation.

Gregory L. Ulmer

1 . 1

William Burroughs left an imperative: storm the reality studio. The reality studio refers in general to all those forces of control in our lives, and in particular to the media institutions that have made us into image addicts. In his iconoclasm Burroughs is a Platonist. The relationship between the reality studio and Plato's allegory of the cave has been noted before, within the context of a critique of the society of the spectacle. From the point of view of grammatology, however, the theoretical hyperbole surrounding debates about the spectacle is a symptom of metaphysical metamorphosis, articulating the emergence of a new apparatus inflected by negation. Academics and intellectuals (like me) cannot be done with Plato or Sophocles and the rest of the Greek heroes for the simple reason that we are their heirs, working the legacy of literacy. To brand this legacy "logocentric" is simply to identify it as relative to an apparatus – a social machine that includes technology, institutions, and identity formations in an irreducible interdependent matrix. The powerful methodologies of critical theory are but one of the features of literacy, whose arguments condemning the spectacle are one of the means by which electracy is being bootstrapped out of literacy.

   A good example of this bootstrapping process by which a new metaphysics rises phoenix-like from the ashes of the old (see how the Greeks hold me in thrall!) is the anthology of essays on the cinematic apparatus edited by Theresa Hak Kyung Cha (1980). The collection of essays by critics and creators – the likes of Barthes, Vertov, Baudry, Deren, Vernet, Straub/Huillet, among others – is framed by a series of citations or extended epigraphs interpolated at regular intervals, in white print on black pages. The first epigraph is a fragment from Plato's *Republic* that begins with Socrates saying, "here is a parable ..." It is the allegory of the cave. The next entry in the series is a paraphrase of the allegory by Diderot, a repetition that sets the theme upon which the remaining entries (by Balzac, Apollinaire, and Henry James) play a variation. This frame evokes in the context of new media the ambivalence of Plato towards the apparatus that he helped to invent. Today we are inventing electracy, rather than literacy, and we cannot help but be ambivalent about a formation that has so little place for the craft that made it possible.

Consider, for example, the table.

1 . 2

The "table," along with the list, formula, and recipe, has been described as one of those "figures of the written word" that are native to literacy (Goody 1977, 17). The heuretic method (using theory for purposes of invention) is to reconsider such claims in terms of the word-thing, to write "table" with its whole set of phrases and shapes. Richard Bolt (Senior Research Scientist in the Perceptual Computing Group at the Media Laboratory, MIT, and author of *The Human Interface*) gave a presentation at Ohio State University in the fall of 1993, part of a series on Technology and Postmodern Culture, in which he argued that "dealing with computers will become less like operating a device and more like conversing with another person." Bolt demonstrated his point by means of a computer program that responded to voice command, gesture (through feedback from a digital glove), and gaze (through feedback from an eye-tracking device). The display contained a chair and a table on which were placed a glass and a pitcher, on a floor covered with black and white tile. Bolt's presentation provides a point of departure for exploring further the nature of choragraphy. I want to look more closely at the two central elements of Bolt's performance: the interface metaphor ("like conversing with another person"), and the demonstration of a virtual (computer-generated) table.

The virtual table brings to mind Plato's three beds (or tables), listed in order of reality from most to least real – the pure form (the idea of the table), the carpenter's table and the picture of a table made by an artist. The virtual table is considered to be a manifestation of simulation, which passes beyond representation to constitute a fourth item on this metaphysical list. How should I take the table Bolt displayed? The very insistence on the table – moving it, raising, rotating, and lowering it before the fascinated gaze of the audience – evokes an allegorical effect. A starting point is the meaning of this table – a hermeneutic question. To move into a heuretic relationship with the table, to learn not what it means but what can be made of it, I put it into a sequence, a series of other tables (the choral table is all tables at once), beginning with the form of which the Periodic Table is the most famous example. In the hybrid circumstances of the webpage, HTML tables show the persistence of literacy during the transition to electracy. Alongside the table as chart is the image of the table with its ability to evoke an associative series.

Bolt's interface metaphor may be contextualized in the history of dialogue, and this metaphor (interacting with a computer is like having a conversation with a person) is a clue that opens up online interactivity to invention. I accept as the terms of my experiment the idea that the future of education in electracy depends upon an ability to extend and adapt the dialogue to computing. What has to be imagined is a dialogue that foregrounds not "communication" but "signifierness" (the "signifiance" of Barthes' obtuse meaning [Barthes 1981]). The first thing that a quick review of the tradition reveals is the fact that the meanings of the terms are unstable and shift from epoch to epoch. Thus, for example, "to converse with a person" means quite different things in an oral civilization and in a literate one. I have to assume that when the technology is electronic rather than print or speech (the different media imply different institutions contextualizing their employment) both the practice of conversing and the nature of personhood will be undergoing a transformation. The task for interface designers is to invent the prototype of an electrate dialogue.

1.3

Plato invented the dialogue as the basic practice for his institutionalization of alphabetic writing in School. As Jan Swearingen has observed:

> When Plato wrote, dialogue was neither an already ancient literary tradition nor the simple transcription of natural conversation. Instead, dialogue was unprecedented and was inaugurated by Plato's hybrid of oral and written conventions, oral genres, and philosophical modes, a blend he termed dialegesthai, not just two but many voices 'crossing speakings' or speaking across one another, or 'spanning' or 'comprehending' each other's statements (Swearingen 1990, 49).

The dialogue actually is part of a collection of interrelated inventions, all of which were designed to take advantage of the material features of alphabetic writing: dialogue, method, and school. Plato's Academy is considered the first school. Method was the practice and dialogue the genre or form in which this practice was expressed in school. The dialogue entitled *Phaedrus* is the first discourse on method in the western world.

Phaedrus is a relay, a point of departure for generating its equivalent for the new apparatus (with its emerging new social machine relating technology, institutional practices, and human subject formation). The dialogue was a hybrid – partly oral, partly written – intended as a way to communicate specific to the new institution (school). What are the elements of the hybrid?

From the old oral culture Plato retained the scene of a face-to-face (f2f) exchange of two speakers; from the new written culture he accepted the abstracting procedures of analysis. When the two are fused in *Phaedrus* (and the other dialogues) the result is "dialectic." The dialogue was a written drama (a narrative) showing speakers performing the dialectical method. In the Academy, students read the dialogues as a basis for further discussion. This discussion was not "conversation" in the form practiced in daily life, but the special way of talking adapted to the work of specialized knowledge. To converse in school, from Plato's day to our own, is as different from conversing in the home as conversing in the home is different from speaking in a ritual ceremony (such as in church).

Oddly enough, this difference, like the technological basis of the practices of schooling, is often forgotten or ignored. To make use of the metaphor of conversing with another person for designing the human-computer interface, it is important to remember that the inventors of modern literacy made use of the same metaphor for the human-book interface. Conversation as represented in the tradition of dialogue must be recognized as a response to the possibilities of a new apparatus, including the demand for new institutional practices and new individual behaviors related to the social assimilation of a new technology of language.

Plato's invention supplies an inventory of what electracy needs: what are the equivalents of school, method (dialectic), and dialogue (genre, mode) in our circumstances? For starters, what are the parts of the hybrid? Literacy is for us what orality was for Plato. From our position at the close of the era of literacy and the opening of the era of electracy, what parts of writing will be retained, and what parts of the new media will be accepted as the elements from which to compose an online educational practice? What will be the result of this fusion, unlike either one of its two component parts? The functionality of this invention should be the electrate equivalent of method (what elsewhere I named "choragraphy," extrapolating from Derrida's updating of *chora* from Plato's *Timaeus* a new practice of "memoria" [Ulmer, 1994]). This discussion of an electrate dialogue is one component of a larger research program engaged with mapping rhetoric onto new media practices. According to this relay, heuretics and choragraphy are the inventio and memoria of electracy. Given the limitations of an article, this status may only be asserted, but not demonstrated nor proven.

1 . 4

Between Plato's time and the present the dialogue underwent a complex evolution, marked by at least three distinct "moments."

> Plato's dialogue transcended earlier 'story,' that of epic and drama, with a protocol for conceptual interlocution that was designed to frustrate the technification of thought. Schleiermacher's translation of Plato's dialogues into German came to be the headwater of a massive reform … that emphasized natural voices, living speech, union between minds rather than understanding of texts, polyphony, deliberately unsystematic philosophizing in spoken and written dialogues. The modern hermeneutic tradition represented by Heidegger, Ricoeur, Gadamer, and Bakhtin has revived and extended the German romantic template to emphasize the irreducibly polyvocal, interlocutionary elements in all language (Swearingen 1990, 68).

The essential shift from Plato to today is the result of the expansion of literacy. Plato's relation to text was pre-hermeneutic, as reflected in Socrates' complaint in *Phaedrus*:

> Writing, you know, Phaedrus, has this strange quality about it, which makes it really like painting: the painter's products stand before us quite as though they were alive; but if you question them, they maintain a solemn silence. So too with written words: you might think they spoke as though they made sense, but if you ask them anything about what they are saying, if you wish an explanation, they go on telling you the same thing, over and over forever (Plato 1956, 69).

Plato feared and warned against the danger of texts being misunderstood in the absence of the author (the text as orphan without its father/author to protect it). Although the methods of interpretation developed in the epoch of literacy proved that the same words could be made to say quite different things, hermeneutics was intended to preserve the presence of the father/author as a spirit accessible by means of the text. "Dialogue persists in evoking the quasi-religious because even when it is not being used in the hermeneutics of sacred texts it invokes the presence of the author" (Swearingen 1990, 69). In other words, the face-to-face encounter of the partners in an oral culture that Plato's hybrid preserved in the written dialogue persisted throughout the entire history of the form, manifesting itself in our own moment as an ethical imperative.

The history of reading in fact may be understood as the exploration of the interface metaphor of conversing with another person. Most of the protocols for reading and writing, developed especially during the era of print, aimed at controlling illocutionary force (to preserve intact the intent of the author during the event of communication) (Olson 1994, 113). The prejudice against the text is reduced, of course, and the condition of having a relationship with a book is accepted. The Socratic question and answer is transferred to this relationship, as in Gadamer's treatment of the text as an answer to a question:

The reconstruction of the question to which the text is presumed to be the answer takes place itself within a process of questioning through which we seek the answer to the question that the text asks us (quoted in Crowell 1990, 345).

Gadamer, that is, transferred "voice" to text in order "to construe the text as a partner in that dialogue constituted on the other side by the interpreter's (reader's) interrogative activity" (343). The ideality of the word and its relationship to truth are not affected by medium or institution in Gadamer's hermeneutics.

1.5

Plato's invention of the dialogue involves the invention of an interface metaphor: writers relate to the written as if they were conversing (communicating) with a person. This metaphor is enforced by means of an ethical imperative whose most recent spokesperson was Levinas.

> Why does Levinas insist on the irreducibility of the face-to-face? The answer lies in his conception of the face as 'expression': 'The face is a living presence; it is expression … The face speaks. The manifestation of the face is already discourse' (Crowell, 1990, 355).

This insistence within the modern moment of dialogue on the oral metaphor of conversation reflects a conception of the human subject that, ironically, came into being as part of the apparatus of literacy. The concept and behaviors of "selfhood," as grammatologists such as Eric Havelock have argued, are as much an invention as is the alphabet or school. An apparatus is a social machine. The ethical problem addressed by the face-to-face encounter is a feature of the separating or alienating of the person from the collective people, a unity that is replaced with a subject-object relationship to the world. The methods of abstraction and the experience of individuality both evolved as part of the matrix of literacy. The invention of the closeup shot in cinema, and the important theoretical discussions devoted to "the face" in media criticism, extend (despite appearances) the communication ideology into electracy.

The ethical dilemma of the relationship between self and other (the problematic of the Other) has dominated and driven the evolution of hermeneutics. The hermeneutic procedure in the modern moment is

> … characterized by a structure of alienation and return, excursion and reunion or, in Bakhtinian terms, of identification and exotopy. As for Gadamer, in the homecoming of Bakhtin's prodigal Self, the Self becomes 'more' than it was before: after returning home from its long journey, the Self is more itself … 'I' can now, upon return in my own unique 'placement' in existence, complete the Other, since I have the Other's vantage point and some extra features to which only I have access (Daelemans and Maranhão 1990, 228).

Part of the difficulty of extending dialogue into the electronic apparatus has to do with the transformation of the apparatus as a whole: not only does the technology change, but so too do the understanding of the person and the behaviors creative of subject formation. Keeping in mind Thomas Kuhn's notion of paradigm shifts, and the incommensurability of the world views of different paradigms, I assume that the ethical dilemma of self/other will not be solved in an electronic apparatus, but simply that it will become irrelevant, just as "appeasing" the gods, which was the problem addressed by ritual, became irrelevant in literacy, even if ritual form – in theater – continued within literacy. These transformations do not happen by themselves, however, nor do they happen in any predetermined way. How might the process of interface design take into account the wisdom of the ethical imperative of the face-to-face conversation in dialogue without losing sight of the emerging conditions of a new apparatus?

DEATH TABLES

L. tabula a flat board, a plank, a board to play on, a writing tablet, a written tablet, a writing, a list, an account, a painted tablet, a painting, a votive tablet, a flat piece of ground, prob. from same root as taberna TAVERN.

2.1

The story of Simonides, the inventor of mnemonics, is well known. As reported by Cicero in *De Oratore*, Simonides left Scopas' banquet hall just before the roof collapsed, killing all those in attendance.

The corpses were so mangled that the relatives who came to take them away for burial were unable to identify them. But Simonides remembered the places at which they had been sitting at the table and was therefore able to indicate to the relatives which were their dead (Yates 1966, 2).

The effectiveness of his recall suggested the method of places and images that allowed rhetors to manage the information explosion created by writing, and to give long speeches from memory.

> [Cicero:] 'He inferred that persons desiring to train this faculty (of memory) must select places and form mental images of things they wish to remember and store those images in the places, so that the order of the places will preserve the order of the things, and we shall employ the places and images respectively as a wax writing-tablet and the letters written on it' (Yates 1966, 2).

Like Simonides, interface designers are in the position of having to invent a practice for the use of a new information technology, with the moment of invention being informed by a catastrophe, by a scene of massive death. My interface design, that is, must work with the Internet (electronic memory technology), which originated in the ARPAnet experiment to support military research:

> ... in particular, research about how to build networks that could withstand partial outages (like bomb attacks) and still function ... In the ARPAnet model, communication always occurs between a source and a destination computer. The network itself is assumed to be unreliable; any portion of the network could disappear at any moment (pick your favorite catastrophe – these days backhoes cutting cables are more of a threat than bombs) (Krol 1992, 11).

The nuclear catastrophe the military researchers had in mind required that the invention of a memory system be accomplished prior to the collapse, rather than after it, as in the case of Simonides. Our new dialogue should take into account and assimilate within itself the anticipation of these conditions of communicating by means of a ruined interchange.

The metaphor for heuretic invention has already been provided by Thomas Erickson, who suggested the evolution of pidgin languages into creole as a model for the evolution of the Macintosh interface.

> The characteristic ways that pidgins evolve into creoles may tell us what properties a linguistic system – or an interface – must have for it to become a powerful communicative device while remaining relatively simple and easy to learn (Erickson 1990, 13).

Erickson does not push the metaphor very far, since the main point of his article is to invite others to use a system of analytical extrapolations from models in order to become more inventive about interface design. The creole model is introduced as an example of how the process might work. Even within the reduced terms of an example, however, the brilliance of Erickson's suggestion is clear and may be appropriated as a "relay" for choragraphy. Pidgin begins as a language for doing business in the absence of a common language among different peoples, and evolves into a creole (that is, into a powerful, full-featured language) in the speech of pidgin-speaking children, once the pidgin has become so common that it is spoken in the home (14).

Since it is not important for his immediate purpose, Erickson only mentions in passing that one of the conditions in which creole developed was "a result of the slave trade in the Caribbean, when slaves from the same areas were deliberately separated to reduce the possibility of uprising" (14). His method of examining the symmetries between two juxtaposed domains (in this case between creole and the Macintosh interface) reveals the fit between the destroyed and dispersed scenario of the Internet design and the conditions of the slave trade that created most creole languages. The choral method, that is, is to consider any figurative juxtaposition in terms of the full semantic domains on either side of the line (bar). The catastrophe in the latter case includes the notorious middle passage in which as many as ten million Africans were forcibly removed from their native lands and dispersed throughout the colonies, a passage which millions of individuals did not survive. A manifesto for creoleness refers to creolization as "the brutal interaction of culturally different populations":

> Generally resting upon a plantation economy, these populations are called to invent the new cultural designs allowing for a relative cohabitation between them. These designs are the result of a nonharmonious (and unfinished therefore nonreductionist) mix of linguistic, religious, cultural, culinary, architectural, medical, etc. practices of the different people in question (Bernabé et al. 1993, 92).

Erickson limits his explicit exploration of the creole interface metaphor to the linguistic dimension (proposing that the Macintosh interface might evolve a greater complexity of syntax, tense, and vocabulary). This analogy may be extended to include Bernabé's, Chamoisseau's, and Confiant's broader understanding of creole as a cultural discourse. Meanwhile, the mnemonic scene supporting and guiding choragraphy portrays the possibilities of surviving the greatest imaginable destruction – of the Internet continuing to function after a nuclear strike; of African culture continuing to function after the diaspora. The creole manifesto makes a good case, moreover, for adopting creole as a model for interaction in the electronic apparatus:

A new humanity will gradually emerge which will have the same characteristics as our Creole humanity: all the complexity of Creoleness. The son or daughter of a German and a Haitian, born and living in Peking, will be torn between several languages, several histories, caught in the torrential ambiguity of a mosaic identity. To present creative depth, one must perceive that identity in all its complexity. He or she will be in the situation of a Creole (112).

2 . 3

The relevance of creole to dialogue has to do with what Bernabé, Chamoisseau, and Confiant describe as the fundamental orality of creole (95). The setting of *Phaedrus* – Socrates and Phaedrus hold their conversation on the banks of the river Ilissos – is one that would immediately evoke in Plato's audience the context of the Eleusinian mystery religion, just as surely as the setting of a frontier town evokes the genre of the western for an American audience.

> At the end of the classical period the philosophical imagination set a higher visio beatifica above the Eleusinian vision, building on this religious experience, known to almost every Athenian, as on an existing, self-evident foundation (Kerényi 1967, 98).

Plato used the mystery religion's ritual practice of moving from a physical to a spiritual seeing in a figurative sense, thus preserving a part of oral practice in his hybrid invention. That the genre of the "dialogue with the dead" evolved in antiquity out of the Socratic dialogue may be due in part to this association of the invention of method with the rituals of the Mysteries, participation in which "offered a guarantee of life without fear of death, of confidence in the face of death" (15).

The dialogue form shares with ritual the function of liminality (theorized in the work of Victor Turner [1982]).

> From its inception in Plato, dialogue has always been and continues to be programmatically liminal: interstructural, between two states or conditions, essentially unstructured rather than structured by contradictions; because of its deliberate avoidance of closure and finality, it serves perpetually as a vehicle for reformulating old elements into new patterns (Swearingen 1990, 47).

The creoleness of the dialogue is evident in Plato's definition of dialogue "as a vast experiment in mixing and mingling discourse conventions in order to bring about optimum understanding, partaking of extant patterns but altering them to fit new objectives. It can, I think, be productively applied to transitional discourses, logics, and literacies today" (67).

The "death" invoked by dialogue includes "the death of one discourse in the birth of another," according to Stephen Tyler.

> Its image is 'X,' the coming-together-of-the crossing-getting-across-crossing-over-crossing-out, the chiasmus within the logos conjuring at one and the same time a bright vision of reason, agreement, consensus, communion, unity, harmony, and mutuality and a darker picture of resistance, disagreement, disensus, opposition, antagonism, and raised voices (quoted in Maranhão 1990, 294).

This "X" is the shape (the *eidos*) guiding the design of an electronic dialogue.

The X eidos plays a foundational role in the African cultures that constitute an essential part of the creole mix. In the Black Atlantic world, for example, the god Eshu is widely honored.

> Eshu came to be regarded as the very embodiment of the crossroads. Eshu-Elegbara is also the messenger of the gods, not only carrying sacrifices, deposited at crucial points of inter-section, to the goddesses and to the gods, but sometimes bearing the crossroads to us in verbal form, in messages that test our wisdom and compassion … He sometimes even wears the crossroads as a cap, colored black on one side, red on the other, provoking in his wake foolish arguments about whether his cap is black or red, wittily insisting by implication that we view a person or a thing from all sides before we form a general judgment (Thompson 1984, 19).

The crossroads, associated with change, were figured in the cosmograms that display the ritual quality of writing in oral culture. According to Wyatt MacGaffey, a scholar of Kongo civilization:

> The simplest ritual space is a Greek cross, marked on the ground, as for oath-taking. One line represents the boundary, the other is ambivalently both the path leading across the boundary, as to the cemetery; and the vertical path of power linking, 'the above' with 'the below.' This relationship, in turn, is polyvalent, since it refers to God and man, God and the dead, and the living and the dead. The person taking the oath stands upon the cross, situating himself between life and death, and invokes the judgment of God and the dead upon himself (quoted in Thompson 1984, 108).

Robert Farris Thompson's *Flash of the Spirit* (1984) inventories the survival and continuation in the diaspora of the different manifestations of the cross from different African cultures into the Atlantic world. All of them have to do with the intersection of two lines associated with the "dia" of dialogue:

> The quartered circle with four eyes – spiritual communication and enlightenment, the core emblem among the signs of the anaforuana corpus – becomes, in Abakuá masking, a Janus with two eyes seen in front and two implied on a disk at the back of the head. And just as the intersection of two lines in nsibidi communicates the intersection of words of one person with words of another, with the same sign the 19th century Abakuá masker visually voices the idea of speech. He crosses broom and wand before his body to indicate a desire to speak of things positive and lasting (262).

The various geometries of the cross were used in practices concerned with mediating the power across worlds, especially between the living and the dead (252). The rituals, that is, served as the methodology for gaining access to the stored wisdom of the ancestors, including a collection of institutional practices such as divination and the making of charms. One of the most elaborate signs in the Abakuá "calligraphy of the dead-the sign of the lifting of the plate," refers to the custom of taking up the plate of a dead person from his table, "for no longer will he use it in this world" (268). The Spanish translation of the ritual cited by Thompson – *levantimiento de plato* – suggests a macaronic pun between "plato" and Plato that signals the nature of the logic of patterning organizing the table series. The purpose of the ritual identified by the ideographic cross of the lifting of the plate, "honoring myriad ancestors simultaneously" (268), is to preserve in memory the cumulative knowledge of the civilization.

Today the function of similar expressions presiding over Kongo graves is the blocking of the disappearance of the talents of the important dead. Lifting up their plates or bottles on trees or saplings also means 'not the end,' 'death will not end our fight,' the renaissance of the talents of the dead that have been stopped, by gleaming glass and elevation, from absorption in the void (Thompson 1984, 144-145).

Plato is one of our ancestors to be honored in this way, even as he is creolized into a new apparatus. Such in any case is the spiritualist setting of table-turnings and levitations evoked by Richard Bolt's virtual table.

TRUTH TABLES

If the letter c is used to represent the compound proposition a truth table can be constructed as follows:

a	b	c
0	0	0
0	1	0
1	0	0
1	1	1

This table is the same as the multiplication table for the numbers 1 and 0. It is also the same as the table for a circuit with two switches in series.

3.1

The convergence of logic with technology that produced the computer is one of those happy moments in the history of invention. The insight that linked the truth table with an electrical switch is to our time perhaps what the conjunction of the theology of God as light with the invention of the pointed arch was to the Gothic period (recognized by the Abbot Suger and realized in Chartres Cathedral). Cyberspace, in any case, has been described as the "cathedral" of the postmodern masses. In the context of interface design, however, the point to keep in mind is the one Raymond Kurzweil makes, tracing the background of this insight from Bertrand Russell's solution to the paradox of the set that is/not a member of itself to the Turing machine.

> The Turing machine can execute any plan to solve a cognitive problem. Its success in doing so will be a function of the validity of the plan. We thus come to the same conclusion that we did with the sea of logic. The Turing machine provides us with a simple and elegant model for the hardware of cybernetic (machine-based) cognition, but not the software. To solve a practical problem the Turing machine needs a program, and each different program constitutes a different Turing machine (Kurzweil 1992, 123).

Here is the crucial point of the invention process – the open relationship between hardware and software. The truth table with its technological embodiment in the computer is the culmination of the history of analytical thinking from Plato to Turing. As this history itself shows, however, analytical logic is only one part of reason.

> Two types of thought processes coexist in our brains. Perhaps most often cited as a uniquely human form of intelligence is the logical process involved in solving problems and playing games. A more ubiquitous form of intelligence that we share with most of the earth's higher

animal species is the ability to recognize patterns from our visual, auditory, and tactile senses (223).

Most of the thinking about computing, from artificial intelligence to video games, has been conducted in terms of the first form of intelligence. The development of dialogue after Aristotle's codification of dialectic into predicational logic reflected this subordination of pattern to analysis, reducing it to the monological recitation of a catechism (Bakhtin 1984). The computational continuation of this tendency may be seen in the expert systems that yoke the dialogue to a diagnostic Q & A. The electronic dialogue, however, must take into account the second form – the intelligence of patterning.

3 . 2

A common way to understand this shift of emphasis from one kind of intelligence to the other is by referring to the shift in Wittgenstein's thought from the *Tractatus* (concerned with abstract truth functions) to *Philosophical Investigations* (concerned with the everyday usage of language). Indeed, the artificial intelligence movement has already made this shift for itself, turning away from its initial attempts to model intelligence in terms of the abstractions of symbolic logic to take up modeling the common sense of everyday life. Such a step is in the right direction, allowing us to think about the implications of communicating with computers by means of the usage of voice, gesture, and gaze (the means Bolt used to manipulate his virtual table) in the context of historical human experience. Thus when considering the gestures to be used in dialoguing with a computer (or of "dialogging on") one might think of the importance Wittgenstein gave to an exchange with his Italian friend, Piero Sraffa, who responded to Wittgenstein's claim that a proposition and what it describes must have the same "grammar" by making the Neapolitan gesture of "the chin flick." Brushing his chin with his fingertips Sraffa asked, "what is the logical form of that?" (Monk 1990, 260-261). The gesture means disinterest, disbelief, a negation.

> It is a symbolic beard-flick, the gesturer flipping his real or imaginary beard upwards and forwards at his companion. As a simple insult, this means 'I point my masculinity at you,' and is associated with verbal messages such as: buzz off, shut up, get lost, don't bother me, or I have had enough of you. But it is also frequently used as a special kind of insult implying boredom (Morris et al. 1979, 170).

Wittgenstein's reliance in his second phase on the metaphor of language games and his concern with the recognition of faces in his use of family resemblance as a way to think about conceptual categories indicate that he was seeking a balance between patterning and analysis. Wittgenstein's adaptation of the Gestaltist duck-rabbit paradoxical drawing late in his career to discuss his own version of the experience of seeing-as, of recognizing an aspect, opens a dimension beyond common sense. The larger importance of this example of pattern may not be recognized until it is put together with the biographer's report that Wittgenstein once revealed to a friend a childhood memory to which he ascribed great significance: he recalled that in the lavatory of his childhood home the plaster had fallen off the wall in such a way as to form a pattern that little Ludwig saw as a duck. The image, which he associated with the monsters in a painting by Hieronymous Bosch, frightened him (Monk 1990, 451). Wittgenstein was thinking with this (irrational) fright.

The details of these anecdotes from Wittgenstein's biography have important implications for interface design in general, and for artificial intelligence in particular (which is why they recur in so many different contexts in these pages). The move to imitate common sense in smart machines goes part of the way towards acknowledging that the mind is in a body. So far, however, the thinking about this body-mind relation has been limited to certain basic physical facts of embodiment (such as the oriented nature of human perception in terms of up-down, front-back, right-left) and certain social facts that support inference (schemas). As important and necessary as this addition of embodied common sense is for communicating with computers, it is still as incomplete in its own way as was the reduction of mentality to the manipulating of formal symbols. Good sense and common sense, according to Deleuze, return everything to the identity of a unified self, and as such are bound up with the circular logic of hermeneutics. The logic of sense theorized by Deleuze addresses instead a subjectivation that is pre-individual and non-personal (the unconscious part of the subject). The second phase of AI is also doomed to failure and must give way to a third phase – the phase of "artificial stupidity" (AS, pronounced "ass," related to ATH) that takes into account unconscious mentality.

3 . 3

Gilles Deleuze provides the Theory of this exploration of choragraphy as a new dialogue, for he makes it possible to state the fundamental insight to be drawn from the X tables. The assumption of interface design is that the computer is a prosthesis augmenting the human mind. That mind is an effect of a brain located in a body. This embodied mind is positioned within a family, in a culture, a society, a historical moment (the "popcycle" of discourse institutions). The person possessed of this mind is sexed, gendered, nationalized, classed, raced, ethnicized, and so on through all the ideological categories used to describe identity. Deleuze provides a theory that suggests a way to take the full scope of this personhood into account in the metaphor of dialogue (to communicate with a computer is like having a conversation with another person). He shows how to include the unconscious in an electrate model of mind. Electracy adds to orality and literacy the possibility of writing the unconscious (and hence of writing with what we don't know – with our stupidity: ATH). The question becomes: in what setting does this conversation occur?

In *The Logic of Sense Deleuze* (1990) shows how to extend method beyond analytical intelligence into the reasoning of pattern, how to go beyond the truth table that brought together logic and the electric switch to solve the problem of hardware, in order to clarify the nature of choragraphy which concerns the logic of software. Partly because he is drawing on the Stoic invention of "sense" (a reversal of Platonism), Deleuze echoes many of the terms of the history of dialogue in his account, including the X or crossing lines of dia.

> The Aion is the ideal player of the game; it is an infused and ramified chance. It is the unique cast from which all throws are qualitatively distinguished. It plays or is played on at least two tables, or at the border of two tables. There, it traces its straight and bisecting line. It gathers together and distributes over its entire length the singularities corresponding to both. The two tables or series are like the sky and the earth, propositions and things, expressions and consumptions. Lewis Carroll would say that they are the multiplication table and the dinner table. The Aion is precisely the border of the two, the straight line which separates them; but it is also the plain surface which connects them (Deleuze 1990, 64).

Here is the key to a choral dialogue – tables for truth and for eating. The interface design needed in order to fully exploit the convergence of technologies in digital hypermedia may be invented out of this convergence in theory of the truth table with the dinner table (producing the series truth table/electric switch/dinner table). Having psychoanalysis in his background the way Socrates had Pythagoreanism, Deleuze formulates the orality of conversation in terms of a new dimension. The mouth participates in more than one circuit – not only the circuit with the ear related to voice, but with the anus related to eating: "speaking will be fashioned out of eating and shitting, language and its univocity will be sculpted out of shit" (193).

A similar insight motivated Michel Serres' demonstration of isotopies across anthropology, biology, and information theory, coordinated (thanks to a puncept in French) by the "parasite":

> In every instance, the parasite disrupts a system of exchange: the disagreeable guest who partakes of a meal and offers only words in return; the organism that physically enters and feeds off of its host; the noise that interrupts messages between two points in an information circuit (Serres 1982, 14).

The condition of invention, Serres points out, is the possibility of interference between systems, with the inventor positioned on the border between message and noise, between the telephone and the feast, in his parable (67). His insight is that, "to succeed, the dialogue needs an excluded third; our logic requires the same thing." He gives us something to think about in the coexistence and rivalry of different economies, different logics: exchange, and gift. "In the logic and economy of the law and of possession, exchange reigns, weighing and measuring, figuring out the balance; in the logic and economy of the freely given, exchange is not there" (30). The riddle about the difference between tourism and revolution is sorted out by situating each in its relevant economy: revolution-Marxism-exchange; tourism-choragraphy-gift.

The interference between these economy-logics occurs in the host-guest-parasite situation – three terms, three positions in any dialogue.

> That is why the relation of exchange is always dangerous, why the gift is always a forfeit, and why the relation can attain catastrophic levels. It always takes place on a mine field. The exchanged things travel in a channel that is already parasited. The balance of exchange is always weighed and measured, calculated, taking into account a relation without exchange, an abusive relation. The term 'abusive' is a term of usage. Abuse doesn't prevent use. The 'abuse value,' complete, irrevocable consummation, precedes use- and exchange-value. Quite simply, it is the arrow with only one direction (Serres 1982, 80).

I am trying to understand, following Serres, the positive or fortunate side of this irreducible abuse, however counter-intuitive it might seem. The response from within the exchange system often is repression, an attempt to exclude the other system, the abuse, the parasite, the gift.

> This couple [noise-message] and their relation are set apart by an observer seated within the system. In a way, he overvalues the message and undervalues the noise if he belongs to the functioning of the system. He represses the parasites in order to send or receive communications better and to make them circulate in a distinct and workable fashion. The repression is also religious excommunication, political imprisonment, the isolation of the sick, garbage collection, public health, the pasteurization of milk, as much as it is repression in the psychoanalytic sense. But it also has to do with a history, the history of science in particular (68).

Invention depends upon the possibility of switching positions, of locating the border, the position from which it is possible to compare inside and outside and to realize that these are both necessary and unstable. The dualism and rivalry of economy-logics is embodied in the Habermas-deconstruction debate. "The language distinction can be described as one between language that 'coordinates action-in-the-world' and language that is 'world-disclosing'" (White 1991, 25). Briefly put, the distinction is between language that works and language that plays (revolution and tourism), understood not as opposites but as supplements (one parasites the other). Habermas favors the former because he fears that the latter renounces responsibility for problem solving in ordinary reality. Chorography as a practice is meant to give the learner the ability to work-play on both sides of this gap. It is not against communication, but for noise as a lever of creativity.

3 . 4

Perhaps it is possible in this context to appreciate the symmetry relating the ruined design of the Internet and the culturally destructive history of creole to the psychoanalytic model of mind. Psychoanalysis, that is, shows in the agencies of repression a mind that must think in conditions of censorship whose effect on thought could be imaged in terms of nuclear strikes and middle passages. All sense, that is, has to be extracted from nonsense (nothing abstract that is not first the body), an aspect of thought that it is impossible to experience or "to think" directly, and which is precisely that part of thought to be augmented and brought into practical service by means of multimedia computing.

> We gave the name 'sublimation' to the operation through which the trace of castration becomes the line of thought, and thus to the operation through which the sexual surface and the rest are projected at the surface of thought (Deleuze 1990, 219).

What occurs between the mouth and brain that allows noise to become speech involves two tables, and the project of design is to model our interface on the frontier of their interaction. This post-structural application of psychoanalytic theory to interface design, however, reveals the misnomer in the terms used to label the experiment: "interFACE." Getting this term right is important because of the intention to extend the tradition of dialogue into the electronic apparatus (to include the unconscious dimension of the embodied mind). What is the status of this "interFACE" with the face-to-face of the dialogic conversation (the model of communication)? The post-structural critique of the philosophy of consciousness supporting the modern moment of hermeneutics is that the face-to-face is an illusion. Or, in psychoanalytic terms, it is a defense mechanism, a denial, a foreclosure of the body.

> Not only does the unconscious constitute an unspoken 'inner speech' that only fragmentarily surfaces in the rhetorical tropes of 'oneiric discourse,' but it also constitutes an Other language of the Self, an ex-centric language that is articulated only through rhetorical repressions and tropological displacements onto other (floating) signifiers whose diffractions seep into the discourse of the consciousness. In other words, the internally dialogic operations of the language(s) of the unconscious speak of a knowledge of the Self, which, however, the Self cannot master (Lacan, paraphrased in Daelemans and Maranhão 1990, 237).

The face-to-face, in short, is a displacement of the encounter of genitals.

The phallus should not penetrate, but rather, like a plowshare applied to the thin fertile layer of the earth, it should trace a line at the surface. This line, emanating from the genital zone, is the line which ties together all the erogenous zones, thus ensuring their connection or 'interfacing,' and bringing all the partial surfaces together into one and the same surface on the body of the child (Deleuze 1990, 201).

The direction of displacement passes from the depth to the surface of the body, the surface organized into erogenous zones as an aggregate of letters, given coherence by the priority of the genital zone.

> And precisely because castration is somewhere between two surfaces, it does not submit to this transmutation without carrying along its share of appurtenance, without folding in a certain manner and projecting the entire corporeal surface of sexuality over the metaphysical surface of thought. The phantasm's formula is this: from the sexual pair to thought via castration (218).

The electrate dialogue must take into account this force of displacement and recognize that the face-to-face has always been a metonym for the interaction of whole bodies. We should keep in mind too that the theory of castration is an account of the mourning process by means of which persons deal with all forms of loss, beginning with the separation from the mother's body. The dead addressed in psychoanalytic dialogue are those figures of the superego, all the authority figures with whom one identifies, introjected to become agencies of unconscious thought. The story of the superego relates how selves (the subjects of literacy) stay in touch with the wisdom of the ancestors.

DRESSING TABLES

Table (Palmistry): The quadrangular space between certain lines in the palm of the hand. Shirley, Love Tricks (1631): In this table Lies your story; 'tis no fable. Not a line within your hand But I easily understand.

4.1

One point of articulation between the two logics needed for an electrate dialogue is Alan Turing, whose personal story offers as much guidance for the software question as his professional work did for hardware design. Turing's biographer reports that in the conjectures attempting to make sense of Turing's suicide on June 7, 1954 (two years after his arrest and conviction for homosexuality) one of the most enigmatic pieces of the puzzle was the incident reported by Turing's (psycho)-analyst, Dr. Greenbaum. In mid-May of 1954, Turing accompanied the Greenbaum family to an amusement park. On an impulse he went into the tent of a gypsy fortune teller to have his palm read. After half an hour he came out of the tent "white as a sheet." Refusing to speak to the family, Turing went home. It was their last contact with him (Hodges 1983, 496).

This incident may be read as an episode in the history of method, understood as a continuous evolution of the attempt to master fortune (or chance) by means of system. Divination is marked negatively as the other of computing in this tale. It is the path not taken at the crossroads, to which choragraphy returns. Turing himself resisted the turn away from system taken by his colleague Wittgenstein (Turing sat in on Wittgenstein's lectures on mathematics), but his fate now may be read as an example of the X formula Deleuze borrowed from Nietzsche: "we must reach a secret point where the anecdote of life and the aphorism of thought amount to one and the same thing. It is like sense which, on one of its sides, is attributed to states of life and, on the other, inheres in propositions of thought" (Deleuze 1990, 128).

This dialogical secret point in Turing's case is the famous Imitation Game he proposed as a way to test whether or not a machine could be considered "Intelligent." The game is a kind of dialogue.

> First he described a parlor game of sorts, the imitation game, to be played by a man, a woman, and a judge. The man and woman are hidden from the judge's view but are able to communicate with the judge by teletype; the judge's task is to guess … which interlocutor is the man and which the woman. The man tries to convince the judge he is the woman, and the woman tries to convince the judge of the truth … Now suppose, Turing said, we replace the man or woman with a computer and give the judge the task of determining which is the human being and which is the computer (Daniel Dennett, in Kurzweil 1992, 48).

The terms of the game are the same ones Serres observed – a couple and an observer – which is to say that those who take the imitation game literally as a straightforward way to determine if a machine is "intelligent" have been tricked by a noisemaker.

In our context the imitation game may be recognized as a survival of the rituals leading to dialogue, such as those associated with the Eleusinian Mysteries.

> Jane Ellen Harrison remarked on the similarities between Plato's presentation of interlocutors as taking on views they did not agree with, or of being forced to say something they didn't mean, and the Eleusinian rites in which males dressed as females. Transformation and transcendence are effected through deliberate reversal. Harrison asserts that 'Plato's whole scheme alike of education and philosophy is but an attempted rationalization of the primitive mysticism of initiation (Swearingen 1990, 65).

The relevance of the Turing test to the new apparatus may be observed in the Internet, which is emerging as the "seat" of electracy. The relevant puncept is "see" in the ecclesiastical sense. A "chora" is a "see," a place or site of authority, displacing the "seeing" or visually dominated sensorium of literacy. The experience of the unified self associated with the assimilation of literacy by civilization is being supplemented by a new experience of a distributed identity forming in the online practices of MUDs and MOOs and the World Wide Web. Impersonation is becoming a motivating force of electronic subject formation.

4.2

What do these tables tell us? What information may be extracted from them relevant to choragraphy? What pattern does their juxtaposition reveal? If Theory instructs me to cross the two logics of analysis and pattern (work and play) to create a hybrid that will be to the electronic apparatus what dialectic was to literacy, it might be helpful to have a scene that condenses and holds the story of pattern as clearly in mind as the Turing test holds the story of analysis. Nietzsche's formula of converging aphorism and anecdote tells us where to look for this story, if it is reduced to the term "vita," whose anagram abbreviates the two intelligences (including their institutions, practices, and subject behaviors) converging in a hypermedial Internet: AI and TV. The figure that is to TV what Turing is to AI is Elvis Presley.

On September 9 [1956] Elvis Presley appeared live, from Los Angeles, on the Ed Sullivan show, performing four songs – 'Don't Be Cruel,' 'Love Me Tender,' 'Ready Teddy,' and 'Hound Dog' – in the more familiar 'Elvis the Pelvis' style (to the delight of screaming fans in the studio) … CBS grew nervous over the then-current wave of Presley detractors and when the singer returned in January for his third and final Sullivan show, the cameramen were instructed to show Elvis only from the waist up. This truncating of Presley inflamed proponents of the new rock'n'roll craze who felt their hero was being unfairly treated (Castleman and Podrazik 1982, 112).

Elvis the Pelvis. This famous pelvis (the vehicle of funk) is the strange attractor that makes sense out of the face-to-face of electronic dialogue, and gives an alternative term for the choral method: interpelvis design. The emphasis of psychoanalytic theory on the phallus marks it as a modern way of dealing with the same concerns addressed by the Mystery religions. The Eleusinian ceremony that Plato turned into a figure involved the display of a concrete thing as the sign of the fertility of nature. The implements and products of the grain harvest (the technology of agriculture) evoked the myth of Demeter and Persephone. The accessories of the Dionysian rites associated with the ceremonies (Dionysus raped Persephone according to the story) were carried in the ritual procession – the phallus and the mask – stored in the winnowing baskets related to those Plato used as a metaphor for chora, the space of generation, in his cosmological dialogue, *Timaeus*. "It is very likely that in the cista mystica [big basket], among the plants that can be seen on representations of the basket, one or more phalluses were hidden" (Kerényi 1967, 66). To be admitted to the site of final vision, the initiates performed a rite meant to be kept secret, of "taking things out of the big basket and putting them in the little basket," and then back into the big basket (66).

This game of "find the phallus" (fetish work) coincides with the history of dialogue at the point of the imitation game, which includes now the phenomenon of Elvis impersonators. In trying to learn how to write and reason with hypermedia I have to bring back together the abstract with the concrete seeing separated by the history of dialectic in literacy. Elvis impersonation, revealed in our context as a kind of pidgin, suggests that television in particular, and the entertainment media in general, are among other things the site of an emerging pidgin discourse capable of creolization when merged with computing. What is the grammar of this new discourse? Its prototype may be witnessed in examples of cultural creole noted by Thompson. The transfer to the Americas of the cult of Osanyin, for example, god of herbalistic medicine in Yoruba culture, included a crossing of the African myth with ready-made materials found in the American setting. An important part of the imagery of the cult is "the equation of bird with head as the seat of power and personal destiny" (Thompson 1984, 45). The sacred metaphor portrayed an "iron bird set upon a single disk of iron surmounting several bells of iron."

By 1954 creole transformations had already occurred. The new forms had absorbed western industrial or cultural fragments – the hubcap of an automobile, a metal rooster from a weathervane or discarded garden furniture, store-bought jingle bells – and invested them with new meaning. The rooster replaced the flattened birds of the elders, the hubcap sometimes became the base, and the jingle bells recalled the agogo gongs (48).

The detachability and remotivation of parts are shared by creole and impersonation. Marjorie Garber has made sense of Elvis impersonation within the general history of transvestism, with cross-dressing marking the X of dialogue.

It is almost as if the word 'impersonator,' in contemporary popular culture, can be modified either by 'female' or by 'Elvis.' Why is 'Elvis,' like 'woman,' that which can be impersonated? From the beginning Elvis is produced and exhibited as parts of a body – detachable (and imitable) parts that have an uncanny life and movement of their own, seemingly independent of their 'owner': the curling lip, the pompadour, the hips, the pelvis (Garber 1992, 372).

The cults formed by fans around certain stars – the likes of Jimmy Dean, Marilyn Monroe, and in general the finite but ever-growing collection of celebrity icons – echo the scene Plato figured in *Phaedrus*, whose setting was the site of an Eleusinian ritual. The best linguistics available to account for fan discourse is based on the psychoanalytic theory of fetishism. The detachable parts that allow the anecdotal life of a historical person such as Elvis to become an aphorism of thought in a language are organized by the logic of fetishism. The detachability that permits transvestism – the deployment of wigs, false breasts, or codpieces – also allows creolization: the switch that turns a hubcap's reflection of the sun into a flash of spirit. In impersonation we have the figure that is to electrate dialogue what dialectic was to *Phaedrus*.

"In the Elvis story the detachable part is not only explicitly and repeatedly described as an artificial phallus but also as a trick, a stage device, and a sham" (Garber 1992, 367). In one of his early performances, Garber relates, Elvis stuffed his pants with the cardboard tube from a roll of toilet paper – an act that we now recognize as writing in pop-pidgin. The logic of the fetish (the doubleness that both sees and does not see the mother's absent phallus) operates by dreamwork, pattern formation, displacement, and condensation. The interchangeability of face and pelvis in psychoanalytic theory makes clear what is involved in the digitizing of dialogue as a face-to-face encounter.

By a familiar mechanism of displacement (upward or downward), which is in fact the logic behind Freud's reading of the Medusa, 'face' and 'penis' become symbolic alternatives for one another (247).

The nickname "Pelvis" (replacing "phallus") suggests how to generalize the sex and gender of this displaced face. Moreover, the pelvis is uni-gendered. By means of dream reason Elvis as icon can perform the work of the unconscious, that is, deal with those anxiety-producing parts of identity formation unthinkable within the subjectivation of selfhood. All the crossings of the borders constitutive of order and defended by taboos – hence denied to the conscious self – are performed by the Greek chorus of pop icons, thus carrying the liminal function of dialogue into entertainment discourse.

I need to argue for an unconscious of transvestism, for transvestism as a language that can be read, and double-read, like a dream, a fantasy, or a slip of the tongue … And that this quality of crossing – which is fundamentally related to other kinds of boundary-crossing in their performances – can be more powerful and seductive than explicit 'female impersonation' (354).

How to creolize the pidgin of pop icons operating in television pidgin? One likely answer lies in the development of interface agents, automated browse and search devices personified with the stereotyped gestures and attributes of cult celebrities. The same link (gram) that makes Elvis available for impersonation also makes him available as an information hieroglyph. The mythological functioning of Elvis in the present information industry may be fused with an analytical functioning by means of a prosthetic unconscious. A goal of choragraphy is to practice this picto-ideo-phonographic writing that makes Elvis and all other manner of impersonation a way to conduct problem-solving, critical thinking, creative innovation, and all the other language practices important to education.

4.3

Garber makes a case for considering the transvestite effect as marking the entry point of the Symbolic order (Lacan's term for that part of the unconscious representing the operations of social institutions). The transvestite, she states, is a good metaphor for writing as such (150). If Garber provides us with a case study, Deleuze supplies us with the grammar of the emerging electrate creole. The doubleness or fuzziness of the fetish structure (a feature of the logic of sense, or of the unconscious dreamwork) forms the same link between pattern and software that the truth table formed between analysis and hardware. If the truth table made logic compatible with the electric switch, the fetish does the same thing for ordinary language. "Each of these words," Deleuze writes, citing Michel Butor, "can act as a switch, and we can move from one to another by means of many passages; hence the idea of a book which does not simply narrate one story, but a whole ocean of stories" (Deleuze 1990, 47).

The electronic analogy is extended to pattern in the operation of a paradoxical word as a switch. This compatibility across electricity, logic, and language is not surprising in itself since Bertrand Russell, inspired apparently by Lewis Carroll's *Alice* books (featured also in *The Logic of Sense*), started the work that led to the Turing machine by sorting out the operations of paradox. Rather than trying to reduce paradox to logic, however, Deleuze pushes paradox to its extreme, treating it as an irreducible logic (or rather dialogic) in its own right.

Deleuze's logic of sense, then, is the postmodern version of the X of dialogical cross-speaking.

> What are the characteristics of this paradoxical entity? It circulates without end in both series and, for this reason, assures their communication. It is a two-sided entity, equally present in the signifying and the signified series. It is the mirror. Thus, it is at once word and thing, name and object, sense and denotatum, expression and designation, etc. It guarantees, therefore, the convergence of the two series which it traverses, but precisely on the condition that it makes them endlessly diverge. It has the property of being always displaced in relation to itself (Deleuze 1990, 40).

In electronic dialogue two semantic domains are juxtaposed, two orders of information are set in motion as two series that is the equivalent in pattern of the step-by-step linear sequence of proof in analysis. The composer's task is to find the entity that produces the X effect:

> Word = X in a series, but at the same time, thing = X in another series; perhaps it is necessary to add to the Aion yet a third aspect, action = X, insofar as the series resonate and communicate and form a 'tangled tale' (67).

Such is the tale of the X tables, with table as the switch – the puncept, the word-thing – that allows at least a glimpse here of a discourse that crosses pattern and analysis. But this glimpse is still only the *myesis*, the lesser mystery of the sort celebrated near the river Ilissos, the setting of *Phaedrus*. The "my" in the verbs *myeo* and *myo* (which imply secrecy) relates these tables to the program of teletheory and the genre of mystory (Ulmer 1989). "The self-evident first object of this verb is the subject itself: he closes himself after the manner of a flower. But a second object is also possible, which must be very close to the subject, his very own possession. Such an object is the secret," which is related to the German terms *Geheimnis*, *heimlich*, and *Heim* (home) (Kerényi 1967, 46); and, we should add, to the *unheimlich*, the uncanny effect of the fetish. The "my" root evokes the heuretic genre of mystory, useful for the beginning of the discovery process, but not yet the *epopteia*, the greater and more hidden mysteries that took place at Eleusis.

Serres announced the possibility of another condition, other than that of the crossroads with its necessary mediator.

> Let us now imagine that any speaker speaks in his own language and that every hearer understands in his own, whatever the language and whatever the location. In that case the relations can be considered to be many-many and the network that describes them is decentered. With neither exchange nor crossroads. Such a graph has never been seen. Hermes agonies along his way – the exchanger has untied his knots (Serres 1982, 42).

Such a graph was not possible in any previous apparatus, but is a vision of what an electrate creole might become. This vision must be undertaken collectively, by the design community – a project of interpelvis (interbody) design for an online polylogue in cyberpidgin.

REFERENCES

Bakhtin, M. 1984. *Problems of Dostoevsky's poetics*. Ed. C. Emerson. Minneapolis: University of Minnesota Press.

Barthes, R. 1981. *Camera lucida. Reflections on photography*. Trans. R. Howard. New York: Hill and Wang.

Bernabé, J., P. Chamoisseau, and R. Confiant. 1993. *Éloge de la Créolité*. Trans. M.B. Taleb-Khyar. Paris: Gallimard.

Castleman, H. and W. J. Podrazik. 1982. *Watching TV: Four decades of American television*. New York: McGraw-Hill.

Cha, Theresa Hak Kyung (ed.). 1980. *Apparatus: Cinematographic apparatus. Selected writings*. New York: Tanam.

Crowell, S. G. 1990. Dialogue and text: Re-marking the difference. In Maranhão 1990.

Daelemans, S. and T. Maranhão. 1990 Psychoanalytic dialogue and the dialogical principle. In Maranhão 1990.

Deleuze, G. 1990. *The logic of sense*. Trans. Mark Lester. New York: Columbia University Press.

Erickson, T. D. 1990. Interface and evolution of pidgins. In Laurel 1990.

Garber, M. 1992. *Vested interests: Cross-dressing and cultural anxiety*. New York: Routledge.

Goody, J. 1977. *The domestication of the savage mind*. New York: Cambridge University Press.

Hodges, A. 1983. *Alan Turing: The enigma*. New York: Simon and Schuster.

Kerényi, C. 1967. *Eleusis: Archetypal image of mother and daughter*. Trans. R. Manheim. Princeton, NJ: Princeton University Press.

Krol, E. 1992. *The whole Internet*. Sebastopol, CA: O'Reilly.

Kurzweil, R. 1992. *The age of intelligent machines*. Cambridge, MA: MIT Press.

Laurel, B. (ed.). 1990. *The art of human-computer interface design*. Reading, MA: Addison-Wesley.

Maranhão, T. (ed.). 1990. *The interpretation of dialogue*. Chicago: University of Chicago Press.

Monk, R. 1990. *Ludwig Wittgenstein: The duty of genius*. New York: The Free Press.

Morris, D. et al. 1979. *Gestures: Their origin and distribution*. New York: Stein and Day.

Olson, D. R. 1994. *The world on paper: The conceptual and cognitive implications of writing and reading*. New York: Cambridge University Press.

Plato. 1956. *Phaedrus*. Trans. W.C. Helmbold and W. G. Rabinowitz. Indianapolis: Bobbs-Merrill.

————— 1971. *Timaeus*. Trans. F. Macdonald Cornford. London: Routledge.

Serres, M. 1982. *The parasite*. Trans. L. R. Schehr. Baltimore: Johns Hopkins University Press.

Swearingen, C. J. 1990. Dialogue and dialectic: The logic of conversation and the interpretation of logic. In Maranhão 1990.

Thompson, R. F. 1984. *Flash of the spirit: African and Afro-American art and philosophy*. New York: Vintage.

Turner, V. 1982. *From ritual to theater: The human seriousness of play*. New York: Performing Arts Journal Publications.

Ulmer, G. L. 1989. *Teletheory. Grammatology in the age of video*. New York: Routlege.

————— 1994. *Heuretics. The logic of invention*. Baltimore: Johns Hopkins University Press.

White, S. K. 1991. *Political theory and postmodernism*. New York: Cambridge University Press.

Yates, F. A. 1966. *The art of memory*. Chicago: University of Chicago Press.

POROUS MEMORY AND
THE COGNITIVE LIFE OF THINGS

John Sutton

130

COGNITIVE SCIENCE AND THE PREHISTORY OF CYBERCULTURE

Recent prehistorians of sound recording have recovered an exotic European fantasy of the early 1630s. A pamphlet called "Le courrier veritable" told Parisians of a strange sponge discovered by a Captain Vosterloch on a voyage to the South Seas. Local people used these sponges to communicate across long distances: a message spoken into one of them would be exactly replayed when the recipient squeezed it appropriately (Marty 1981, 10; Levin 1995; Draaisma 2000, 85-86).

These wondrous sponges, then, were unique cognitive tools, soaking up sound, embodying particular acoustic signals in an unusually porous medium. They are strange objects to have had this cognitive and cultural role, even in an imaginary space of early modern European fantasy. As a cognitive artifact, the sponge was more commonly a figure for the effacing of memory, so that Confession, for example, could be described as "that happy Spunge, that wipeth out all the blottes and blurres of our lives."[1]

Sociologists and historians describe for us the complex social life of things, the peculiar fetishized or mundane biographies which certain objects accumulate (Appadurai 1986).[2] But some things, natural as well as artificial, also have a *cognitive* life. In use, these sponges were to act as what Merlin Donald (1991) calls "exograms," objects which embody memories and which combine in many different ways with the brain's distributed, context-ridden "engrams." In this short paper, I frame two historical examples in the bare outline of a framework for understanding the multiplicity of relations between engrams and exograms. Darren Tofts' kind of "prehistory," with its "plausible narratives which make links between disparate, achronological moments," and

---

1   The quotation is from John Trapp's commentary on the book of Ezra (1657), in the O.E.D. s.v. "sponge," 4b.

2   On early modern objects see also de Grazia, Quilligan, and Stallybrass 1996; Jones and Stallybrass 2000.

which recognize "fusions between the past and the present, between the *present* and the present" (Tofts and McKeich 1998, 10, original emphasis), is a natural part of the project. Brief forays into the history of theories and practices of memory can, I hope, be both justified and improved by attention to provocative and puzzling points of contact between new media theory and recent dynamical cognitive science.

The delicious story of message transmission by sponge brings home the difficulty we have in remembering just how magical it is, in a world of flux and mixture, that information can ever be enduringly stored, transmitted without distortion, and precisely reproduced. Our lives are irretrievably tangled with artificial systems which keep their contents ordered and immune from melding, and we trust that our computers won't creatively blend our files overnight. The media we use to fix, transmit, and reformat information, and to shift or transform representations from one context to another, are often more stable, less porous, than these sponges. But durable information storage is a cultural and psychological achievement, not a given, and it depends on the construction and exploitation of social and technological resources. In writing a paper, for example, I toggle between yesterday's handwritten scribbles, printed notes from months back, a few words which I jotted down during a phone conversation with a colleague this morning, and the crisp on-screen fonts in which I churn out, obliterate, and rework each version. The process involves multiple feedback loops as I rely on external jottings of yesterday's work, jottings which are more enduring and less context-sensitive than any traces overlaid on others in my brain. My brain and body can temporarily couple with external tools or media, forming an integrated cognitive system with capacities, characteristics, and idiosyncrasies quite different from those of the naked brain (Clark and Chalmers 1998).

In art, science, and ordinary life we construct, lean on, parasitize, and transform artifacts and external symbol systems. And in turn our bodies and brains are inflected and contaminated by the material supplements and cognitive prostheses which we incessantly internalize. Marius Kwint, recently urging us to address the sensuous and physiological dimensions of embodied memory, puts the point thus:

> [H]uman memory has undergone a mutual evolution with the objects that inform it ... the relationship between them is dialectical. Not only does the material environment influence the structure and contents of the mind, but the environment must also have been shaped along the lines of what persists in the mind's eye (Kwint 1999, 4).

Kwint acknowledges that the attempt to fathom such loops must be specific, historically anchored, and insistently interdisciplinary.

So the cognitive life of things takes shape not only in their roles as storage aids or tools at the capricious or agonized disposal of the creator. An abstract artist, to take another example, may work incessantly with a sketchpad, because imagining an artwork "in the mind's eye" will not successfully allow the perception, creation, recognition, and transformation of the kind of hidden patterns which support surplus structure in the work. Just as the aesthetic appreciation of the layered meanings in the finished artwork may take prolonged interactive viewing, so the initial creation of such hidden regularities may have to be an iterative process of sketching and perceptually (not imaginatively) re-encountering the forms. The sketchpad here isn't just a convenient storage bin for pre-existing visual images: the ongoing externalizing and re-perceiving is an intrinsic part of artistic cognition itself. The artist and the sketchpad may be so tightly coupled that it's possible to see them more as a single temporarily integrated system than as an agent operating on a distinct passive medium (Van Leeuwen, Verstijnen, and Hekkert 1999; Clark 2001, 147-150).

In attending to this dynamic interplay between brains and world, we don't need to *identify* internal with external resources. Post-connectionist cognitive scientists like Andy Clark argue, on the basis of everyday cases like these, that the brain is a leaky associative engine, good at pattern-matching and pattern-transformation, but poor (in isolation) at the permanent storage or logical manipulation of individual items (Clark 1993; 1997, 53-69). The classical neuroscientific search for the engram failed because there are no enduring single memories stored alone at local fixed addresses (one neuron for my grandmother, one for my grandfather). Since brain traces are dynamic, we often leave information out in the environment, using the world as its own best representation (Brooks 1991). Brains don't replicate, but rather complement, the alien formats and media of external resources. It's just because representations in the brain are partial and action-oriented that external cognitive scaffolding and tools of many varieties supplement our relatively unstable internal memories. As Clark puts it, "our brains make the world smart so that we can be dumb in peace" (1997, 180). In a range of couplings with other people, instruments, machines, and objects, bodies come into what Clark calls relations of "continuous reciprocal causation" (1997, 163-166). In dance, improvisational music, interactive sport, and ordinary conversation, or in working, feeling, and thinking with cars, computers, airplanes, and sketchpads, there can emerge a mutually modulatory dynamics. Each component in the larger system is continuously responsive to the activity of the other components, and at the same time feeds back its own influences into the web of causal complexity (Haugeland 1998).

These concerns may seem remote from cultural history and theory. But the methodological revolution implied by these new sciences of the interface, which must combine cognitive science and media theory, is far-reaching. Cognitive systems can genuinely extend across brain, body, and world, and are potentially smeared across the natural, technological, and social environment. As Clark says:

> The cash value of the emphasis on extended systems (comprising multiple heterogeneous elements) is thus that it forces us to attend to the interactions themselves: to see that much of what matters about human-level intelligence is hidden not in the brain, nor in the technology, but in the complex and integrated interactions and collaborations between the two ... The pay-off, however, could be spectacular: nothing less than a new kind of cognitive scientific collaboration involving neuroscience, physiology and social, cultural, and technological studies in about equal measure (2001, 153-154).

This means that the particular histories of cryptograms and codes, perspective, autobiographical genres, tattoos, roads, diagrams and graphs, photography, artificial memory techniques, laboratory practices, maps, clothes, and religious ritual (to name just a few) now become an integral part of a historical and comparative cognitive science, rather than mere humanistic curiosities. Careful analysis of historical theories and practices with a cognitive-scientific eye, then, may find problems in the past unlike those perceptible by the more cautious historian.

Some straightforwardly exegetical accounts of Augustine's philosophy of memory, to take one emblematic prehistorical example, focus on his attitude to Platonic doctrines of reminiscence and his efforts to parallel the Holy Trinity with the psychological triad of memory, understanding, and will (Teske 2001). Others examine the role of memory in his account of our awareness of time as a "distention of the mind," or investigate the significance of his theory of memory in light of the autobiographical structure of the *Confessions* (Krell 1990, 52-55; Lloyd 1993, 14-22; Mendelson 2000). These are important and fascinating projects, but they do not rule out a different, more present-centered kind of cognitive-cybercultural history.[3]

The Renaissance arts of memory so richly described in Frances Yates' *The Art of Memory* (1966) and Mary Carruthers' *The Book of Memory* (1990) form one of two prehistorical case studies I sketch in more detail below. Augustine's attitude to the classical *ars memoria* is unclear (Yates 1966, 61-62; O'Donnell 1992, 177-178). But his notoriously spatial images of memory in Book 10 of the *Confessions* as a field, a palace, or a storehouse drive a less commonly noticed demand that in memory "everything is preserved separately."[4] Items channeled through the different senses, for Augustine, are captured by the (personified) memory which swiftly "stores them away in its wonderful system of compartments." Augustine knows that control over the contents of memory is not always easy, for sometimes unwanted past experiences "come spilling from the memory, thrusting themselves upon us when what we want is something quite different." But his regulatory ideal is to be able at will to gather (*cogitare*) the scattered items in memory from the "most remote cells [and] … old lairs," so that a mind which "has the freedom of them all" can "glide from one to the other," effortlessly surfing this strange virtual inner place which is yet not a place.

I want now to quiz two other sets of historical doctrines and practices of memory for their answers to these distinct questions we already glimpse in the case of Augustine, questions about subjective control of memory, and about the format of the vehicles or medium of "storage." Firstly I suggest that it's not unique to any modern crisis of memory to think of the inner components of memory systems as porous and active, rather than fixed in archives. Then I go back to one great era of the cognitive use of artifacts and imaginal places. These analyses only begin to apply the cybercultural grid which the historical material warrants, but they bring some tantalizing topics in the history of memory to the attention of new media theorists.

THE CARTESIAN PHILOSOPHY OF THE BRAIN

Just at the time the story of the sponge was circulating in France, a little-known anatomist and proponent of the new mechanical philosophy was completing a strange book on the philosophy of the body. After an extensive program of dissection in which he had opened "the heads of various animals [to] explain what imagination, memory, etc. consist in," René Descartes described the brain as "a rather dense or compact net or mesh," composed of "tissues" or flexible filaments with "pores or intervals" between them (Descartes 1996, 1.263, 11.171).[5] Through these pores or conduits flow nervous fluids, the fleeting "animal spirits," which "trace figures in these gaps," patterned traces which somehow represent remembered objects and events (1996, 11.178).[6]

---

3 Among treatments more sensitive to questions about memory and media, see also especially Brian Stock's inventive account of Augustine's "theory of reading," which takes him as "the Western originator of the notion of autobiographical memory" (1996, 13, 212-220), and James O'Donnell's discussion of "an online Augustine" (1998, 124-143), which draws on his pathbreaking Internet Augustine seminars of 1994 and 1995: see http://ccat.sas.upenn.edu/jod/augustine/. One different extended attempt (which grew out of those seminars) to link Augustine's accounts of memory with contemporary debates is Katz n.d.

4 All quotations from Augustine are from R. S. Pine-Coffin's translation of the *Confessions* (Augustine 1979), Book 10 chapters 8-17. See also O'Daly 1987, 131-151; Coleman 1992, 90-100.

5 References are to volume and page numbers of the 1996 Vrin reprint of Adam and Tannery's edition of Descartes. Some of the passages cited are translated in Descartes 1985 and 1991. *L'homme* (Descartes 1996, 11:119-202) was not fully translated until 1972, and only brief extracts are included in the standard translation (Descartes 1985). For more on this remarkable work see Krell 1990, 62-73 and Gaukroger 1995, 269-290.

6 Descartes was fascinated by the forms of encoding and decoding found in anamorphosis and visual play, not just in aesthetics but as a model for the psychology of imagination and distortion. See Decyk 2000.

These animal spirits, which are the medium of memory and the passions, are derived from blood.[7] Their particular state – their agitation, abundance, and purity – depends on the balance of bodily fluids (blood, semen, spirits, humors, sweat, tears, milk, fat) in the individual's internal environment; and this balance in turn depends on a ceaseless cosmobiological exchange of vapors between body and world (Descartes 1996, 11.167-170; compare Carter 1983). Descartes' body-machines, animated statues that dream, imagine, feel, and remember (Descartes 1996, 11.120, 201-202), are embedded in the same fluid dynamics which drive the whirling vortices of Cartesian cosmology (Gaukroger 1995, 249-256; Sutton 1998, 82-90). The body, like the cosmos, is full, so that every motion is inevitably coupled with other motions, in a physics of circulation and displacement, which is quite unlike a system of isolated atoms colliding in a void.

Through the blood and animal spirits, as Descartes' follower Malebranche ([1674] 1980, 341-342) wrote, after the Fall we are all "to some extent joined to the entire universe," for each man is linked "through his body to his relatives, friends, city, prince, country, clothes, house, land, horse, dog, to this entire earth, the sun, the stars, to all the heavens."

So the Cartesian body is not rigid and dull, its behavior "automatic" in the sense of endlessly repeatable. Rather, "with its interactive openness," it is the means by which difference is introduced into the human compound (Foti 1986, 76; compare Rorty 1992, Reiss 1996, Sutton 2000a). External parameters like diet, climate, social interactions, and stress, which change at a relatively slow rate, directly affect the fast dynamics of the internal state variables of blood and spirits. But because the spirits are the medium of perception, passion, memory, and imagination, and thus cause our behavior, changes in those external parameters are themselves partly caused by the internal processes with which they are coupled (for the terminology compare Van Gelder and Port 1995, 23-25).

Every act of remembering, then, as the reconstruction of patterns of flow in the animal spirits roiling through the pores of the brain, is context-dependent and causally holistic. Several different figures, Descartes notes, are usually "traced in [the] same region of the brain" (1996: 11.185), so that every recomposed memory pattern is composite, just as every sensation dangerously carries the perceptual history of the perceiver. A single "fold of the brain" can "supply" many of the things we remember: Descartes thus dismisses any worries about the problem of finding room in the brain for all our memories, as they are "stored" only superpositionally and implicitly (Descartes 1991, 143, 148; Sutton 1998, 57-66).

Misassociation and imagination are thus intrinsic to the fluid dynamics of the Cartesian brain. Order is not built in to memory. Descartes hoped nevertheless to enforce clear distinctions between memory and imagination externally, by recourse to the guidance of reason. But few contemporaries found this at all plausible, and Descartes' theory of memory was thus one of the most fiercely criticized strands of his natural philosophy in the second half of the 17th century. English natural philosophers in particular complained that Descartes couldn't *guarantee* personal control over the preservation of the personal past. On Descartes' view of memories as motions, argued Joseph Glanvill, remembering anything would at once "put all the other Images into a disorderly floating, and so raise a little *Chaos* of confusion, where Nature requires the exactest order" ([1661] 1970, 36). The 1650s and 1660s, on either side of the Restoration of the monarchy after the regicide and Commonwealth, saw a terrible crisis of public memory in England,

---

7   On the prehistory of animal spirits see Sutton 1998, 25-49. For Descartes, these nervous fluids are "merely bodies," which "never stop for a single moment in any place" (Descartes 1996, 11.335, 129). So it's not quite true, as Carolyn Merchant suggested, that in Descartes' work "all spirits were effectively removed from nature" (1980, 204).

reflected in neglected yet obsessive debates among natural philosophers about the neuro-physiology of individual memory. After the uncontrolled multiplicity of opinion allowed free rein in the Interregnum, unity had to be imposed not only in worship, dress, and conduct, but in narratives of the personal and political past. "Memory is a slippery thing," wrote a preacher in 1657 (quoted in Cressy 1994, 68), and the reception of Descartes' physiological psychology in England was driven, in a sense, by the desire not to slip (Sutton 1998, 129-148).

Because the mere roaming of fickle fluids and spirits through the brain's networks would allow memories to interfere and blend with each other, the English instead constructed systems of internal fixity. In a lecture of 1682, the Royal Society technician Robert Hooke, for example, saw each item in memory as separately stored in order on physical coils of memory in the brain, spirals down which the soul could radiate its attention in calculating the temporal sequence of past experiences (Hooke [1705] 1971, 140). Descartes' innards, then, were too wet, his brain too porous for the English, making memory hostage to fluid animal spirits which, complained Henry More, are "nothing else but matter very thin and liquid." The brain can't reconstruct motions by itself, as Descartes' theory required, since it is just a "loose Pulp" of "a laxe consistence" which is no more fit to perform our noble cognitive operations than is "a Cake of Sewet or a Bowl of Curds" (More [1653] 1978, 33-34).

The naturalizing of localist or archival models of memory was thus a wishful resistance to Cartesian confusion. For Hooke, individual memory ideas must be "in themselves distinct," so that "not two of them can be in the same space, but that they are actually different and separate one from another" (Hooke [1705] 1971, 142). Even though Hooke himself used external aids to memory remorselessly and was an inveterate list-maker, recording the weather, his health, and his every orgasm (Mulligan 1992), his theory of human memory also imposed pure, "cleansed" order on our internal "Repository." If memory traces were active patterned motions, as Descartes argued, loss of control was inevitable: Glanvill complained that "one motion would cross and destroy another … and there would be nothing within us, but Ataxy and disorder" ([1661] 1970, 39). Far safer, thought the English, for ideas in memory to be themselves passive and independent, to leave it up to the soul or the will to read, decode, and manipulate them (Sutton 1998, 135-156). Kenelm Digby ([1644] 1978) was the first to argue in English that Descartes' philosophy of the brain could not explain "how thinges are conserved in the memory" (282). Digby wanted every memory idea, on its entry into the brain, to "find some vacant cell, in which they keep their rankes and files, in great quiett and order; all such sticking together, and keeping company with one an other, that entered in together: and there they lie still and are at rest, untill they be stirred up" by appetite or by the will (284-285). But the task of the cognitive agent in raiding the spongy brain's caches and compartments is not easy: when it has trouble recollecting some particular idea from memory:

> [I]t shaketh again the liquid medium they all floate in, and rooseth every species lurking in remotest corners, and runneth over the whole beaderoule of them; and continueth this inquisition and motion, till eyther it be satisfied with retriving at length what it required, or that it be grown weary with tossing about the multitude of litle inhabitants in its numerous empire, and so giveth away the search, unwillingly and displeasedly (Digby 1978, 285-286).

Prone to boredom and petulance, lost in its own archive, Digby's soul is unable to navigate its own liquid empire. Disputes between dynamic and static accounts of memory traces are political as well as empirical, the historical distance afforded by examining these quaint and alien 17th century debates revealing just the issues about control of the personal past which may still animate brain theory.

THE ARTS OF MEMORY

One reason prehistory is useful is that it's often difficult to see the mutual contaminations operating between brains and technology in the present. Historical case studies offer a better grasp on the ways in which machines like us are naturally cultural, flexibly soaking up and hooking up with a variety of norms and artifacts. The medieval and Renaissance arts of memory, way back beyond Descartes' porous memory, offer an example of the way humans can freeze their thoughts, interiorizing relatively stable forms of scaffolding in the quest for self-mastery. But where the English critics of Cartesian confusion tried to believe that stability was natural to the brain's storage systems, these earlier practitioners were sensitive to the inevitability of prosthetic supplements in anchoring human memory. The monks, scholars, and magi described by Yates and Carruthers can be seen as laboriously disciplining their brains by the use of specific inner objects.

Cybercultural theorists return zealously to these early forms of intelligence augmentation. They may celebrate the memory artists' architectonic immersion in an array of virtual inner data spaces, strange interactive habitats of the imagination (Davis 1994), or query new media hype by carefully teasing apart analogies and disanalogies between old and new forms of artificial memory (Tofts and McKeich 1998, 62-82). My brief remarks here seek to link this kind of "interiorization of the artefactual" (Scarry 1988, 95-96, 101-102) with my prehistorical topics, porous memory and the mnemonic role of things, as I try to historicize what the historians themselves tend to see only as "certain enduring requirements of human recollection" (Carruthers 1990, 130; compare Coleman 1992, 600-614).[8]

The techniques of local or place memory involve the internalization of a memory architecture, most simply a set of palace corridors with rooms on each side, but alternatively grids, theaters, bestiaries, alphabets, and wheels. I must insert a permanent set of memory locations or niches – two, perhaps, in each memory room – on which I will mentally place items when I'm learning a speech or a set of instructions. Then in recall I mentally walk down the corridor, entering each room in my chosen sequence and reading off whatever is stored at each address. Then I can erase this set of items and store new ones in the same locations for future use.

Initially the process seems to double the cognitive load: what's the point of remembering this memory palace as well as having to remember your speech? But the system is highly flexible: once locations are built in to my own memory architecture, I can use them for any purpose. The art of memory allows me to construct, or to turn my mind into, a random access memory system (Carruthers 1990, 7).[9] Items are kept rigidly ordered by their location, to be inspected and manipulated only at will. Whether "stored" as images or as text, pictorially or linguistically, the key to success is the rigidity and the static nature of the format. Even when the images used to chunk encoded information were strikingly affective, bloody and violent, each atomistic item was to remain independent of all others, isolated at encoding. These are not *external* objects, yet they

---

8 For a more detailed treatment of these themes, in particular on the neglected role of physiology in Renaissance accounts of mental representation, see Sutton 2000b. There are intriguing essays on early modern information storage and retrieval in Rhodes and Sawday 2000.

9 Carruthers here equates "rigid order" with "easily reconstructible order." But these methods are not genuinely reconstructive, for after careful local encoding all the memory images are always already there, waiting; they have only to be found by an active, searching consciousness or subject, and do not (like dynamic or distributed memory representations) have themselves to be recreated anew each time. This is what gives local representations their characteristic context-independence.

are clearly artifacts, interiorized prostheses intended to revise the brain to render it susceptible to voluntary control. The desire is to trap intensity in the memory rooms. So the system has no intrinsic dynamics: the point is to eliminate the activity endemic in what was called "natural" memory, because it leads inevitably to the confusion of items. Semantic stability is thus built in, to allow only *deliberate* combination and recombination. These men's fantasy is of *totally* voluntary memory. So the Renaissance arts of memory were not wild proto-hypertextual schemes for the free flow of information, but the disciplined purging of what St Bernard called "filthy traces" (see Coleman 1992, 182-191) from the past. Adepts imposed (an approximation of) rigidity and inflexibility on their own mental representations. By freezing the contents of memory, monks and scholars sought to tame and recalibrate their minds. The control of items in memory was to be guaranteed by separating data from process, memory from executive self. Artifice was required just because of corruption, the result of sin or of embodiment, where one effect of the Fall was loss of control over the personal past. We are immersed in matter and in time: where angels constantly have in view the whole scene of their former actions, humans need to scramble for the past in the face of oblivion (Locke [1690] 1975, bk. 2, chap. 10). Hamlet assures the Ghost that he'll wipe away all trivial records from the table of his memory, vowing that the Ghost's urgent command "all alone shall live / Within the book and volume of my brain, / Unmix'd with baser matter" (*Hamlet* 1.5.102-104). The arts of memory were a moral quest, so that the true memory artist would never be haunted by reminiscence and the intrusion of unwanted thoughts. Escaping the murky forests of natural memory, the artist aspires to the angelical, using his artificial memories to resist the crowding, interfering, and overlapping of traces in the brain (compare Tofts and McKeich 1998, 80-82).

MATERIAL MEMORIES AND EXTENDED MINDS

But of course the branding of morality on the memory was always wishful. Hamlet fails to flatten his past out, to eradicate affect from memory, to act as a free or sovereign executive. Volition in memory is a vanishing goal, for the putative autonomous memory artist is already caught up in a vast and uneven world of objects inside and outside the skin (compare de Grazia, Quilligan, and Stallybrass 1996). As I've argued, we can't avoid leaning on artificial systems whether inside or outside of skull and skin. There's a continuum between the relatively mindless tidying of the local world in which most animals engage and these highly socialized and morally-charged quests for mastery of the self by the self. Many civilizing processes require a kind of self-oppression, in which control of the brain involves the assimilation of symbolic props and pivots. As Derrida argues, it's not as if evicting every such "prosthesis of the inside" would leave subjective reminiscence as "spontaneous, alive and internal experience" (1996, 11, 19; compare Wills 1995).

To celebrate the Internet as "a chaotic memory system" (Locke 2000, 30) is to be overimpressed by the decentralizing of authority, and to forget how familiar is the Net's primary localist mode of information storage. As Tofts notes, "sites are simply 'there,' located at a particular address" (Tofts and McKeich 1998, 115). Not only does the hype confuse issues about control with quite distinct issues about the activity of the bearers or vehicles of information,[10] it also takes our attention from the deep contingency of the dynamic historical and developmental processes by which we extend our minds with various forms of external scaffolding.

---

10 The Internet is in fact an interesting case here, demonstrating that ease of control doesn't automatically follow from the nature of the vehicles of representation. Items in the virtual environments of the *ars memoria* were rendered passive specifically in order to aid the quest for mastery. And it's because of the intrinsic activity, or tendency to confusion, of the distributed post-connectionist vectors or animal spirit patterns in our brains that creatures like us must come to terms with the limits of choice and will. So far, the more dynamic the

Just as infants learn to walk by leaning on objects and by holding others' hands, until they achieve some fragile motor autonomy, so our cognitive skills require scaffolding. The development of autobiographical memory exemplifies the process (Sutton 2002a). Children learn to remember in company, with their initial narratives of experienced episodes being prompted and heavily guided by parental intervention and shared reminiscence. This scaffolding doesn't then simply disappear with the inevitable triggering of a blueprint for autobiographical memory. Instead, the parental scaffolding is internalized, often in some idiosyncratic detail. Developmental studies show that the particular emotional tone, and the elaborative or pragmatic style of talk about the past in the child's local narrative environment, influences not just the expression but the contents of the child's own memories (Nelson and Fivush 2000). A child's autobiographical memory, then, isn't the product of an automatic unfolding of autonomous capacities; rather it's already sculpted by and embedded in specific and uneven narrative worlds.

Questions about the *location* of the cognitive technology in this kind of scaffolding thus become less pressing, for there just may not be constant or determinate interfaces between brain, body, and world (Haugeland 1998). More interesting are the idiosyncratic cognitive trajectories along which our particular cultural and institutional learning aids allow us to go. We can understand the old arts of memory as one culturally-anchored way to "minimize contextuality" (Clark 1997, 210). Clark's description of the cognitive function of the reusable, relatively stable linguistic media in which we learn to fix our mental representations could be applied equally well to the special fixed pictorial images with which the Renaissance memory artists sought to order their minds:

> [B]y 'freezing' our thoughts in the memorable, context-resistant, modality-transcending format of a sentence, we thus create a special kind of mental object – an object that is amenable to scrutiny from multiple cognitive angles, [and] is not doomed to change or alter every time we are exposed to new inputs or information (Clark 1997, 210).

The biggest challenge, then, in constructing a genuinely dynamical framework to analyze the cognitive life of things in memory, is to acknowledge the diversity of feedback relations between objects and embodied brain. Just as architects can occasionally be too confident that buildings or monuments can act as simple analogues or substitutes for memory (Forty 1999), so cognitive anthropologists and psychologists can too easily neglect the sheer variety of the forms of media and exograms which humans have developed since the Palaeolithic emergence of notations and external symbol systems. Merlin Donald's initial classification, for instance, strongly contrasts the fading, constantly-moving contents of biological working memory with the enduring, unlimited, supramodal, context-independent, and reformattable nature of exograms (1991, 314-319). Certain formats *do* freeze information, allowing it to be held up to multiple scrutiny in future, transmitted more widely across a variety of networks, altered and then re-entered into storage; and these properties of exograms have had essential roles in the development of artistic and theoretic culture. But of course different external media hold information in quite different ways, on quite different timescales, and interact quite differently with individual memories.

---

representation, the more decentralized the access and executive processes. But the Internet's much-vaunted resistance to global control is a counterexample to this neat equation: at least at present, what some critics lament as its "regrettable lack of organization, uniformity, and strategic planning" (Floridi 1999, 85) nonetheless coexists with a thoroughly atomistic, localist, page-by-page representational format. The distinction between local and distributed representation, though, is notoriously hard to draw (Sutton 1998, 149-156), and (paradoxically) the increased control permitted by new digital Web technologies may encourage the use of increasingly dynamic bearers of information.

Information in notebooks, sketchpads, and word-processing systems, whether really external or interiorized, may normally sit passively on call, awaiting mobilization. But other kinds of memory objects are themselves dynamic, like pets and landscapes and cars and friends and ghosts, or will themselves decay or fade or break, like films and knots and bowls and buildings and unreliable machines. Information and emotional memory are held also in rituals which occur only once, or in the dynamic singularity of a group performance, or in other human minds, unpredictable and fragile. It's just because our bodies and brains are porous, our memory thus opened up to time, sensation, and pain, that objects don't just trigger and unlock memory retrieval, but can also stagger it, halt it, haphazardly twist it, and leave it in disarray.

The desire thus to attend to artifacts, media, and brains all at once does not require a unitary view of memory along classical reductionist lines; rather, the idea is the construction of parts of a partial but potentially integrated framework within which different memory-related phenomena might be understood (Sutton 2002b). Memory may have to be studied in both natural and human sciences, while such institutional distinctions remain; but nature is as patchy and idiosyncratic as culture, and the social and technological products of human cognition and action in turn "have direct effects upon individual cognition" (Donald 1991, 10). I suggest that, in the bewilderingly interdisciplinary future of the sciences of memory, from neurobiology to narrative theory, from the computational to the cross-cultural, historical and prehistorical studies should play a significant role.[11]

11 My thanks to Belinda Barnet, Tony Bond, Chris Chesher, Andy Clark, Charles Green, Adam Holland, Doris McIlwain, Andrew Murphie, Will Sutton, Darren Tofts, Maria Trochatos, Mitchell Whitelaw, and Elizabeth Wilson for help and for comments on earlier versions of this material.

REFERENCES

Appadurai, A. (ed.). 1986. *The social life of things: Commodities in cultural perspective*. Cambridge: Cambridge University Press.

Augustine 1979. *Confessions*. Trans. R. S. Pine-Coffin. London: Penguin.

Brooks, R. 1991. Intelligence without representation. *Artificial Intelligence* 47:139-159.

Carruthers, M. 1990. *The book of memory*. Cambridge: Cambridge University Press.

Carter, R. B. 1983. *Descartes' medical philosophy: The organic solution to the mind-body problem*. Baltimore: Johns Hopkins University Press.

Clark, A. 1993. *Associative engines*. Cambridge, MA: MIT Press.

————— 1997. *Being there: Putting brain, body, and world together again*. Cambridge, MA: MIT Press.

————— 2001. *Mindware: An introduction to the philosophy of cognitive science*. Oxford: Oxford University Press.

Clark, A. and D. Chalmers. 1998. The extended mind. *Analysis* 58:7-19.

Coleman, J. 1992. *Ancient and medieval memories*. Cambridge: Cambridge University Press.

Cressy, D. 1994. National memory in early modern England. In *Commemorations: The politics of national identity*, ed. J. R. Gillis. Princeton NJ: Princeton University Press.

Davis, E. 1994. Techgnosis, magic, memory, and the angels of information. In *Flame wars: The discourse of cyberculture*, ed. M. Dery. Durham, NC: Duke University Press.

de Grazia, M., M. Quilligan, and P. Stallybrass. 1996. *Subject and object in Renaissance culture*. Cambridge: Cambridge University Press.

Decyk, B. N. 2000. Cartesian imagination and perspectival art. In *Descartes' natural philosophy*, ed. S. Gaukroger, J. Schuster, and J. Sutton. London: Routledge.

Derrida, J. 1996. *Archive fever: A Freudian impression*. Trans. E. Prenowitz. Chicago: University of Chicago Press.

Descartes, R. 1985. *The philosophical writings of Descartes, volume 1*. Trans. J. Cottingham, R. Stoothoff, D. Murdoch. Cambridge: Cambridge University Press.

————— 1991. *The philosophical writings of Descartes, volume 3: Correspondence*. Trans. J. Cottingham, R. Stoothoff, D. Murdoch, A. Kenny. Cambridge: Cambridge University Press.

————— 1996. *Oeuvres de Descartes*. Eds. C. Adam, P. Tannery. Paris: Vrin.

Digby, K. [1644] 1978. *Two treatises*. New York and London: Garland Publishing.

Donald, M. 1991. *Origins of the modern mind*. Cambridge, MA: Harvard University Press.

Draaisma, D. 2000. *Metaphors of memory: A history of ideas about the mind*. Cambridge: Cambridge University Press.

Floridi, L. 1999. *Philosophy and computing: An introduction*. London: Routledge.

Forty, A. 1999. Introduction. In *The art of forgetting*, ed. A. Forty and S. Kuchler. Oxford: Berg.

Foti, V. 1986. Presence and memory: Derrida, Freud, Plato, Descartes. *The Graduate Faculty Philosophy Journal* (New York: New School for Social Research) 11:67-81.

Gaukroger, S. 1995. *Descartes: An intellectual biography*. Oxford: Clarendon Press.

Glanvill, J. [1661] 1970. *The vanity of dogmatizing*. Brighton: Harvester Press.

Haugeland, J. 1998. Mind embodied and embedded. In *Having thought: Essays in the metaphysics of mind*, ed. J. Haugeland. Cambridge, MA: Harvard University Press.

Hooke, R. [1705] 1971. Lectures of light. In *The posthumous works of Robert Hooke*, ed. R. Waller. London: Frank Cass and Co.

Jones, A. and P. Stallybrass. 2000. *Renaissance clothing and the materials of memory*. Cambridge: Cambridge University Press.

Katz, S. n.d. Memory and mind: An introduction to Augustine's epistemology. At http://ccat.sas.upenn.edu/jod/augustine/sheri (accessed April 22, 2002).

Krell, D. F. 1990. *Of memory, reminiscence, and writing: On the verge*. Bloomington, IN: Indiana University Press.

Kwint, M. 1999. Introduction: The physical past. In *Material Memories*, ed. M. Kwint, C. Breward, and J. Aynsley. Oxford: Berg.

Levin, T. Y. 1995. Before the beep: A short history of voice mail. *Essays in Sound 2: Technophonia*.

Lloyd, G. 1993. *Being in time: Selves and narrators in philosophy and literature*. London: Routledge.

Locke, C. 2000. Digital memory and the problem of forgetting. In *Memory and methodology*, ed. S. Radstone. Oxford: Berg.

Locke, J. [1690] 1975. *An essay concerning human understanding*. Ed. P.H. Nidditch. Oxford: Clarendon Press.

Malebranche, N. [1674] 1980. *The search after truth*. Trans. T. M. Lennon, P. J. Olscamp. Columbus, OH: Ohio State University Press.

Marty, D. 1981. *The illustrated history of phonographs*. Trans. D. Tubbs. New York: Dorset Press.

Mendelson, M. 2000. Venter animi/distentio animi: Memory and temporality in Augustine's *Confessions*. *Augustinian Studies* 31:137-163.

Merchant, C. 1980. *The death of nature: Women, ecology, and the scientific revolution*. New York: Harper and Row.

Mulligan, L. 1992. Robert Hooke's "Memoranda": Memory and natural history. *Annals of Science* 49:47-61.

More, H. [1653] 1978. An antidote against atheism. In *A collection of several philosophical writings, volume 1*. New York and London: Garland Publishing.

Nelson, K. and R. Fivush. 2000. Socialization of memory. In *The Oxford Handbook of Memory*, ed. E. Tulving and F.I.M. Craik. Oxford: Oxford University Press.

O'Daly, G. 1987. *Augustine's philosophy of mind*. Berkeley: University of California Press.

O'Donnell, J. J. 1992. *Augustine, Confessions: Introduction and text, volume 3*. Oxford: Clarendon Press.

————— 1998. *Avatars of the word: from papyrus to cyberspace*. Cambridge, MA: Harvard University Press.

Reiss, T. J. 1996. Denying the body? Memory and the dilemmas of history in Descartes. *Journal of the History of Ideas* 57:587-607.

Rhodes, N. and J. Sawday. 2000. *The Renaissance computer: Knowledge technology in the first age of print*. London: Routledge.

Rorty, A. 1992. Descartes on thinking with the body. In *The Cambridge companion to Descartes*, ed. J. Cottingham. Cambridge: Cambridge University Press.

Scarry, E. 1988. Donne: 'but yet the body is his booke.' In *Literature and the body*, ed. E. Scarry. Baltimore: Johns Hopkins University Press.

Stock, B. 1996. *Augustine the reader*. Cambridge, MA: Harvard University Press.

Sutton, J. 1998. *Philosophy and memory traces: Descartes to connectionism*. Cambridge: Cambridge University Press.

———— 2000a. The body and the brain. In *Descartes' natural philosophy*, ed. S. Gaukroger, J. Schuster, J. Sutton. London: Routledge.

———— 2000b. Body, mind, and order: Local memory and the control of mental representations in medieval and Renaissance sciences of self. In *1543 and all that: Word and image in the proto-scientific revolution*, ed. G. Freeland and A. Corones. Dordrecht: Kluwer.

———— 2002a. Cognitive conceptions of language and the development of autobiographical memory, *Language and Communication* 22.

———— 2002b. Representation, reduction, and interdisciplinarity in the sciences of memory. Forthcoming in *Representation in mind*, ed. H. Clapin, P. Staines, and P. Slezak. Westport, CT: Greenwood Publishers.

Teske, R. 2001. Augustine's philosophy of memory. In *The Cambridge companion to Augustine*, ed. E. Stump and N. Kretzmann. Cambridge: Cambridge University Press.

Tofts, D. and M. McKeich. 1998. *Memory trade: A prehistory of cyberculture*. Sydney: Interface Books.

Van Gelder, T. and R. R. Port. 1995. It's about time: An overview of the dynamical approach to cognition. In *Mind as motion: Explorations in the dynamics of cognition*, ed. R. F. Port and T. Van Gelder. Cambridge, MA: MIT Press.

Van Leeuwen, C., I. Verstijnen, and P. Hekkert. 1999. Common unconscious dynamics underlie uncommon conscious effects: a case study in the interaction of perception and creation. In *Modeling consciousness across the disciplines*, ed. J. Jordan. Lanhan, MD: University Press of America.

Wills, D. 1995. *Prosthesis*. Stanford: Stanford University Press.

Yates, F. 1966. *The art of memory*. London: Routledge and Kegan Paul.

BECOMING IMMEDIA :
THE INVOLUTION OF DIGITAL CONVERGENCE

Donald F. Theall

"Now, to the degree that ... the human elements infiltrated more and more into each other, their minds (mysterious coincidence) were mutually stimulated by proximity. And as though dilated upon themselves, they each extended little by little the radius of their influence upon this earth which, by the same token, shrank steadily. What, in fact, do we see over and over again? Through the discovery yesterday of the railway, the motor car, and the aeroplane, the physical influence of each man, formerly restricted to a few miles, now extends to hundreds of leagues or more. Better still: thanks to the prodigious biological event represented by the discovery of electromagnetic waves, each individual finds himself henceforth (actively and passively) simultaneously present, over land and sea, in every corner of the earth" (Teilhard de Chardin 1959, 240).

The publication in the 1950s and 1960s of Pierre Teilhard de Chardin's writings on the evolution of mind and consciousness received considerable attention because the works marked what Stephen Toulmin called "The Return to Cosmology," and since the Jesuit paleontologist and philosopher had seemed to marry the theory of evolution to a natural theology. Strongly praised by Thomas Huxley, Teilhard's theory of the emergence of the "noosphere" seemed also to have anticipated the emergence of a new cybernetic age. Teilhard, who encountered objections to his work from the Catholic Church and the Jesuit order, had obviously been strongly influenced by the turn of the century philosophical writings of Henri Bergson. But the scientifically and theologically trained Teilhard, with his theory that humanity is evolving, mentally and socially, to a final spiritual unity, did not have the knowledge of the perceptual power of the arts and the conceptual power of philosophy, which in a gradually developing counter-tradition, would evolve a cybernetic "chaosmology." That chaosmology would have even stronger intuitions of the rapidly developing digiculture with its cyberspace and virtual (or artificial) realities.

James Joyce had already, before 1940, introduced the term "chaosmology" which was to be further conceptualized by McLuhan in the 1960s and later in the writings of Gilles Deleuze and

Félix Guattari. However, the rise of McLuhanism at the opening of the 1960s provided a strong impetus to Teilhard's theories as possible precursors for a cybernetic cosmology grounded in a natural theology. In 1962, when Marshall McLuhan discussed his concepts of the externalization of the senses and the global village in his groundbreaking book, *The Gutenberg Galaxy*, he identified them with Teilhard's concept of the "noosphere" – or as McLuhan glossed it, the "technological brain" (described by Teilhard as "the 'thinking layer,' which … has spread over and above the world of plants and animals, and thus worlds beyond the biosphere" (Teilhard de Chardin 1959, iii. i. 182).

While references to Teilhard's writing appear frequently in McLuhan's proclamation of the end of the era of the primacy of the printing press (not the death of the book), the first time McLuhan cites a quotation from Teilhard – the opening quote in this essay – is in his discussion of the outering of the senses. That passage concludes with the paleontologist noting that the discovery of electromagnetism is the most recent and far-reaching extension of the radius of influence of the "technological brain" achieved through social evolution. McLuhan further observes that Teilhard has seen into the heart of contemporary panic, since by 1962 the world, instead of tending towards a huge Alexandrian library, has become "a computer, an electronic brain" (McLuhan 1962, 32). Viewed from the year 2001, the concerns of McLuhan and Teilhard with a world that has become an electromagnetically produced "technological brain" appear somewhat anachronistic, for there is now not only a potential for merging the Alexandrian library with the electronic brain, but still further a potential for their eventually becoming intertwined with a multisensory "virtual reality."

In raising this image of Alexandria, McLuhan ironically, but also unconsciously, echoes the intimations of a future which Borges satirically anticipated in the "The Total Library" (Borges 1981, 35). McLuhan frequently revisited such visions of the future in his various writings, which have now led to his being denominated one of the prime prophets of cyberspace, the wired world, and/or digiculture. While Teilhard envisioned the evolution of mind and its impact on the processes of contemporary society – the emergence of the ultrahuman from the compression and connectedness of people's minds in an electromagnetic shrinking world – he did not, and probably could not, foresee the full extent to which the electromagnetic era would lead to the metamorphosis and transformation of the book. Nor could he have foreseen the para-electronic "library" as information databank being able to encompass the book within a totally new set of technologies, just as the book in its origins had been transformed from ogham and runes to papyrus to parchment to paper to print to mass-produced print and images. While McLuhan in the 1950s intuited this process, he only gradually and then partially began to perceive that electromagnetic transformations of the book, through convergence of media and the accompanying merging of the senses, could, in the long term, offset the potential of the postelectric media to suppress the "Gutenberg" effect; the forestalling of the end of the book became transformation in the evolution of the book.

In the 1950s the roots of such a visionary understanding of the long-ranging implications of the electromagnetic and then the digital revolutions came from a diverse group of sources. These sources were primarily associated with the creative or critical practice in the arts and humanistic studies rooted in cultural history:

(a) radical and avant-garde modernist art and poetry subsequent to 1850, most specifically French symbolism and such European art movements from 1900 as vorticism, constructivism, dadaism and surrealism;

(b) the particular critical integration of theories of human evolution, of the evolution of creative intuition and expression, including the book and the consciousness of the unconscious through the dream, as realized in contemporary poetry and art such as the vision of James Joyce's *Ulysses* and *Finnegans Wake*;

(c) an increased knowledge of the history of the humanistic arts of communication and their transmission (that is, the history of the trivium – grammar, logic, rhetoric – and the history of early education) from the classical world until the enlightenment;

(d) the new appreciation after 1850 of the history of the alchemical, the occult, and the gnostic, which led to observations such as W. B. Yeats saying that "[t]he visible world is no longer a reality and the unseen world is no longer a dream" (quoted in McLuhan 1964, 35).

McLuhan, who built his own prophecies from quixotic interpretations of all of the above sources, linked his own views to Yeats' observation of the merging of the seen and unseen worlds being paralleled by a merging of reality and dream. These views concerned the contemporary reversal, under the impact of electronic media, of the real world into science fiction and the accompanying reversal by which the western world is going eastern, as the east goes western. Played off against the science, epistemology, and technology of the new cyberneticians (Turing, Wiener, Shannon, Weaver, and the Macy seminars; the ecology of mind of Gregory Bateson and the Palo Alto group; and the autopoiesis of Varela and Maturana), the aforementioned history of the arts and of the humanistic tradition could provide contexts for partially understanding the nature of the future through knowledge of past transformations of modes of communication and expression and of ways of storing and processing data. Simultaneously, it could also provide insight into the significance of the ongoing metamorphoses of relationships between matter and consciousness. Because Teilhard did not have a mastery of this specific body of artistic and humanistic knowledge, his account of the technological brain was for McLuhan the "the lyrical testimony of a very romantic biologist" which while "perceptive" and "prophetic," yet had "a shrill vehemence" accompanied by "an uncritical enthusiasm for the cosmic membrane that has snapped around the globe by the electric dilation of our various senses" (McLuhan 1962, 32).

McLuhan, denying he was either an optimist or a pessimist, while asserting he was an apocalyptic, claimed to discover a more complex, critical, and historically contextualized focus for his understanding of the electromagnetic and its outering of the senses from his "partial" readings of the work of symbolist poets, avant-garde artists of the first half of the 20th century and the later works of the anti-apocalyptic Joyce – *Ulysses* and *Finnegans Wake*. From these sources, and particularly Joyce's work, McLuhan launched an intuitive, perceptual, and artistic critique of the emergence through electromagnetism and technology of cyberculture or digiculture. Of particular interest at the moment is the way in which he positioned himself as a balance between the optimism of Teilhard and what he took to be the pessimistic "gnosticism" of the cyberneticists (Wiener, etc.). Simultaneously he seems to have positioned Joyce as a balance between the optimism of the Bergsonian *élan vital* (the basis of creative evolution) and what seemed to be the devolutionary eternal return of Nietzsche. Probably through his suspicion that Joyce was of the "devil's party" (along with Milton and Blake), McLuhan ultimately favored the Viconian Joyce, declaring Giambattista Vico, with his progressive cyclicity, to be a Baconian and the last pre-electric grammarian.[1]

---

1 In the posthumous work McLuhan and McLuhan 1988, Vico is mentioned more frequently than Joyce and Bacon's work is stressed. See the discussion in Theall 2001. For the importance of Bacon to McLuhan long before he started his career as a media guru, see the posthumous (McLuhan 1991, 7-27), a version of a paper written for the Modern Language Association in 1942.

If Bergson was a more important precursor than Darwin of Teilhard's "natural theology," it was Joyce who produced a vision of electromagnetic and digital evolutionary involution. Building on Bergson and Whitehead and revising their cosmologies as a "chaosmo[logy]," Joyce developed a comedically modified interpretation of Vico's *Scienza Nuova*, beginning from the "increasing, livivorous, feelful thinkamalinks; luxuriotiating everywhencewithersoever" (FW 613.19-20) (Toulmin 1985, 119, 123).[2] As his antihero, HCE, falls into an inebriated sleep, he is imagined to be the "last pre-electric king of Ireland" (FW 380.12), a protoptypical King Roderick O'Connor. Joyce's dream, nearly a quarter of a century earlier than McLuhan and years before Teilhard, explores the transformation of mind through the transformation of electromagnetic technologies. The immediately preceding episode in which HCE's school-age children are studying their lessons ("Triv and Quad"), examines the relations of mind, memory, meaning, and cultural production in which, "After sound, light and heat, memory, will and understanding ... (the memories framed from walls are minding) ... " (FW 266.18-21).

By relating fundamental matter as implicit in sound, light, and heat to mind and memory, Joyce, in his post-Bergsonian vision, explores the roots in gesture – for "in the beginning was the gest" (FW 468.5) – of the evolution of postelectric consciousness. As they are maturing, the children's learning is guided by "Mimosa Multimimetica, the maymeaminning of maimoomeining!"; that is, "the meaning of meaning" through a multiplicity of modes of mimesis and mimicry (FW 267.1-3).[3] This is a quest that produces a virtual world of "Singalingalying. Storiella as she is syung. Whence followeup with end-speaking nots for yestures" (FW 267.7-10) that can be realized through meaningful signs which are equally words and speech tending to postelectric codes – white light, spectral color, and motion: "Belisha beacon, beckon bright! Usherette, unmesh us! That grene ray of earong it waves us to yonder as the red, blue and yellow flogs time on the domisole, with a blewy blow and a windigo. Where flash becomes word and silents selfloud" (FW 267.12-16).[4]

Here the convergence of media is projected into the remote historical past. From the mnemonic and its encounter with "end-speaking nots" and "yestures" emerges the basis of that semiotic foundation of the "meaning of meaning" that Ogden expounded from Malinowski, Peirce, and others ("Multimimetica, the maymeaminning of maimoomeining!"). When postelectric mimesis (virtual realization) takes place in an electromagnetic context, the original multimedia nature of communication, recognized by the early grammarians, rhetoricians, and their commentaries on poetry, will be seen as essential to constructing virtual realities. Joyce thus involved all of the previously mentioned artistic, humanistic, and historic strands, embedding them within his "critico-satiric" play with a helical theory of evolutionary history derived from Vico in interface with pre-psychoanalytic and other alternative theories of dream and the emergence of consciousness of the unconscious that counterpointed the emergence of psychoanalytic theory.

Such a vision at the same time explored, anticipated, and queried the extreme, uncritical optimism Teilhard later associated with his account of the noosphere, while still providing a foundation for McLuhan's theories. Vico's history is relevant to the issue at hand, since his "new science" was based on the history of language rather than etymology. Further, Vico's "new science" does not strictly depend on a four-part structure because his account consists of three

---

2   Joyce's affinities with the "time philosophy" of Bergson and Whitehead were extensively discussed by Wyndham Lewis and resulted in the well-known counterattack on Lewis as Professor Jones by Joyce (FW I.q 11).

3   For a discussion of this section and of Joyce's interest in Ogden see Theall 1997, 10, 93, 132.

4   It should be noted that a "Belisha beacon" is a traffic crossing sign.

ages (Gods, heroes, men) plus a ricorso which is quite distinct from the three ages that precede it and each ricorso, as it introduces the next cycle of ages, does not imply going back to a totally new beginning. Each Viconian ricorso, as Joyce interprets it, returns the course of historical life from a decadent age of men to begin the "seim anew" (FW 215.23) in an age of gods, which retains in memory the previous cycle. This is why Vico's vision is helical in structure rather than an ever-repeating cyclical one. The *Wake* both utilized and satirized Vico. *The New Science*'s emphasis on a poetic wisdom had strong affinities with Joyce; his helical theory of history provided an appropriate counterpoint to progressive evolutionary theories, and even allowed for a possible vision of an a*Wake*ning into a new cycle based on the postelectric age. However, Joyce wished to move beyond Vico's Homeric poetic wisdom to a post-Homeric, post-Cartesian, techno-scientific poetic vision confronting the emergence of a new cognitive-sensory hypertextuality.

In the *Wake* Joyce recognized the apparent paradox of his preoccupation with time, mind, and memory in relation to a machinic world that is part of a "chaosmos" presided over by "blankde-blank, god of all machineries" (FW 253-33).[5] This is a delirious (*délire*) deity whose dreams are crafted by each and every person (i.e., Everybody), each of whom is a "harmonic condenser enginium" (FW 310.1) and enunciated by poets and artists, who assemble abstract machines. So Joyce considered himself to be "the greatest engineer." But as Joyce's self-directed, comic comments on his role as engineer indicate, the engines he assembled are the "abstract machines" of the "musicmaker" or the "philosophe."[6] Possibly his awareness that he was assembling such machines grew naturally out of his early realization, in 1923, of the radical complexity of his work (he described it as "complex," "duplex," "perplex," "stuplex," etc.) and beginning in 1927 from his thinking of himself as a great engineer, so that by the late 1930s the conscious recognition of the interplay of machinic motion, complexification, and chaos led to his inserting into the evolving text the words "gossip will cry it from the housetops no surelier than the writing on the wall will hue it to the mod of men ... every person, place and thing in the chaosmos of Alle anyway connected ... [is] moving and changing every part of the time" (FW 118.20-22).

In describing his dream action or the poetic action of a (or more specifically *his*) manuscript discovered by a hen in a midden heap, Joyce, moving beyond the interplay of the cosmos and random chance, sees the experimentation with language in the *Wake* as directly related to the technoscientific thrust towards convergence of media. This is a convergence that Joyce – along with many avant-garde contemporaries such as Le Corbusier, the Bauhaus group, and the Giedions – realized had been intuited in early multimedia aspects of the medieval cathedrals.

That this was an important foundation for Joyce's composition of the *Wake* is confirmed in that one of his earliest fragments or "vignettes" (1923) composed in the design of the *Wake* – the debate between the Archdruid (Berkeleyan philosophy and idealism) and the Saint (matter and medieval realism) – eventually, as the process of composition progresses over the years, comes to take place in a setting mediated by television cameras, newspaper coverage, and film journalism, and gradually comes to be more and more associated with discussions of the nature of light, the spectrum, and optics. From its earliest moments there was an intuitive sense of the *Wake* uniting the poetico-artistic and the technoscientific in the midst of late 19th and early 20th century developments leading to the convergence of media and ultimately to the emergence of a world "beyond media."

---

5   "Blanc de Blanc" is a French appellation for a white wine. For suggested relationships between delirium, dream, and intoxication and the "chaosmos," see Guattari 1995, and Deleuze and Guattari 1994.

6   James Joyce to Harriet Shaw Weaver, April 16, 1927 (Joyce 1957, 251).

All of these concerns are intrinsically involved with problems of virtuality and actuality, of the possible and the real, and of the monism of mind and body that were implied in the philosophy of Berkeley and his theories of vision, which resurfaced in the interest of contemporary cosmologists such as Alfred North Whitehead. What Joyce undertakes through parody and comedy is to transform Vico's equating of memory, understanding, and imagination, as well as the transformation of his helical, evolutionary history of humankind into a pre-postmodern, poetic vision of the interplay of the future and the past in the evolution of mind, resulting from the emergence of the postelectric, electromagnetic and electrochemical world. Joyce's hero, HCE, Here Comes Everybody, prefigures an evolutionary process of becoming "machinic" ("Harmonic Condenser Enginium") (FW 310.1). Teilhard and McLuhan see invention as one of the key elements at the heart of this processual evolving of consciousness. Joyce locates that invention in the very nature of the "language" he is inventing – developed from the nature of language as code and gesture, for in a "mock" portrait of himself as Shem, the sham poet, it is said:

> … and him, the cribibber like an ambitrickster, aspiring like the decan's, fast aslooped in the intrance to his polthronechair with his sixth finger between his cats eye and the index, making his pillgrimace of Childe Horrid, engrossing to his ganderpan what the *idioglossary he invented* … (FW 423.5-9 [italics mine]).

By approaching his subject as an "ambitrickster" and learned satirist such as Swift ("the decan's fast aslooped …"), Pope (in whose *Dunciad*, the description of Dean Swift as asleep in his easy chair) or Byron (who wrote "Childe Harold's Pilgrimage"), Joyce – avoiding the tendency towards linearity in Teilhard – is able to encompass semiotic ambivalence and the complexity of the helical-cyclical nature of the cognitive-sensory evolution of consciousness, from which the convergence of modes of expression emerge. While Teilhard sees humanity moving into a posthistorical period of the ultrahuman, Joyce's vision is only partly reflected in McLuhan's assertion that:

> … we live in posthistory in the sense that all pasts that ever were are now present to our consciousness and that all the futures that will be are here now. In that sense we are posthistory and timeless. Instant awareness of all the varieties of human expression reconstitutes the mythic type of consciousness, of once-upon-a-time-ness, which means all-time, out of time.

> It is possible that our new technologies can bypass verbalizing. There is nothing impossible about the computer's … extending consciousness itself, as a universal environment. In a sense, the surround of information that we now experience electrically is an extension of consciousness itself (McLuhan 1999, 88).

Joyce, unlike Teilhard and McLuhan, confronts the "ambiviolent" nature of the "chaosmos" and simultaneously sees that the physical and somatic bodies are involved in this moving beyond. The process of being more parahuman than posthuman is constituted in a "pre-posthistory" that still involves the historical, for in time there is no present, since the present is always already immediately past at the moment we contemplate it and the future necessarily implicates the past.

Grasping the complexity of the "chaosmos" underlies the necessity of such phenomena as the Joycean "collideorscape" (FW 143.28), which is a medley of verbal, gestural, visual, harmonic, and optical elements that a "fargazer seems[s] to seemself to seem seming of" (FW 143.26-27). This is important since the identity of invention-imagination and memory within the time-space world of the virtuality of the cultural product is essential to moving beyond the book. As Joyce's dream action moves towards its conclusion it is described as "Our wholemole millwheeling vico-ciclometer, a tetradomational gazebocroticon" (FW 614.27-28) of which it can be said as of a dream, a speech act, a miming gesture, or a traditional book:

What has gone? How it ends?
Begin to forget it. It will remember itself from every sides,
with all gestures, in each our word. Today's truth, tomorrow's trend.
Forget, remember! (FW 614.19-22).

Such a machine composed through matter, mind, and memory, in whatever medium it may be assembled has the "sameold gamebold adomic structure" which must ultimately have an electro-magnetic foundation "as highly charged with electrons as hophazards can effective it" (FW 615.6-8).

Although Teilhard intuited the significance of the emergence of the contemporary electro-magnetic world through memory and mind becoming conscious of itself (unlike McLuhan, and even more so Joyce, Deleuze, or Guattari), he does not examine the metamorphoses of modes of communication, which seem to be an essential component of the electromagnetic evolution of consciousness. Joyce partly shares with Deleuze and Guattari a world in which metamorphoses are accompanied by a fluidity of signifiers, since becomings are states of virtualization prior to actualization: man can become woman and woman, man; where Shem can become Shaun and Shaun, Shem; where wo/man can become insect (i.e., earwig) or insect, wo/man; where the tree can become stone and the stone, tree; where the river can become mountain and mountain, river. The fluidity and complexity of Joyce's language is indispensable for crafting this verbal poetic vision of the "chaosmotic" evolution of the digital universe. Guattari, in his *Chaosmosis*, offers an explanation for the importance of such linguistic experimentation as Joyce's language or the semiotic experimentation of avant-garde artists such as Duchamp and Klee:

[they reverse] subjectivity['s being] standardized through communication which evacuates as much as possible trans-semiotic and amodal enunciative compositions [by] slip[ping] towards the progressive effacement of polysemy, prosody, gesture, mimicry, and posture to the profit of a language rigorously subjected to scriptural machines and their mass media avatars (Guattari 1995, 104).

Through its transversality as a satire of "subjectivity standardized through communication," Joyce's *Wake* re-establishes polysemy, prosody, gesture, mimicry, and posture. In a sense it revisits the concerns of the "Oxen of the Sun" section of *Ulysses* and thus retraces an evolutionary semiology in which the imaginary simulations of poetry and the arts are transformed, as different modes and media converge in a electromagnetic, technological simulation of cyberspace. Brenda Laurel, in her *Computers as Theater*, argued that cyberspace, as the simulation of virtual worlds, should be construed in terms of a theory of dramatic mimesis (Laurel 1991). Having posited dream, hallucination, inebriation, and delirium as generators of virtual worlds, Joyce in his "Phoenix Playhouse" (or children's nurseryroom) section of the *Wake* (FW II.i) relates his work as the "Feenichts Playhouse" (FW 219.1) to Bergsonian motifs of time, duration, memory, and creative evolution with a specific allusion to Bergson's *élan vital*:

Time: the pressant. With futurist onehorse balletbattle pictures and the Pageant of Past History worked up with animal variations amid ever-glaning mangrovemazes and beorbtracktors by Messrs Thud and Blunder. Shadows by the film folk, masses by the good people. Promptings by

Elanio Vitale. Longshots, upcloses, outblacks and stagetolets by Hexenschuss, Coachmaher, Incubone and Rocknarrag. Creations tastefully designed by Madame Berthe Delamode. Dances arranged by Harley Quinn and Coollimbeina. Jests, jokes, jigs and jorums for the Wake lent from the properties of the late cemented Mr T. M. Finnegan R.I.C. (FW 221.17-27).

The full significance of Joyce's satiric counterattack against Lewis's attack on his Bergsonism only becomes clear when it is recognized how Bergson's discussions of memory, duration, the embodied nature of the brain, and creative evolution contributed to an early understanding of the semiological evolution taking place in tandem with the technological evolution of the post-electric era. If Teilhard's work appears to have a relevance to the chaosmology of a world of complexity, it is partly because of Bergson's influence. In this excerpt from the *Wake*, there is an interplay between a theory of time – how the present which is always already past makes urgent (*pressant*, Fr. = urgent) the future – and the avant-garde poet's or artist's technological vision, arising from past history. In "The Mime of Mick, Nick and the Maggies" this is realized through the intermixture of artistic modes ("the onehorse balletbattle pictures and Pageant"), the cinematic (the "film folk") and the *élan vital* ("promptings by Elanio Vitale"). The imaginary dream theater of the children's nurseryroom is the *Wake*, as this excerpt asserts (since it asserts that the "Feenichts theatre" is the *Wake* for which "the jests, jokes, jigs and jorums" are those of Finnegan).

The Joycean vision of the transformation of the primitive into a complex and chaotic techno-culture is a prophecy of cyberspace which foresees, but in a more pre-postmodernist mode, the affiliation between the theater and cyberspace which Laurel developed, arguing that Aristotelian poetics provided an account of drama that was the most appropriate way to understand the human-computer interface and its generation of the virtuality of cyberspace.[7] The *Wake* (being also a "mime") is a drama, "wordloosed over seven seas crowdblast in cellelleneteutoslavzend-latinsoundscript. In four tubbloids" (FW 219.16-18). The fluidity of its "language" parallels the fluidity of electromagnetic media such as cinema and its extended cinematic successors, subsequently moving towards the convergence of media in digiculture. If Bergson influenced Teilhard and Whitehead and ultimately Bateson, the major Bergsonian discovery in *Matter and Memory* in 1896 of a movement-image in the physical reality of the external world and a time-image in the psychic reality of consciousness, influences the construction of the language and structure of *Ulysses* and particularly of *Finnegans Wake* (Deleuze 1986, passim).

Bergson had come to see that duration was virtual coexistence, as Deleuze points out in his study of Bergsonism (Deleuze 1988). By the time of his writing of the *Wake*, Joyce had developed a "synthetic" language which, within itself, encompasses past-present-future in the virtual coexistence of a remembered dream. At various moments he links this very specifically to electromagnetic media, such as the conclusion of a TV battle between two figures – Butt and Taff:

> [*The pump and pipe pingers are ideally reconstituted … All the presents are determining as regards for the future the howabouts of their past absences which they might see on at hearing could they once smell of tastes from touch. To ought find a values for. The must overlistingness. When ex what is ungiven. As ad where. Stillhead. Blunk.*] (FW 355.1-7).

The virtual (that is, the ideally reconstituted) dwells in duration ("the presents are determining as regards for the future the howabouts of their past absences") and is embedded in the reality,

7   See Rice 1997; I also discuss Joyce's attitudes toward chaos and complexity in Theall 1997 (passim).

but not the actuality, of the senses and synesthesia. As the "probapossible polegomena" to his "ideareal history" (FW 262.R1-2 and 5-6) illustrates, the *Ideal Present Alone Produces Real Future* (FW 303.L.12-4).

As a sidereal history (a virtual chaosmology) as well as an "ideareal" (ideal + real, i.e., a virtual history) Joyce critiques, transforms, and contemporizes the poetic science of a Viconian "ideal eternal history." This is an "ideareal history" realized only through the mythico-mathematical language that is "an autocratic writings of paraboles of famellicurbs and meddles muddlingisms, thee faroots of cullchaw" (FW 303.19-21).

Joyce's poetic history is a history moving through the virtual of the fictional via dream, inebriation, hallucination and delirium, to the wo/man-made electromagnetic virtualities converging in cyberspace. Such satirico-comedic Viconian histories are ones in which an "alshemist" poet can write "over every square inch of the only foolscap available, his own body, till by its corrosive sublimation one continuous present tense integument slowly unfolded all marryvoising mood-moulded cyclewheeling history ..." (FW 185.35-186.2). They both illustrate the ongoing cultural production of consciousness and yet allow for localized regressions of decadent consciousness through a ricorso or eternal return. This is "History as her is harped ... Mere man's mime. God hath jest. The old order changeth and lasts like the first" (FW 486.6-10). The "ambiviolence" of an electromagnetic world of virtuality, its implication by the Bergsonian *durée*, its reflection in language and convergent modes of cultural production, and its links to Joycean poetic history, are the subject of another one of the original vignettes – the earliest fragments of the composition of *Work in Progress* from which the *Wake* evolved – an episode entitled "Mamalujo" (FW II.4).

These four comic annalist-historian-evangelist-psychoanalysts (*Matt* Gregory, *Marcus* Lyons, *Luke* Tarpey, and *Johnny* MacDougall [FW 384]), in one aspect the tellers of the *Wake* as a Swiftian "Tale of a Tub,"

> ... used to give the grandest gloriaspanquost universal howldmoutherhibbert lectures on anaxarquy out of doxarchology ... according to the pictures postcard, with sexson grammaticals, in the Latimer Roman history ... (FW 388.28-32).

The four, the "mamalujo," as eternal historians (the four evangelists) also being evolutionary historians as well as "howldmoutherhibbert" lecturers (the Hibbert lectures being devoted primarily to philosophy and theology) are a satiric anticipation of the natural theology of Teilhardian evolution that Joyce could have derived from a combination of Bergson with H. G. Wells and/or Huxley. Theirs are the "grandest gynecollege histories ... on which purposeth by spirit of nature as difinely developed in time by psadatepholomy, the past and present ... and present and absent and past and present and perfect" (FW 389.9-17). A history told through "picture postcards" and developed by pseudo-telephony is then a mode of virtual electromagnetic history, even if it is a parodic one. In fact, this is the very "ambiviolence" of the Joycean vision regarding the simultaneity of the evolutionary and the cyclical modes of Viconian "science" as a helical theory of history applied to the emergence of the contemporary world. Throughout the *Wake* the complex play with compounds of tele- and their relation to the convergence of media builds up to its culmination in the climactic merging of TV, film, cathedrals, and dream, with visions of the work itself as an imaginary electromagnetic machine: "the vicociclometer, a "tetradomational gazebocroticon."[8]

8 See Theall 1997, chap. 4.

The Joycean dream as a "wake," as well as a prelude to an awakening, anticipates how Guattari, half a century later, speaks of "a-waking" as a way of understanding the presence of multiple heterogeneity of media in chaosmosis:

> The heterogeneity of components (verbal, corporeal, spatial …) engenders an ontological heterogenesis all the more vertiginous when combined, as it is today, with a proliferation of new materials, new electronic representations, and with a shrinking of distances and an enlargement of points of view. Informatic subjectivity distances us at high speed from old scriptural linearity. The time has come for hypertext in every genre, and even for a new cognitive and sensory writing that Pierre Lévy describes as 'dynamic ideography' (Guattari 1995, 96-97).

This is the merging of media that Joyce intuited which had led to the "junction of informatics, telematics, and the audiovisual moving ultimately in the direction of interactivity, towards a post-media era" (Guattari 1995, 97). Eventually through combinations of speech, gesture and tactility – a movement beyond the orality/literacy principle – communication with machines will be initiated through multisensory, intuitive and cognitive machines.[9] Joyce, like Deleuze and Guattari, did not speak of evolution when dealing with this becoming that produces a secular approximation of what Teilhard has described as his vision of the emerging noosphere.

Joyce intuited what Deleuze, specifically with respect to his philosophy of cinema, was to conceptualize as the importance of the progression (a becoming or involution) in modern media from the movement-image to the time-image and beyond to "immedia" (i.e., a state beyond media or of "paramedia"). Since Deleuze marks this point of transition primarily in terms of the pre- and postwar films – considering directors such as Hitchcock and Welles to be transitional moments – the publication of *Finnegans Wake* at the outbreak of World War II marks not only the poetic recognition of that transition, but the anticipation of its movement beyond media (i.e., to "immedia"). In a sense, even earlier, one could speak of *Ulysses* as developing from the mode of the movement-image with its interest in space-time relationships where the movement and the time are linked to a state of affect and action, to gradually moving beyond to a psychological time-image ("Sirens," "Cyclops," "Oxen of the Sun," and "Circe") and then to a mental state of relations ("Ithaca") to be returned through Molly's soliloquy ("Penelope") to an eternal return which eludes the ever-insistent, unfolding time-images and relational states that have led towards the virtuality of Molly's dream.

These chapters of *Ulysses*, which were a result of Joyce's revisions, occurred almost simultaneously with the rise of dadaism, vorticism, constructivism, and abstract expressionism, and the early days of the cinema, mark the beginnings of the development from the early narrative *Ulysses* to the polysemic complexity of the *Wake*. This development interprets modernity in the impact of the past in the generation of the future. It evidences Joyce's realization that the machines of cultural production are revivifying for an electromagnetic era, "the hieroglyphs of engined egypsians" (FW 355.23) – a theme which had appeared in the question and answer catechism of Bloom and Stephen's encounter in the "Ithaca" section of *Ulysses*:

---

9 Although Guattari continues that "correlatively, an acceleration of the machinic return of orality" will occur since speech will replace the keyboard, as "it is through speech that dialogue with machines will be initiated – not just with technical machines, but with machines of thought, sensation and consultation …" (Guattari: 1995, 97) that mode of entry will both permit a new total sensorial participation-communication achieving the understanding of moving beyond the orality/literacy principle and be affected by the same total gestural activity.

In what common study did their mutual reflections merge? The increasing simplification traceable from the Egyptian epigraphic hieroglyphs to the Greek and Roman alphabets and the anticipation of modern stenography and telegraphic code in the cuneiform inscriptions (Semitic) and the virgular quinquecostate ogham writing (Celtic) (U 17.769-773).

What Joyce is playing with in both *Ulysses* and *Finnegans Wake* is unique and yet central to understanding the convergent evolution (i.e., involution or "becoming immedia") of an electromagnetic, virtual semiotic machine that is accompanying the multisensory becoming mind and understanding. It has echoes of, yet essential differences from, Teilhard in that he presents this process as implicit in a historical movement going back to the originary moment of mind and memory, for "In the beginning there was the gest." This prehistory of cyberspace and digiculture has been discussed elsewhere, but essentially it is summarized in taps, tips, types, topes (topoi, [topics] and tropes) contributing not only to the multiple, ambiviolent "typtopsical raidings" of the alphabets of printed texts, but of all modes that "typtopsical" codes produce.[10] The crucial ground of this then is the intersection of memory and human history – Joyce's "ambiviolent" revision of Bergsonian creative evolution – which requires the complex treatment of space, time, movement, and "information" that has led to the vertiginous ontological heterogenesis of "hypertext in every genre" and a "dynamic ideography."

In contradistinction to Bergson, Vico, Teilhard, and even McLuhan, the Joycean (and for that matter the early avant-garde artists') perception of digiculture and its virtual realities is not only grounded in the grammatico-rhetorical view of art as techné, and drama as the pinnacle of the poetic, but also in the long historical process of becoming conscious of the inter-relationships of mind and unconsciousness within the interplay of matter, memory and mind. In Joyce's vision Teilhard's noosphere becomes a far more complex interplay of mind, matter, memory, and history, which ultimately leads to the technoscientific discovery of the electromagnetic and intuits the significance of its emergence for establishing new modes of sense and meaning through the convergence of multiple means of communication and expression.

---

10 See Theall 1997, chap. 4.

REFERENCES

Borges, J. L. 1981. The total library. In *Borges: A reader*, ed. E. R. Monegal and A. Reid. 94-96. New York: E. P. Dutton.

Deleuze, Gilles. 1986. *Cinema 1: The movement-image*. Trans. Hugh Tomlinson and Barbara Habberjam. Minneapolis: University of Minnesota Press.

————— 1988. *Bergsonism*. Trans. Hugh Tomlinson and Barbara Habberjam. New York: Zone Books.

————— 1989. *Cinema 2: The time-image*. Trans. Hugh Tomlinson and Robert Galeta. Minneapolis: University of Minnesota Press.

Deleuze, Giles and Félix Guattari. 1994. *What is philosophy?*. Trans. H. Tomlinson and G. Burchell. New York: Columbia University Press.

Guattari, Félix. 1995. *Chaosmosis: An ethico-aesthetic paradigm*. Trans. Paul Bains and Julian Pefanis. Bloomington: Indiana University Press.

Joyce, James. 1957. *Letters of James Joyce*. Ed. Stuart Gilbert. New York: The Viking Press.

————— *Finnegans Wake* (FW). At http://www.trentu.ca/jjoyce/

————— 1986. *Ulysses*. Ed. Hans Walter Gabler et al. Harmondsworth and New York: Penguin Books/Bodley Head.

Laurel, Brenda. 1991. *Computers as theater*. Reading, MA: Addison-Wesley.

McLuhan, Marshall. 1962. *The Gutenberg galaxy: The making of typographic man*. Toronto: University of Toronto Press.

————— 1964. *Understanding media: The extentions of man*. New York: McGraw Hill.

————— 1991. Francis Bacon's patristic inheritance. *McLuhan Studies* 1:7-27.

————— 1999. *The medium and the light: Reflections on religion*. Toronto: Stoddart Publishing Company.

McLuhan, Marshall and Eric McLuhan. 1988. *The laws of media*. Toronto: University of Toronto Press.

Ogden, C. K. and I. A. Richards. 1947. *The meaning of meaning: A study of the influence of language upon thought and of the science of symbolism*. 8th ed. (1st ed. 1926). New York: Harcourt Brace.

Rice, Thomas J. 1997. *Joyce, chaos and complexity*. Urbana: University of Illinois Press.

Teilhard de Chardin, Pierre. 1959. *The phenomenon of man*. Introduction by Julian Huxley. Trans. Bernard Wall. New York: Harper.

————— 1964. *The future of man*. Trans. Norman Denny. London: Collins.

Theall, Donald. 1997. *James Joyce's techno-poetics*. Toronto: University of Toronto Press.

————— 2001. *The virtual Marshall McLuhan*. Montreal: McGill-Queen's University Press.

Toulmin, Stephen. 1985. *The return to cosmology: Postmodern science and the theology of nature*. Berkeley: University of California Press.

TOO REAL

McKenzie Wark

If virtual reality is mimesis, then mimesis of what? This is the burden of Ray Bradbury's short story "The Veldt," collected in *The Illustrated Man* (1952).[1] The device Bradbury uses to link his stories in that collection is the conceit that the Illustrated Man's tattoos "come to life" when any passing stranger looks at them.[2] "The colours burned in three dimensions. They were windows looking in upon fiery reality" (2).

The Illustrated Man leads a solitary life, trying to discourage people from looking into his tattoos, in which, eventually, inevitably, they see their own death. Children follow him. As he says to the book's narrator, with a striking fatalism:

> Sometimes at night I can feel them, the pictures, like ants, crawling on my skin. Then I know they're doing what they have to do. I never look at them any more. I just try to rest. I don't sleep much. Don't you look at them either, I warn you. Turn the other way and go back to sleep (4).

Of course, the narrator does look at the illustrations, the first of which is the story of "The Veldt." The last is a story not told in much detail, but it reprises "The Veldt" in a curious way:

---

1 All page references to *The illustrated man* are to this edition.

2 Most of Bradbury's best work takes the form of linked short stories. Ray Bradbury was born in Waukegan, Illinois in 1920, but has lived most of his life in Los Angeles. He quit his job selling newspapers in 1943 and became a full time writer, mostly of short stories for the then-flourishing 'pulp' fiction magazines. These stories also appeared in book form. Apart from *The illustrated man*, Bradbury's best known story collection is *The martian chronicles* (1950). Those stories express a range of American "atomic age" anxieties through the struggle by Earth to colonise Mars, an effort thwarted repeatedly by the gentle and elusive Martians. Bradbury's best known novel, *Fahrenheit 451* (1953), which also grew out of a series of stories, presents a world in which firemen do not put out fires, but start them. Their job is to burn books. That work is a prescient imagining of the media's transformation of the public sphere. Many of Bradbury's stories were adapted for television in shows such as The Twilight Zone and the Ray Bradbury Theater. With director John Huston, Bradbury adapted *Moby Dick* for a film version released in 1956. The best film made from Bradbury's work is certainly *Fahrenheit 451*, directed by Francois Truffaut, released in 1966. There is also a 1969 film version of *The illustrated man*, starring Rod Steiger. Bradbury's brand of cerebral, often melancholic fantasy has had a viral effect on the culture, turning up in all sorts of strange places. Elton John's song 'Rocket Man' describes a character from a Bradbury story of the same name. There is even a crater on the moon named after a collection of his stories.

It was very hazy. I saw only enough of the Illustration to leap up. I stood there in the moonlight, afraid that the wind or the stars might move and wake the monstrous gallery at my feet. But he slept on, quietly. The picture on his back showed the Illustrated Man himself, with his fingers around my neck, choking me to death (186).

The first thing to be said is that the Illustrated Man is not entirely truthful. He does sleep, and soundly, contrary to his word. The second is to note Bradbury's association, in "The Veldt" and the framing narrative, of sleep, dream, media, future – and death. If media is mimesis, it is not mimesis of what is real, but of what is "too real" (8) – the fantasy of death, the death of fantasy.

The Illustrated Man is an outcast, an outsider, an itinerant wanderer – but his illustrations still address the social. Or what will become of the social, perhaps, after its demise. They speak of a space between bodies, after the social, a space yet to come. The first story of *The Illustrated Man*, "The Veldt," does not so much represent something to come, as offer a primal scene for its arrival. The arrival of a world no longer social, no longer based on the intersubjective, but perhaps based on the interobjective. A world no longer social, but perhaps still a republic. The public thing, or public reality, is very much at issue in "The Veldt," no matter how much it may appear to concern a private home and domestic matters.

This postsocial world is a world away, if not a world after, the world of representations exchanged between subjects. A representation must be different from what it represents. Otherwise, it is what it mimics. The representation may fall short of what it represents. This is the fate of the copy. Or it might be more than what it represents: "too real," as Bradbury says. The too real is what the Illustrated Man warns against, even as he knows nobody will heed the warning.

There is something remarkable about such a popular author warning us not to read his stories, for surely both the narrator and the Illustrated Man are Bradbury himself, in his dreaming and lucid aspects. Or perhaps he is warning himself not to write the stories in the first place – a warning that, like all warnings in stories, must go unheeded. Nothing holds back the narrator from gazing into the Illustrated Man's sleeping skin. "Primarily my eyes focused upon a scene, a large house with two people in it. I saw a flight of vultures on a blazing flesh sky, I saw yellow lions, and I heard voices. The first Illustration quivered and came to life …" (5).

The house is a Happylife Home, owned by George and Lydia Hadley, inhabited by them and their two children, Peter and Wendy. It's a Buckminster Fuller contraption, "this house which clothed and fed and rocked them to sleep and played and sang and was good to them." It is a wonder of "soft automacity" (7), a "a miracle of efficiency" (8). The question being, as we shall see, efficiency of what?

The house seems, for narrative purposes, to have three spaces: the kitchen, center of rational discussion and the orderly preparation of food, the sleeping areas above, reached by the whimsical device of an air-lift, and the nursery. Dreams hatched up aloft, discussed in the kitchen, have their theater of operations in the nursery. We never see the sleeping quarters. They are the real absent space in the story.

The nursery is a miraculous theater of virtual reality, which, as it will turn out, is also, necessarily, a theater of cruelty. It is composed out of science fiction superlatives: "dimensional superactionary supersensitive colour film and mental tape film behind glass screens. It's all odorphonics and sonics" (9). Its effect is to realize fantasies in the form of sensory experience: "when you felt like a quick jaunt to a foreign land, a quick change of scenery. Well, here it was!" (8).

Something is not right in this technological Eden. George inspects the nursery, curious, expectant. "The nursery was silent. As empty as a jungle glade at high noon" (7). At first, the jungle appears as a metaphor for the nursery's ominous silence. But a jungle is what the Hadley children, Wendy and Peter, have been creating in the nursery, as George discovers: "An African veldt appeared, in three dimensions; on all sides, in colours reproduced to the final pebble and bit of straw" (7).

George has the old problem of "Africa in your parlour," when what he wants his children to play in is a "green, lovely forest" (9). Jungle equals Africa equals the dark side of fantasy. Forest equals Europe equals sweetness and light. A Eurocentric tic worth noting, but which need not detain us here.

What is more interesting is what Bradbury does with this material. The nursery externalizes the unconscious of those who play in it. Or as Bradbury phrases it: "the nursery caught the telepathic emanations of the children's minds and created life to fill their every desire" (10) and so "whatever you thought would appear" (11). What it expresses is the colonial unconscious, Africa as other.

Curiously, Bradbury expresses George's sensation of discovering the secret Africa of his children's minds – lions in the jungle – in a pair of metaphors: "your mouth was stuffed with the dusty upholstery smell of their heated pelts, and the yellow of them was in your eyes like the yellow of an exquisite French tapestry" (8). This Africa is "so real, so feverishly and startlingly real" (8) and yet it makes its appearance via two moments of metaphoric reversal. The nursery in its blank but ominous state is like the jungle at noon, but then the jungle at noon is like an over-stuffed European sitting room. Stuck with the problem of conjuring up, in words, an expression of something that exceeds the mimetic power of words, he creates a space in which this inexpressible experience of the too real might be present in its absence, somewhere between two moments of metaphor.

George doesn't want his children playing in an Africa stalked by carnivorous beasts. He wants Alice, Aladdin, Dr. Doolittle "all the delightful contraptions of a make-believe world" (11). He decides to turn off the nursery, turn off the Happylife Home. In a neat reversal, he suggests to his wife that the family take "a little vacation from the fantasy" (11). If everyday life is a fantasy, then the "real world" appears as a break from its labors, an other place and time in which to clean up one's act.

Lydia has divined the problem that besets George. It's something she feels herself: "You're beginning to feel unnecessary too" (10). The work of everyday life does not press its necessities on them. The house shines your shoes and ties them. Every aspect of the body is scrubbed, buffed, bathed, and clad mechanically. Even the unconscious gets a workout, in the nursery. The result is not utopia but superfluousness.

This liberation of the body from suffering necessity and from necessary suffering does not lead to the good life. As Schopenhauer put it:

> Work, worry, toil and trouble are indeed the lot of almost all men their whole life long. And yet if every desire were satisfied as soon as it arose how would men occupy their lives, how would they pass the time? Imagine this race transported to a Utopia where everything grows of its own accord and turkeys fly around ready-roasted, where lovers find one another without any delay and keep one another without any difficulty: in such a place some men would die of boredom or hang themselves … (Schopenhauer 1970, 43).

The Hadley parents may suffer from the loss of suffering; the Hadley children know better. As we shall see, they create a whole new world of it.

Wendy and Peter are less than happy with their meddling parents and their crisis of utility. As Peter says, "I don't want to do anything but look, listen and smell; what else *is* there to do?" (14). Africa expresses their rebellion. George consults a psychologist friend, David McClean. The doctor's explanation is that the nursery is "supposed to help them work off their neuroses" (13) – but it might do the reverse. Interestingly, access to media, education, and the nursery combined has produced in Peter and Wendy children with an elaborate relationship to the world that is not mediated by their parents. "They treat us as if we are their offspring," lament George and Lydia (13).

McClean's arrival is the cue for some prescient moralizing about an emerging consumer society. "You've let this room and this house replace you and your wife in your children's affections," McClean chides (16). "Like too many others, you've built your life around creature comforts" (16). George and Lydia agree. "We've been contemplating our mechanical, electronic navels for too long" (17). This lament will take a tragic turn: "What prompted us to buy a nightmare?" "Pride, money, foolishness" (18). All timely stuff for a story written in 1950, and published in collected form in 1951.

If that were all there were to it, "The Veldt" would be an interesting curiosity, a story that anticipates consumerism, youth rebellion, and the impact of mass media on subjectivity. One could admire the way Bradbury makes the impossible appear at the limit of textuality with a few carefully controlled metaphors. Or one could get out the critical scalpel and dissect the stereotypes of colonialism and gender that mark it as being too much of its time. But Bradbury has some untimely surprises for us yet.

Curiously, the more serious trouble to come in the nursery is signaled by three instances, which, within the story itself, are not signs but indexes. George is standing outside the nursery. "He looked at the door and saw it tremble as if something had jumped against it from the other side" (10). Then, George finds in the nursery an old wallet with "blood smears on both sides" (13). The next index that turns up is Lydia's "bloody scarf" (17). Where signs have merely conventional relations to materiality, indexes are, paradoxically, only signs to the extent that they are material traces of cause and effect. It is not at the level of the sign, but the index, that the too real will manifest its "realness."

George slowly starts to get the picture. He asks McClean about the lions stalking the nursery: "I don't suppose there is any way ... that they could *become* real?" (16). The reader knows what the Hadley parents cannot see, that it is already too late. The children lock their parents in the nursery, where they find "the lions on three sides of them, in the yellow veldt grass, padding through the dry straw, rumbling and roaring in their throats" (18). No more metaphors for the lions. They are what they are, or rather, what they have become.

Bradbury's story is told from the point of view of the parents, not that of the Machiavellian children, but it is the parents who die. They have to die, for something new is entering the world. Where once the relationship children could entertain with the world passed through the mediation of parents, the machinery of media usurps their role. It is not just these parents who must pass, in Bradbury's story, but a whole media regime in which the parent still plays a role in mediating the child's access to the world.

The story of "The Veldt" is a story that falls short of the object of representation, but which gestures to the other possibility, an exceeding of representation. Romanticism might be a good name for this longing for the excess, this desire for an other side to representation. But in romanticism, this other side is an ineffable, sublime – impossible – void.[3] The too real is the being of death and the death of being. *The Illustrated Man* plays with the romantic figure of the too real of death, particularly in "The Veldt," and yet "The Veldt" also points to another possibility.

Representations that are too real, ants crawling upon the skin, are phenomena that haunt literature, that haunt the book. A recent and striking example might be David Foster Wallace's *Infinite Jest* (1997). The Infinite Jest is (or might be), like the illustrations, a kind of virtual reality experience fatal to its user. It's striking how books validate the moral authority of the book in general, time and again, by warning of the fatal dangers of the too real. A too real to which the romantic in literature longingly desires as much as it hysterically fears.

There may be another kind of too real, other than the romantic conception of it as an exceeding of representation to the point of a beyond of representation, to the point of death. "The Veldt" points to the consequences of this other side of mimesis, this more tangible exceeding of the real. The author, like the parent, fears the loss of authority over the mediated experience. Let loose in the funhouse of the too real, the childish desire to become more than a mere passive consumer of the less than real, but a willing participant in the more than real, might take hold. Wendy and Peter are not just readers, or even "users" of their too real world. They are gamers. They "set up" their parents, for they know what their parents do not. This is not a representation of a prior world, but a model for a world yet to come. A model in which the parent (or author) as arbiter of representation as a gateway to the world it represents is no more.

While the work of art invariably falls short of the real, the work of the artist exceeds it in its power to create a world. What literature both desires and fears is the democratization of this power. Interestingly, Bradbury anticipates the cultural domain within which this possibility of democratization might indeed occur. The nursery only looks like a children's book "come to life," or made more mimetic. It is not a text space at all; it is a game space. The inability to conceive of what they see as being something other than a representation will be what proves fatal to the Hadley parents. The Hadley children know too well that what they have at their disposal is the space of a game.

Games usually exist within the constraint of boundary that separates a game world from a nongame world (Sutton-Smith 1998). The backgammon pieces, their rules of engagement, and their iterations of interaction, exist within the space of the board. Any game with logical rules exists as a thing apart, a world unto itself. Are games a representation? Of formal relations, yes; of surfaces and appearances, no.

In "The Veldt" the parents think that what is at stake is surfaces and appearances. They see lions; they would rather see fairylands. Yet it is not the surfaces that matter, but the formal properties of this world. George and Lydia Hadley pay scant attention to the no doubt elaborate logic by which the nursery produces its representations. They see the surface, not the logic; they see appearances, not form. Yet it is through the formal properties of the nursery that Peter and Wendy trap their parents. They do not necessarily know the computational processes by which the nursery produces its effects, but by trial and error they have clearly learned how to exploit the machine's capacity to produce worlds. Through the practice of gaming they become programmers of their own game of cruelty.

They are, as indicated earlier, not readers, but gamers. They do not scan the data to evoke representations, as if it were a book, they interact with the data and through experiment and

---

3 See Lucy 1997.

experience, set the game to work – in this case, to work as a trap. This is the difference between the less than real of representation and the more than real of the interactive. The former is a palimpsest of this world; the latter is a working model for another world. The nursery is a game engine, not just a game. It is an engine for making the real itself conform to the structure of a game.

The mistake, when thinking about virtual reality, is to put too much stress on mimesis. Representation is never the same as the represented. It either falls short of, or exceeds, what is represented. But it is not in terms of how it *appears* that the too real manifests itself as too real. It is in terms of how it *works*. The too real is a machine for producing new worlds, not for reproducing prior worlds. The Hadley parents are fatally deceived by the almost real appearances of lions in the jungle, and miss entirely the endgame of a more than real world constructed with no end in sight but their death.

It is not just the parent and the author who pass away in "The Veldt." This is the "Perfect Crime," as Jean Baudrillard might put it. "The Perfect Crime no longer involves God, but Reality" (Baudrillard 2000, 61). Not the extermination of Reality, but of its former opposite, Illusion. Dreams, fantasies, utopias are eradicated by being immediately realized. Or in other words, as the gap between representation and the real, held open by the inadequacy of representation, closes, both poles disappear into that closing gap.

What is interesting about "The Veldt" is that it does not stop at this denunciation of the closure of difference and representation, but provides the primal scene, or the crime scene, for a whole new logic of appearances. The scene of this crime is intimate, whereas in Baudrillard it is abstract. The point of transition from one media regime to the other takes place in an everyday scene conveniently imagined as in the future, but which is clearly already taking place mid 20th century. The less than real gives way to the more than real; the media of appearances gives way to the media of modeling; the authority of the author gives way to that of the programmer.

Appearances do not dissolve into Baudrillardian melancholia. On the contrary, a new world is born, with a new possibility for negotiating between image and reality, albeit negotiating in reverse. Reality will now lag behind, be not quite adequate, to the modeling of its game-like architecture. There may even be a new space for ethics here, albeit one founded in the murder of the past rather than just in the birth of the new. An ethics of existence, as Elspeth Probyn puts it, of "ways of living informed by both the rawness of the visceral engagement with the world, and a sense of restraint in the face of the excess" (Probyn 2000, 3). How might the power of the game world, of the power to play and program the too real world as a precursor of the real, be checked by a negotiated process?

In the light of the heroic and triumphalist rhetorics and narratives about the coming of cyberspace, of virtual reality, of hypertext, of multimedia, there's something salutary about Bradbury's primal scene, in which the new media regime is ushered in with an act of matricide and parricide. The death of the old is the reverse side of the euphoric tales of the birth of the new.

It is not the reality in virtual reality that matters. It is the virtual. What is virtual in virtual reality? Not its mimetic qualities, but its potential to pass through and go beyond mimesis. The possibility of creating new worlds. Think "virtual" and the expression "virtual reality" quickly comes to mind, describing a technology that creates an immersive, three-dimensional environment within which a "user" can move and look at things.[4] This is a rather odd use of a very ancient word, which almost by accident brings together three quite different meanings.

---

4 See Rheingold 1992 and 1993.

Virtual has its roots in the word *vir*, or "man." From ancient times it is a word that has come to identify not just a person in general but the best qualities of a person – virtue. And so at various times it has meant valor, righteousness, influence, excellence. It can point to a moral quality, or an aesthetic one. When the 18th century novelist Samuel Richardson speaks of an "object of virtue," he means a work of art. Virtue has valued different qualities in different times and places as the best of what it is to be human.

Virtue has also designated different qualities in men from those it nominates for women. A man of virtue in Nicolo Machiavelli's Renaissance Florence was a man of boldness and cunning.[5] A virtuous woman in Richardson's England was chaste and modest. There is no particular quality that is virtue, but virtue is always a quality of people, rather than of institutions. In other words, western culture has for a long time recognized the need for the concept of virtue, even though the particular values attached to the concept change across time and place.

I mention Machiavelli because it was he who first began to inquire into the qualities of institutions and of people that are required for a regime to last, through good times and bad. He thought a lot about institutions, and influenced a whole line of English-language thinkers on the topic. What "The Veldt" and the Illustrated Man stories as a whole direct our attention to is the question of what might constitute the qualities of a media regime, as a particular kind of institution.

Freed from necessity, to do nothing but look, listen, and smell – what else to do? What kind of possibility exists in the too real of the game, besides that of death? What public world, for example, might be created by a republic of programmers? If, as I have suggested, the death of the Hadley parents signals a none too ethical but highly effective end to a media regime of mediated authority – then what might take its place? Without falling into the utopian celebration of the emergent, perhaps it is nevertheless possible to think about what might replace the patriarchal regime of information – a "virtual republic."

Machiavelli's republic was a small city-state where citizens freed from necessity could easily meet in a public place. Contemporary republics can cover vast territories. Railways and the telegraph made it possible to create a republic across so vast a space, yet still dependent on the intimate, everyday media regime that reaches into the home. They are the tools by which the conversation moves through the people, and the people move freely through the space of the country. Railway and telegraph were useful tools, just as the explosion of new means of communication in the late 20th century provided useful tools.[6] But the institutions of the new media regime do not in themselves make for a conversation of any remarkable quality.[7]

I don't think it is appropriate any more to think about the virtue of a free people as consisting of any one particular quality. We don't all need to meet the same exacting standard of boldness, morality, or chastity to embody a virtue of value to the republic. What is troubling in a contemporary republic is how to combine and relate the particular and very widely varied qualities that people have in useful, fair, and creative ways. The word *virtue* can refer to a quality all people are supposed to have, but it is also the root of the word *virtuoso*, which refers to a special, highly developed quality of a particular kind. In the overdeveloped world of the west, we are almost all virtuosos at something. Our jobs and the cultures that come with them are far more highly

---

5   A particularly readable account of the life of this key republican thinker is Sebastian de Grazia's *Machiavelli in hell* (1989).

6   I may appear to be unduly neglecting the vectoral revolution of the 19th century, the railway, on which see Schivelbusch 1986.

7   It seems ironic that the Internet should spawn a vast literature in book form, but it has. Two particularly interesting works are Stone 1995, chap. 3, and Mitchell 1995.

specialized than in the days when the republican idea first entered the heads of people whose passion was to govern themselves.

I'm holding a pen in my hand, a Schwan Stabilo 188 – standard government issue from the university. I hold the pen and I look at it. My eyes relay to my brain, via the optic nerve, an image of this pen I can feel in my hand. I hold the pen up before a mirror. I look in the mirror, and I see the pen in my hand. Once again my eyes relay to my brain, via the optic nerve, an image of this same pen. Only it's not quite the same image. It is what is known in optics as a virtual image. By extension, any image I have that comes via a mediating source, not just a mirror but a screen or a speaker, might also be called virtual.

From the same word that means the best quality of a person, we get this word that means a reflected image of a thing. I don't really know why, but it's a happy accident for this essay because I want to float the idea of a virtual republic, which combines these senses of the virtual. A republic composed of institutions that produce a conversation, which all citizens are entitled to join, where we come to know ourselves by bringing to it our particular qualities. A republic that can only converse in a virtual space composed of dimensional superactionary, supersensitive color film, odorphonics, and sonics, rather than in the public square of the old Renaissance republics. It is this virtual conversation that produces our sense of the public thing. It is through the virtual that we discuss and confirm what is real to us.

And so, the virtual republic, the whole point of which is to create a people aware of its capacity to produce itself as a people. Not a people bound to any fixed idea of itself, but which knows something of the many pasts from which it descends to the present. A people aware of its potential, of the things it can make of itself, the things it can do and be. This is a third sense of the word *virtual*, and the hardest to grasp. We oscillate between the euphoria of thinking ourselves absolutely free and capable of anything; and the pessimism of thinking our lives absolutely determined, ground between the wheels of inexorable laws of history or nature. Or in short, between the worldview of the Hadley children and the Hadley parents, the worldview of executioner or victim.

Why does Africa stand in for the too real, in "The Veldt"? What place does the periphery play in the Eurocentric world of Bradbury's characters? While not wanting to fall into a critique of representation, there is something to be said about the way communication between center and periphery creates, for the center, the possibility of exploiting the periphery. The periphery appears as a territory to be known, catalogued, valued, and exploited as a resource. It also appears as the nightmare of what always exceeds the capacity of the colonial project to incorporate anything and everything in its media regime.

For a long time postcolonial culture has manifested a desperate attempt to fix a few things in consciousness between two great abstract terrains of movement. The first is the sea. The sea, as the philosopher Hegel says, "gives us the idea of the indefinite, the unlimited, and infinite: and in feeling his own infinite in that Infinite, man is stimulated and emboldened to stretch beyond the limited: the sea invites man to conquest and to piratical plunder, but also to honest gain and to commerce" (Hegel 1991, 90). Thus, ambivalently, did this first tentative vector traverse the sea.

This word *vector* has traveled a bit, from language to language, discourse to discourse, meaning to meaning. I'm very fond of it (Wark 1994, 11-14). Its roots mingle with those of the word *way* – the way: the road, the course of movement, the path of life. Also tangled up in there is the sense of "to carry." The vectors traced by these old English, Dutch, and German senses cross with the Latin *via*, and with the sense of "to weigh." From there it's a short path to specialized technical meanings. In geometry a vector is a line of fixed length but no fixed position; in physics, a quantity having direction as well as magnitude; in biology, the means of transmission of an infection.

The sense I give to the term traces a line through all of those senses. To me, a vector is a technology that moves something from somewhere to somewhere else, at a given speed and cost and under certain specified conditions. They come in two kinds: those that move mostly physical objects about the place, and those that move only information. Transport and communication were once one and the same thing. Now communication moves at a faster rate, and is able to model and coordinate movements of ever more intricate design over great distances.

"Cyberspace," people call it, this emergent terrain of information vectors. The novelist William Gibson popularized the term, and it caught on, spreading over the vector, naming the world the vector makes.[8] Cyberspace is the second great abstract terrain of movement, after the sea. It began with the telegraph, but speeds up and proliferates in the late 20th century. Like the navigation of the sea, it gives us an idea of the unlimited: digital! cyber! hyper! multi! inter! data! space! media! active! Like the sea: plunder and conquest; commerce and honest gain – and Africa in the parlor. Immersed in cyberspace, people now experience the three kinds of relation that people once felt about the sea.

First, there are *imaginary relations* to the other. The vector connects one to an elsewhere, but rather than think about this as *relating* formerly separate things together and making of them a third and different thing, people become preoccupied by the difference of the other place, and forget about what relates them. In other words, rather than seeing the relations passing *between* places, one sees only the borders that *separate* them. Rather than seeing the way different qualities mix and combine into a whole new *type* of space, one sees only what is strange, what is other. Which is what the Hadley parents see in the nursery veldt.

Secondly, there is the world of *potential relations*, lurking within the vector: the world of honest gain and piratical plunder. A vector can connect anywhere to anywhere, within the limits of what is technically feasible at a given time. So it has the potential to make connections of a certain kind, which in turn can form the basis for producing something out of what is related. Which is what the Hadley children make out of the nursery's technics – a way of exploiting the potentials, of the nursery, and of Africa.

But thirdly, there is the *virtual dimension* to the vector. An imaginary relation projects a fantasy of how different the other place is, and forgets about what passes to and fro. It is about hanging onto an old identity, by distinguishing it from that with which it mingles. A potential relation makes a fetish out of what passes to and fro, and deals with differences only in quantities – expenses, wages, quantities of goods and their prices; pounds, shillings, and pence. It is about making things, but always making more of the same. A virtual relation is about the differences between places *and* about what passes between them. It is about how places differ without forgetting they are connected, and about how they are connected without forgetting that they differ. The virtual side of a vector is all the things that might happen across the terrain it creates that are singular, unique, unrepeatable events – experiences that exceed all categories.

The virtual dimensions to the Hadley's nursery go unrealized in the story, but they are lurking there in the text of the "The Veldt" nonetheless, legible at least to a virtual reading. This virtual reality might not be about representation, either individual or social at all, but about the virtual republic – a space of possible concatenation of elements that traverse both objects and subjects. This virtual republic would be a vectoral republic, in which brute space maintains less and less

8   See Gibson 1984, 1988a, 1988b; Bukatman 1993; Dery 1996.

of a gap between the geographical locus of what is represented and its representation. A vectoral reality emerges, in which the capacity to construct relations is the engine of power and of change itself. There is nothing outside the vector.

Here then is one more way of thinking about what culture is, and why it matters: culture is the virtual lurking in the vectoral. Every vector has its virtual side. Heidegger worried about the way in which what I would call vectoral technologies of perception produce the world as if it were a series of pictures, *framed* as if they were meant for us, brought into *proximity* as if they were meant to be – just for us. The combination of vectors of mapping and movement produce the earth itself as if it were the most natural thing in the world that it respond to our passions (Heidegger 1977).[9]

This new experience of difference is an experience of an active trajectory between places, identities, formations, rather than a drawing of borders, be they of the self or place. This is what I would call antipodality. Antipodality is the cultural difference created by the vector. The acceleration of the vectors of transnational communication makes this antipodality more common. With satellite TV beaming into every part of the globe that can afford it, with the Internet spreading from west to east, many people are experiencing it. In the overdeveloped world, both the culture of everyday life and the culture of scholarly thinking about the present seem to me to betray traces of unease if not downright paranoia about antipodality. Yet it is the emergent axis of technocultural conflict.

What would things be like if the vector were perfected? What if there were no blinds to keep out the light? Imagine; but imagine carefully. Don't think utopia, the best of all possible worlds. Don't think dystopia, which is just a utopian dream turned upside-down. Think all the consequences and possibilities at once. Think of the future as a heterotopia, a mix of different kinds of space. As Deleuze and Guattari (1984) say, "Perhaps we have not become abstract enough" (321). What would it mean to become more abstract, ever more abstracted from the boundedness of territory and subjectivity? One can imagine a delirious future, beyond cyberspace. Not the future of Marx's communism: from each according to their abilities, to each according to their needs. Rather, the future of the abstract, virtual space of the vector made actual: where third nature is not just a space of resentful imaginings of the other, nor of feverish gambling on potentials that promise only more of the same, but a zone of indifference for free creation.

"We no longer have roots, we have aerials" (Wark 1994, 64). When one is implicated in a network of vectors, chance cuts across the past, but leaves its traces on the bodies of all those it cuts. The orders of the real and of representation may have imploded. So much the worse for them. But there still remains means by which to triangulate one's location, to "orient" oneself in the matrix. Only it is no longer a question of a hierarchy of orders, of the real and the less than real, as if these were two completely separable but homogenous planes. Rather, it is a matter of geometries, of heterogeneous vectors, each composed of unknown alloys of the real and the too real. Geometries of potential relation, the virtual dimension of which offers to free chance from necessity, and open postsocial spaces in which to really dream, and dream really.

---

9 I pursue this more fully in Wark 1994, 158-164.

REFERENCES

Baudrillard, J. 2000. *The viral illusion*. New York: Columbia University Press.

Bradbury, Ray. 1952. *The illustrated man*. London: Rupert Hart-Davis.

Bukatman, S. 1993. *Terminal identity: The virtual subject in postmodern science fiction*. Durham, NC: Duke University Press.

Deleuze G. and F. Guattari. 1984. *Anti-Oedipus: Capitalism and schizophrenia*. Trans. Robert Hurley, Mark Seem and Helen Lane. London: Athlone Press.

Dery, M. 1996. *Escape velocity: Cyberculture at the end of the century*. New York: Grove Press.

Gibson, W. 1984. *Neuromancer*. New York: Ace Books.

————— 1988a. *Count zero*. London: Victor Gollancz.

————— 1988b. *Mona Lisa overdrive*. New York: Bantam.

Grazia, Sebastian de. 1989. *Machiavelli in hell*, Princeton, NJ: Princeton University Press.

Hegel, G. W. F. 1991. *The philosophy of history*. Trans. J. Sibree. Buffalo, NY: Prometheus Books.

Heidegger, M. 1977. The age of the world picture. In *The question concerning technology and other essays*. Trans. W. Lovitt. New York: Harper.

Lucy, N. 1997. *Postmodern literary theory*. Oxford: Blackwell.

Mitchell, W. J. 1995. *City of bits: Space, place and the infobahn*. Cambridge, MA: MIT Press.

Probyn, E. 2000. *Carnal appetites: Food, sex, identities*. New York: Routledge.

Rheingold, H. 1992. *Virtual reality*. London: Mandarin.

————— 1993. *Virtual community*. Reading, MA: Addison Wesley.

Schivelbusch, W. 1986. *The railway journey: The industrialization of time and space in the nineteenth century*. Berkeley: University of California Press.

Schopenhauer, A. 1970. *Essays and aphorisms*. Trans. R. J. Hollingdale. Harmondsworth: Penguin.

Stone, A. R. (Sandy). 1995. *The war of desire and technology at the close of the mechanical age*. Cambridge, MA: MIT Press.

Sutton-Smith, B. 1998. *The ambiguity of play*. Cambridge, MA: Harvard University Press.

Wallace, David Foster. 1997. *Infinite jest*. New York: Little Brown.

Wark, M. 1994. *Virtual geography: Living with global media events*. Bloomington: Indiana University Press.

Virtuality : Webworlds and Cyberspaces

SPACE FOR RENT IN THE LAST SUBURB

Scott McQuire

While it's still too early to say whether William Gibson's *Neuromancer* will eventually transcend the chrysalis of the buzz-word it spawned, to re-emerge, like Orwell's *1984* or Heller's *Catch-22*, as a book which is still read and written about decades later, I suspect it will always be best known as the book which introduced the term "cyberspace" to our lexicon.

Mike Davis has placed this moment as an epiphany belonging to a glowing lineage, suggesting that:

> … the opening section of *Neuromancer*, with its introduction of cyberspace, is the kind of revelation – of a possible but previously unimagined future – that occurs perhaps once a generation. Charles Babbage's and Ada Lovelace's anticipation of a programmable computer in the 1820s, Friedrich Engels' 1880s prophecy of a mechanized world war and H. G. Wells' prevision of the atomic bomb in 1900 are comparable examples (Davis 1993, 10).

Other writers and critics have waxed even more enthusiastic about the historical significance of Gibson's vision. At the beginning of the 1990s, Fred Jameson lionized Gibson, and the cyberpunk movement he initiated, as "the supreme *literary* expression if not of postmodernism, then of late capitalism itself" (Jameson 1991, 419 n. 1), while avid cyber-convert Timothy Leary (who later bookmarked his place in Internet history as the first man to die online), pronounced that Gibson "has produced nothing less than the underlying myth, the core legend, of the next stage of human evolution" (quoted in Kellner 1995, 298).

At a more terrestrial level, social theorists have celebrated Gibson's work as the interface between fiction and the future: Doug Kellner suggests that Gibson can be read "as a sort of social theory" which grabbed the initiative from those such as Baudrillard who had "dropped the theoretical ball" (Kellner 1995, 299, 327), while Sandy Stone argues that Gibson's work transcends differences between literary and scientific discourse:

[It] is a massive textual presence not only in other literary publications ... but in technical publications, conference topics, hardware design, and scientific and technological discourses in the large (Stone 1991, 95).

There is a palpable irony in the thought of Gibson's inventive novel influencing developments in "real" science and technology: as is well known, Gibsonian cyberspace, which precociously fused the emergent global information network of the Internet with the immersive capabilities of virtual reality, sidestepping the need for screen, projection or HMD (head-mounted display) with the sci-fi staple of direct neural stimulation, was, precisely, an invention. A far cry from the seasoned computer programmers he clothed in the new romantic garb of outlaw hackers and console cow-boys, the Gibson who wrote *Neuromancer* – using a typewriter no less – was a technical neophyte with little understanding of computing:

It wasn't until I could finally afford a computer of my own that I found out that there's a drive mechanism inside – this little thing that spins around. I'd been expecting an exotic crys-talline thing, a cyberspace deck or something, and what I got was a little piece of a Victorian engine that made noises like a scratchy old record player. That noise took away some of the mystique for me ... (McCaffrey 1991, 270).[1]

However, rather than discrediting the author for lacking an adequate technical base, Gibson's admission should focus attention on the complex interactions of theory and fiction, particularly when both are concerned with the social effects of new technology. Rather than asking whether *Neuromancer* was technically well-founded, or how accurately it prefigured future developments in computing and information technology, it seems more instructive to examine the itineraries of desire it cathected. In other words, to read the popular success and critical influence of the book as symptomatic of a complex of social changes and cultural anxieties which is still far from exhausted. In choosing this approach, I find some comfort in Gibson's comment (cited in Leary 1989, 59) on the much-celebrated "anticipatory" aspects of *Neuromancer*:

It's not really about an imagined future. It's a way of trying to come to terms with me about the awe and terror inspired in me by the world in which we lived.

In what follows, I want to (re)place *Neuromancer* in the world in which we live, or rather, the world in which it was written: the early 1980s, on the cusp of the mass production of personal computers, when hacker counter-culture began to merge with big business; a moment when it seemed to many North Americans that what *Time* founder Henry Luce had dubbed the "American Century" was fast giving way, after barely five decades, to a Japanese future. (It is these cross-cultural elements which are most obscured by Leary's premature elevation of *Neuromancer* to the status of universal myth.)

In fact, the shock of finding that "American" culture was no longer an exclusively national possession – in the 1980s Japanese interests acquired prime U.S. icons such as the RCA Building and Columbia Studios, while Detroit's car makers surrendered to the likes of Toyota, Honda and Mitsubishi – was merely the tip of the much bigger iceberg of globalization. Cross-cultural anxiety towards "Asia" overlaid a series of disappointments "at home": most notably the fading of the suburban vision of the 1950s in the urban dysfunctions of the 1970s, and the decline of the

---

1 Scott Bukatman (1994) has interpreted Gibson's use of a typewriter as symptomatic of his methodology of using the paradigm of industrial technology to give form to post-industrial technological developments.

traditional rural and manufacturing sectors. In their place – but scarcely a replacement for those rendered economically and socially redundant – rose the shimmering image of the nascent information and biotechnology industries which, for the privileged, hold the promise of becoming dominant forces in the knowledge-based economies of the 21st century.

In retrospect, it is clear that the impact of *Neuromancer* belongs to the extent to which its fantasy of the cyberspace matrix was able to touch all these nerves simultaneously. In particular, it registered the cumulative impact of a century of communication technologies in dissolving old certainties about space, place and physical boundaries, depicting instead the de-territorialized cultural flows which are the everyday experience of post-televisual generations; it also resonated the ambiguous mix of fascination and horror surrounding biotechnologies such as IVF and genetic engineering which were fast redefining the parameters of human identity. From the first line of the novel, which evokes a sky "the color of television tuned to a dead channel," to its casual references to "vatgrown" flesh and posthuman bodies transformed by implants, both surgical and software, in an "age of affordable beauty" and simulated experience, it is clear that the parameters of neither nature nor self are what they were long thought to be.

Rather than rehash the novel's plot, which is fairly staple sci-fi fare about the machinations of a computer wanting to become sentient, in what follows I want to explore the underpinnings of Gibsonian cyberspace: in particular, the way it is imaged and imagined, and the social implications of the investments it supports.

DATA CITY

Neuromancer offers a compact definition of cyberspace which is fast rivaling André Breton's 1924 gloss on surrealism for frequency of quotation. Less remarked is the fact that the "definition" is coded in the novel as one supplied by a machine; Borgesian fragments lifted from an imaginary database which the protagonist Case immediately qualifies as a "kid's show":

> Cyberspace. A consensual hallucination experienced daily by billions of legitimate operators, in every nation, by children being taught mathematical concepts ... A graphic representation of data abstracted from every computer in the human system. Unthinkable complexity. Lines of light ranged in the nonspace of the mind, clusters and constellations of data. Like city lights, receding ... (Gibson 1995, 67; all ellipses in original).

Both the fragmentary nature of the definition and Case's dismissive response immediately align cyberspace with the sublime, a domain of "unthinkable complexity" which eludes the nets of representation.[2] Nevertheless, in order to function in the narrative, the matrix still has to be figured in some way. Of all the metaphors Gibson uses to evoke cyberspace, including experiences of sex and drugs, the image of the modern city proves most persistent. As Scott Bukatman has noted:

> If cyberspace is a 'consensual hallucination' that enables users to make sense of both their actions and the circulation of information, then that hallucination works by continually referencing the kinetic urban landscapes of machine-age modernity (Bukatman 1994, 86).[3]

---

2 This alignment is significant in considering the extent to which Gibsonian cyberspace functions as a site for the projection of traditional gendered fantasies of transcendence. I will return to this point below.

3 Elsewhere (Bukatman 1989, 48) he suggests: "Perhaps we can begin to learn about cyberspace by learning from Las Vegas or Times Square or Tokyo for, on one level, cyberspace only represents an extension (and implosion) of the urban topography located at the junction of postmodernism and science fiction."

More precisely, it is the image of city lights which provides Gibson with the most potent frame of reference for the "non-space" of the matrix, in which information is arrayed in luminous grids or radiant lattices stretched across an indeterminate void. Overlaying the post-industrial wasteland of massive urban sprawl and generalized urban decay which forms the novel's setting, intermingled with the neo-orientalist collage of western caricatures drawn from 1980s Tokyo (the stackable hotel rooms Gibson dubs "coffins," the depiction of relentlessly alien salary-men in their "corporate arcologies" and tattooed yakuza in their equally corporate crime worlds), the modern synergy between bright lights and big city exerts a gravitational pull in rendering the new datascape.

Walter Benjamin long ago noted the atavistic appearances of early elements of industrial culture, such as iron cast in arboreal forms, which he suggested represented an "attempt to master the new experiences of the city" and industrial technology nostalgically, by positioning them "in the frame of the old, traditional [forms] of nature" (cited in Buck-Morss 1991, 145, 111). For Gibson, it is not organic nature, but the city itself which has become the new ground to be mined for metaphors to imagine the future. Recalling the World Bank forecast that the year 2000 will mark the point at which more than half of the world's population will live in cities (with estimates suggesting 85 percent by 2025), the fact that Gibsonian cyberspace can be convincingly represented by an image of the modern city asks us to consider what has happened to the city in modernity. The issue is complicated, not least by the way that dramatic changes in architecture and urban form have been counterpointed at every turn by equally dramatic transformations in technologies of representation.

In fact, the relation that Gibson posits between urban space and data space is overlapping, as the image of each consistently modifies the conception of the other. While cyberspace resembles the city seen from a distance at night, Gibson's cities are themselves indexed as a collection of discrete, discontinuous spaces, framed by neon signs and giant holograms, and woven together by invisible networks and abstract linkages. This reciprocity directs attention to the dual impact of artificial light in shaping the modern metropolis, transforming the external appearances of the city along one axis, while simultaneously underwriting new means for seizing and transmitting the urban milieu *as image* along another. In short, the electrification of the streetscape, begun in the late 19th century, established the metropolis as a spectacular space for its increasingly mobile citizen-spectators, while new image technologies such as cinema and later television, themselves dependent on electric light, lifted the process of dematerialization to another plane by uncoupling the city's image from any and all material constraints. It's worth briefly tracing this trajectory, insofar as it forms one of the key cultural reservoirs on which Gibson draws.

Architectural historian Rayner Banham has claimed that electrification was "the greatest environmental revolution in human history since the domestication of fire" (Banham 1969, 64). Electrical energy not only powered many of the new forms of movement which characterized the industrial city, but profoundly changed the visual experience of the city, charging the urban landscape with the qualities of spectacle and display previously reserved for specialized show-places such as the theatre, diorama and amusement park. While there are innumerable testaments to the revolutionary social impact of electrification (one of Lenin's key slogans was "Soviets + Electrification = Communism"), the extraordinary *perceptual* effects of Edison's bulb on the human sensorium are registered in a passage from Thea von Harbou's novel *Metropolis* (which formed the basis for husband Fritz Lang's epic film):

The workman No. 11811, the man who lived in a prison-like house, under the underground railway of Metropolis, who knew no other way than that from the hole in which he slept to the machine and from the machine back to the hole – this man saw, for the first time in his life, the wonder of the world, which was Metropolis: the city, by night shining under millions and millions of lights.

He saw the ocean of light which filled the endless trails of streets with a silver, flashing luster. He saw the will-o'-the-wisp sparkle of the electric advertisements, lavishing themselves inexhaustibly in an ecstasy of brightness. He saw towers projecting, built up of blocks of light, feeling himself seized, over-powered to a state of complete impotence by this intoxication of light, feeling this sparkling ocean with its hundreds and thousands of spraying waves, to reach out for him, to take the breath from his mouth, to pierce him, suffocate him … (Harbou 1927, 50-51).

In von Harbou's florid prose, infinite towers of light form the modern architecture for a metaphysical struggle between the forces of light and darkness, democracy and despotism, head and hand. The sense that electrified urban space was not only intoxicating but also potentially capable of overwhelming the modern subject, is equally evident in Sergei Eisenstein's first impressions of late 1920s New York, which are couched in his characteristically cinematic language:

All sense of perspective and of realistic depth is washed away by a nocturnal sea of electric advertising. Far and near, small (in the *foreground*) and large (in the *background*), soaring aloft and dying away, racing and circling, bursting and vanishing – these lights tend to abolish all sense of real space, finally melting into a single plane of colored light points and neon lines moving over a surface of black velvet sky. It was thus that people used to picture stars – as glittering nails hammered into the sky. Headlights on speeding cars, highlights on receding rails, shimmering reflections on wet pavements – all mirrored in puddles which destroy our sense of direction (which is top? which is bottom?), supplementing the mirage above with the mirage beneath us, and rushing between these two worlds of electric signs, we see them no longer on a single plane, but as a system of theatre wings, suspended in the air, through which the night flood of traffic lights is streaming (Eisenstein 1963, 83).

Despite the ambitions of modern architects to rationalize urban space through systematic planning, the modern city-machine, powered by electricity and charged with dynamic movement, constituted an environment of visual and sensory excess, which no longer provided a stable framework against which the dimensions of space and time could be measured in traditional terms. The historical function of the city as a map of sociopolitical hierarchies and a repository of collective memory defined in the stability of sites, monuments, meeting places and neighborhoods began to give way to a new urban topography, in which the coordinates of home, self and community would have to be plotted in new ways.

This liquidation of the old city was frequently celebrated by the modernist avant-garde. In a 1927 essay defending Eisenstein's *Battleship Potemkin* (1925), Walter Benjamin premiered arguments (later repeated in both his "Artwork" essay and his unfinished "Arcades" project), extolling cinema's capacity to reveal the quotidian secrets of urban space:

To put it in a nutshell, film is the prism in which the spaces of the immediate environment – the spaces in which people live, pursue their avocations, and enjoy their leisure – are laid open before their eyes in a comprehensible, meaningful and passionate way. In themselves these offices, furnished rooms, salons, big-city streets, stations, and factories are ugly, incomprehensible, and hopelessly sad. Or rather, they were and seemed to be, until the advent of film. The cinema then exploded this entire prison-world with its dynamite of fractions of a second, so that now we can take the extended journeys of adventure between their widely scattered ruins. The vicinity of a house, a room, can include dozens of the most unexpected stations, and the most astonishing station names. It is not so much the constant stream of images as the sudden change of place that overcomes a milieu which has resisted every other attempt to unlock its secret, and succeeds in extracting from a petty-bourgeois dwelling the same beauty we admire in an Alfa Romeo. And so far, so good (Benjamin 1999, 17).

It is important to register the cautionary note sounded by the closing sentence. On the one hand, the modern city demanded more abstract maps – such as those offered by cinema, collage and photomontage – in order to penetrate the increasingly complex patterns of movement, assembly, circulation and social interaction driven by the invisible force lines of industrial capitalism. But the politics underlying the increasing abstraction of the image manifest in the modern logic of montage were always intensely ambiguous. If cinema's novel perceptual apparatus was to supply a new socio-political map of urban experience, it depended not only on a new cinematic "language" (such as Eisenstein's "intellectual montage"), but demanded an audience capable of reconnecting fragmented and fragmentary images to material structures and historical events. Even in the 1920s, writers such as Benjamin and Siegfried Kracauer saw that, in the absence of significant political change, the "explosive" experience of cinema could install itself as part of a new mythologized "nature," regimenting viewers to the mechanical drill of the city as commodified spectacle.

AT HOME IN CYBERSPACE

Where does Gibson's *Neuromancer* intersect this legacy? Its setting is characterized by a state of excess which affects both the boundaries of urbanism and the means to represent it:

> Home. Home was BAMA, the Sprawl, the Boston-Atlanta Metropolitan axis. Program a map to display frequency of data exchange, every thousand megabytes a single pixel on a very large screen. Manhattan and Atlanta burn solid white. Then they start to pulse, the rate of traffic threatening to overload your simulation. Your map is about to go nova. Up your scale. Each pixel a million megabytes. At a hundred million megabytes per second, you begin to make out certain blocks in mid-town Manhattan, outlines of hundred year old industrial parks ringing the old core of Atlanta … (Gibson 1995, 57).

Exponential urban sprawl renders "home" susceptible to a crisis of representation, demanding a new, even more abstract, form of mapping to succeed the camera as a visual solution. And, in a world where all experiences can be simulated and everything, including body parts, has a market price, being "at home" takes on increasingly alien connotations. Surveillance has become so omnipotent that privacy has become the ultimate commodity. (It is not the State – Orwell's "Big Brother" – but Big Business which is listening and watching. When Molly needs to talk to Case she takes him to a place protected by sophisticated electronic circuitry for which she pays by the second: "This is as private as I can afford," she tells him.) More so than in the Sprawl – or any *place* – Case feels at home wherever he can access the matrix: "He knew this kind of room, this kind of building; the tenants would operate in the interzone where art wasn't quite crime, crime not quite art. He was home" (Gibson 1995, 58).

When Case finally "jacks in" to the matrix for the first time, after an operation to repair the neural damage inflicted on him by some employers he crossed, he experiences cathartic release:

And in the bloodlit dark behind his eyes, silver phosphenes boiling in from the edge of space, hypnagogic images jerking past like film compiled from random frames. Symbols, figures, faces, a blurred, fragmented mandala of visual information … And flowed, flowered for him, fluid neon origami trick, the unfolding of his distanceless home, his country, transparent 3D chessboard extending to infinity. Inner eye opening to the stepped scarlet pyramid of the Eastern Seaboard Fission Authority burning beyond the green cubes of the Mitsubishi Bank of America, and high and very far away he saw the spiral arms of military systems, forever beyond his reach (Gibson 1995, 68-69).

The sensation that the dimensions of time and space have been altered, if not annihilated, has been a consistent reaction to the deployment of new generations of transport and communication technologies, from railways to airplanes and cinema to the Internet, animating writers from Vertov to McLuhan and Virilio. If the "distanceless home" of cyberspace echoes the perceptual discontinuity of cinema ("like film compiled from random frames"), it also holds out the promise of transcending the cinematic to establish a new perceptual order.[4] While Gibson uses a dystopian image of the modern city grown monstrous, accentuating its electronic dazzlements and labyrinthine and ex-centric forms which depart from all traditions of "form," in order to depict the "unthinkable complexity" of cyberspace, he also preserves the modern ambition of using a new medium to construct a legible image of social reality. While the Night City enclave of Chiba is described in quasi-cinematic terms as "a deranged experiment in social Darwinism, designed by a bored researcher who kept one thumb permanently on the fast-forward button," Case reflects that it was also "possible to see Ninsei as a field of data" (Gibson 1995, 14, 26).

When Case finally encounters the artificial intelligence (AI) which gives the novel its name, it speaks to him "… of the patterns you sometimes imagined you could detect in the dance of the street," adding: "Those patterns are real. I am complex enough, in my narrow ways, to read those dances" (Gibson 1995, 305). This echoes the investment that an earlier generation of theorists such as Kracauer and Benjamin once placed in cinema as the means for interpreting the random movements of urban crowds flowing through the new sites of ephemeral social encounters such as the train station, hotel lobby and street. The desire once invested in cinema's capacity to map the modern machine-city has been transferred to the computer's capacity to map the postmodern information-city, even as differences between film and computer become less relevant in the digital era.

For adepts, cyberspace holds the promise of deciphering the complex real and instilling order in a social and architectural milieu where it is felt to be lacking. This investment becomes even clearer in Gibson's *Mona Lisa Overdrive*, where the matrix is described as "all the data in the world stacked up like one big neon city, so you could cruise around and have a kind of grip on it, visually anyway, because if you didn't, it was too complicated, trying to find your way to a particular

---

4   *Neuromancer* contains several other "cinematic" references to cyberspace, most notably when the AI Wintermute takes Case on a virtual tour of the Villa Straylight. The experience echoes contemporary ride-film simulators adapted from defence equipment for high-end theme parks: "The walls blurred. Dizzying sensation of headlong movement, whipping around corners and through narrow corridors. They seemed at one point to pass through several meters of solid wall, a flash of pitch darkness" (Gibson 1995, 205-6). In his postscript (319) Gibson gives a more pragmatic rationale for his literary recourse to the space-time compression of cinematic form, noting that "much of the cyberspace technology so beloved of VR enthusiasts arose from my impatience with figuring out how to write physical transitions."

piece of data you needed" (Gibson 1988, 19). In *Neuromancer*, Gibson grafts the "infinite" dimensions of cyberspace onto a striated crystalline universe made up, like the early suprematist art of Malevich and Lissitsky, of basic shapes such as pyramids and cubes, arranged in a spatial hierarchy akin to the medieval cosmology of spheres.[5] It is this abstract space of urban building blocks occupying a space (like Lissitsky's *prouns*), somewhere between art and architecture, that the cyberspace jockeys navigate, if not at will, then with greater agency than they enjoy in the physical universe.

The parallel Gibson draws between electrified urban space and electronic data space is perhaps starkest at the novel's conclusion, at the moment when Case, assisted by the code-breaking program Kuang, penetrates the database of the Tessier-Ashpool corporation in order to "liberate" the AI code-named Wintermute:

> 'Christ,' Case said, awestruck, as Kuang twisted and banked above the horizonless fields of the Tessier-Ashpool cores, an endless neon cityscape, complexity that cut the eye, jewel bright, sharp as razors (Gibson 1995, 303).

In contrast to his bodily inertia, Case has become the virtual pilot of a data-vehicle flying through the spatialized information order. The heart of the Tessier-Ashpool database is represented by one of the icons of architectural modernism: the RCA building topping the "city within a city" of New York's Rockefeller Center: "The Kuang program dived past the gleaming spires of a dozen identical towers of data, each one a blue neon replica of the Manhattan skyscraper" (Gibson 1995, 302-303). As Scott Bukatman has argued persuasively, it is via these urban-kinetic metaphors that Gibson annexes the abstract dimensions of cyberspace to a narrative of human agency.[6] Such agency is inevitably riven with contradictions. Even as it heightens feelings of personal autonomy, the matrix is also an addiction, a dependency. At home in cyberspace, Case is often at a loss elsewhere. In the baroque, feminine labyrinth of the Villa Straylight, maternal "hive" of the Tessier-Ashpool clan which counterpoints the geometric lattices of cyberspace, Case experiences vertigo: "He was now thoroughly lost: spatial disorientation held a peculiar horror for cowboys" (Gibson 1995, 249).

The significance of Gibson's identification of Tessier-Ashpool and the RCA building goes further than some kind of nostalgia. It seems no accident that Gibson has chosen a complex funded by one of the best-known capitalist dynasties of the 20th century (the Rockefellers) to represent the "atavistic" structure of the Tessier-Ashpool "clan."[7] And here, of all places, Gibson's evocation of cyberspace in the guise of Manhattan registers the extent to which the relation

---

5   Gibson's image of cyberspace comprised of pyramids, grids, cubes and spirals resembles that adopted by the "Cyber City" website developed in Germany by ART + COM, simulating a model of the Berlin National Museum as conceived by Mies van der Rohe, and housing four contemporary thinkers representing the concepts of adventure, hope, utopia and catastrophe. Christine Boyer (1996, 67) notes: "A red pyramid, symbol of fire, represents adventure and is controlled by Vilem Fusser. A green cube symbolizes the house of hope; Joseph Weizenbaum is assigned to it. Marvin Minsky occupies a hexahedra of blue water, which is assumed to be a utopian place. And Paul Virilio is housed in a sphere of yellow air – the house of catastrophe."

6   Bukatman (1989, 47) writes: "Works such as TRON and *Neuromancer*, most obviously, have 'simply' rendered the invisible visible, reconstituting the terminal spaces of the datascape into new arenas susceptible to human experience, perception and control ... Science fiction stands as a significant attempt to visualize and dramatize this bewildering and disembodied environment, thereby to endow it with the weight of a phenomenal familiarity". Elsewhere Bukatman (1993, 259) argues that cyberpunk is in fact restorative of the humanism that it ostensibly displaces: "In one way or another, every work of cyberpunk produces the same radical and reactionary formation."

7   "Power, in Case's world, meant corporate power. The zaibatsus, the multinationals that shaped the course of human history, had transcended old barriers. Viewed as organisms, they had obtained a kind of immortality. You couldn't kill a zaibatsu by assassinating a dozen key executives; there were others waiting to step up the ladder, assume the vacated position, access the vast banks of corporate memory. But Tessier-Ashpool wasn't like that, and he sensed the difference in the death of its founder. T-A was an atavism, a clan" (Gibson 1995, 242).

between architectural space and social interaction in the modern city became increasingly ambivalent. From inception, the base of the RCA building was designed to be occupied by the studios and offices of the fledgling NBC TV network. As Rem Koolhaas has noted, this meant that, even in the 1930s, the complex was conceived in terms of a different relation between architecture and space:

> In anticipation of the imminent application of TV technology, NBC conceives of the entire block (insofar as it is not punctured by RCA's columns) as a single electronic arena that can transmit itself via airwaves into the home of every citizen of the world – the nerve center of an electronic community that would congregate at Rockefeller Center without actually being there. *Rockefeller Center is the first architecture that can be broadcast* (Koolhaas 1994, 200).

Whether or not one agrees with Koolhaas' attribution of this particular point of origin (what of the Eiffel Tower, which found its first "functional" use as a radio transmitter?[8]), his comment situates the new relation between the built environment and data-space which is a key to understanding the social dynamics hinging modernity to postmodernity. The generalization of electronic communities which "congregate … without actually being there," which is the fictional conceit underlying Gibsonian cyberspace, represents an *epochal* change in human modes of inhabiting space and relating to place. The overlapping and interpenetration of the different spatio-temporal regimes sustained by communication technologies, from telegraph and telephone to television and Internet, has gradually diminished the primacy of social relations based on physical proximity, as the pressure of the absent begins to displace the customary metaphysical coordinates of "presence."

THE ELECTRONIC POLIS

At the conclusion of both the Great War in 1918 and the second world war in 1945, movie cameras recorded the crowds which rushed into city streets across the world in collective acts of celebration. Faced with events of similar historical magnitude today, the orientation has changed: it is not outward, into the streets, but inward, to the home, and the electronic screen which has become its hearth, that many turn. (The latest in a long line of examples was the 24-hour worldwide telecast to celebrate the "new millennium." Even those who still gathered in city centers found themselves spectators as much as actors, watching similar groups on the giant screens that have become so popular in contemporary urban public spaces.)

Heidegger once identified the *polis* as "the historical place, the *there* in which, *out* of which and *for* which history happens" (quoted in Lacoue-Labarthes 1990, 17). In an era in which it has become increasingly persuasive to argue that this symbolic function has been annexed by the ubiquity of the electronic screen, it is clear that the notion of community and the dynamics of social bonding are subject to new exigencies. Internet enthusiasts such as Howard Rheingold long ago proclaimed the emergence of virtual communities as the practical antidote to the demise of older forms of communality, riven by economic dislocation and urban dysfunction.[9]

---

8 As Benjamin noted in one of his own radio broadcasts: "At the time of its creation, the Eiffel Tower was not conceived for any use; it was a mere emblem, one of the world's wonders, as the saying goes. But then radio transmission was invented, and all of a sudden the edifice had a meaning. Today the Eiffel Tower is Paris's transmitter" (quoted in Mehlman 1993, 14).

9 Rheingold (1994, 12) wrote of "the need for rebuilding community in the face of America's loss of a sense of social commons." Michael Heim (1991, 61) gives the investment in computers an existential edge: "Our love affair with computers, computer graphics, and computer networks runs deeper than aesthetic fascination and deeper than the play of the senses. We are searching for a home for the mind and the heart."

Others are less sanguine about the often uncritical displacement of utopic aspirations from physical to virtual space. Paul Virilio argues that electronic networks, with their new forms of access which replace traditional portals, have limits which are just "as real, constraining and segregating" as the doors and windows of antiquity (Virilio 1986, 20). As much as they are able to unite spatially disconnected social fractions, electronic media have also established a new architecture splitting the city into two societies:

> One is a society of cocoons and home offices where people hide away at home, linked into communication networks, inert. Call it home automation or what you will. The other is a society of the ultra-crowded megalopolis and of urban nomadism (Virilio 1993, 75).

Gibson's novel resides within this disjunction, offering an uneasy blend of the dystopic urbanism characteristic of postmodern fiction, manifested in his visions of urban wasteland and the violent, chaotic "dance of the street," coupled to a fantasy envisaging technological transcendence of body and street alike. Such investment registers the shift from the traditions of geopolitics to what Virilio has called chrono-politics, highlighting the importance of "real time" transactions to both political economy and cultural politics in the age of electronic media. As the screen interface becomes increasingly ubiquitous, old spatial divisions, such as those between center and periphery, or public and private, become subject to new exigencies. If, as Virilio suggests, the city is no longer a geographical entity, but is in the process of becoming "a world-city," according to a trajectory which empties rural regions and agrarian societies of their sociability, the flip side of the same process is the increasing provincialization of the metropolitan center: "The suburb is in the process of becoming the norm, the center is becoming the last suburb" (Virilio 1993, 78, 72). One might also add that the home is becoming a key node in the electronic marketplace, as Internet and interactive television converge in domestic "point of sale" technology.

The pervasive blurring and profound instability of geographical, cultural and conceptual boundaries which lies at the heart of postmodernity can itself be related to the media-induced crises of reference (where to find the real in an era of simulation?) and dimension (what happens to here and there, near and far, when distances collapse and we are all potentially "everywhere" and "nowhere" at once?).

The difficulty of answering such questions in traditional terms is intimately connected to the displacement of embodied experience as an authoritative measure. This displacement is due to two processes which are tightly interlaced without being entirely coextensive: the increasingly abstract social relations which characterize capitalist society on the one hand, and the growing primacy of what McLuhan once dubbed the "extensions of man" on the other. Far from passports to the global village, Virilio interprets the effects of electronic media on the human sensorium as the infrastructure for a new sort of incarceration:

> The urbanization of real time is in fact first the urbanization of *one's own body* plugged into various interfaces (keyboard, cathode screen, DataGlove or DataSuit), prostheses that make the super-equipped able bodied person almost the exact equivalent of the motorized and wired disabled person … Having been first *mobile*, then *motorized*, man will thus become *motile*, deliberately limiting his body's area of influence to a few gestures, a few impulses like channel surfing (Virilio 1997, 11, 16).

This transformation of bodily experience, and the concomitant disjunction between visual space and the space of bodily action, represents a particularly acute problem for architecture and urbanism. The long reign of the Vitruvian body as a humanist matrix for harmony, proportion and order (a model which persisted into modern architecture), today finds itself doubly displaced, on the one hand by the fragmented and fragmentary body of postmodern theory,[10] and, on the other, by the immaterial architectures of media networks. How, then, do we reconcile the body to the uncertain spaces in which it now dwells, without giving in to the lures of either technological transcendence or nostalgia for a pre-technological existence?

While *Neuromancer* leans heavily towards the first option, the way it does this should not be simply equated with the viewpoint of Case, "who'd lived for the bodiless exultation of cyberspace," who experiences his surgical eviction as "the Fall," and continually describes the body as "meat" and the flesh as a "prison." Such sentiments are themselves set within a narrative in which the body reasserts its primacy as the site of genuine knowledge and real experience at critical moments.[11] Rather, it is the plot's metaphysical investment in the technological mastery of space and time which most seriously undermines any phenomenological security. This becomes clearest at the conclusion of the novel, in the context of a discussion about the spatio-temporal dimensions of "being." At one point, Case asks Wintermute: "I mean, where, exactly, are all our asses gonna be, we cut you loose from the hardwiring?" Wintermute replies: "I'm gonna *be* part of something bigger. Much bigger." At the novel's conclusion, the conversation is resumed: "I'm the matrix, Case … Where's that get you? Nowhere. Everywhere" (Gibson 1995, 246, 315).

Wintermute's success in attaining god-like status is only possible because of a similar transformation which first occurs to Case. In the final run into the Tessier-Ashpool "cores," Case undergoes an extreme experience paralleling the traditional quest, in which the (male) hero experiences dissolution of ego as the pre-condition for spiritual rebirth:

> And when he was nothing, compressed at the heart of all that dark, there came a point when the dark could be no *more*, and something tore … Case's consciousness divided like beads of mercury, arcing above an endless beach the color of dark silver clouds. His vision was spherical, as though a single retina lined the inner surface of a globe that contained all things, if all things could be counted (Gibson 1995, 304).

The alchemical references to mercury (the *prima materium* of transubstantiation) and omniscient vision (a quality historically attributed to gods of all persuasions, including the new techno-gods of camera and computer) returns a few pages later:

> And then – old alchemy of the brain and its vast pharmacy – his hate flowed into his hands. In the instant before he drove Kuang's sting through the base of the first tower, he attained a level of proficiency exceeding anything he'd known or imagined. Beyond ego, beyond personality, beyond awareness … (Gibson 1995, 309).

Rather than his oft-cited references to the body as "meat," which I would argue are framed by character and couched in ambivalent, if not negative terms, it is Gibson's recourse to this most traditional myth of metaphysical transcendence which seems most retrograde in his vision of cyberspace. As Donna Haraway notes:

---

10 See, for example, Vidler 1990.

11 Most intriguingly, when Case makes love to Linda Lee – stimulated only by the memories accessed and animated by the AI Rio (a.k.a. Neuromancer) – he finds the experience still shakes him: "It was a vast thing, beyond knowing, a sea of information coded in spiral and pheromone, infinite intricacy that only the body, in its strong blind way, could ever read" (Gibson 1995, 285; see also 148, 171-72 for other examples).

Any transcendentalist move is deadly; it produces death through fear of it. These holistic, transcendentalist moves promise a way out of history, a way of participating in the God trick. A way of denying mortality (cited in Penley and Ross 1991, 20).

As vital as it is to criticize the transcendental fantasies which have been so regularly projected onto technology, it is equally imperative to avoid doing so by giving way to nostalgia, particularly for the putative certainty of old cultural boundaries. Occasionally, even commentators as perceptive as Virilio can veer close to right-wing populism:

> Sociality is in doubt because the nation no longer has any cultural standards. The countries of Europe have opened their borders and lost their specificities (Virilio 1993, 79).

The problem is not so much the "opening" of national borders to foreigners and foreign ideas, but rather the historical failure of the bourgeois public sphere, which proclaimed its universality but always operated by a system of strategic exclusions. The point today is to recognize this *structural* failing, which has been exposed by a century of mass migration and mass media, in order to develop a more genuinely inclusive and reflexive public sphere. This demands the formation of cultural institutions and the inculcation of cultural values capable of expressing plurality and difference, rather than blaming migrants and foreigners for cultural de-territorialization. To argue that such "leveling" effects are more properly laid at the door of commodity capitalism is also to argue for the importance of unstitching the synonymity of marketplace and media. If the Internet is fast becoming a new suburb in the world-city, it is one increasingly dominated by the values of commerce rather than communality. Like radio and television before it, it has become a primary avenue for introducing advertising to the home (and the school), rendering domestic space another node of global capitalism. This insertion of commodity relations into the private realm parallels the extensive privatization of public assets and the public sphere which has occurred over the last two decades.

The point is not to simply reject the increasing penetration of media technologies into the fabric of everyday life, but to more rigorously debate the socio-political framework underpinning such extensions. Virilio's conclusion to the interview cited above offers a more thoughtful way forward: "The bonds will have to be reinvented … Will we rediscover the religious bond, and so re-establish sociability? Are there new, as yet unimaginable bonds?" (Virilio 1993, 80). This comes close to the point Donna Haraway made long ago in her celebrated "Manifesto for Cyborgs," where she argues "for pleasure in the confusion of boundaries and for responsibility in their construction" (Haraway 1985, 66-67). At present many seem content to dwell on the first part of this statement to the detriment of thinking about the second, which is ultimately a question of politics. In lieu of broaching this question as one of collective action, Gibson is inevitably drawn towards the recapitulation of fictions of individual mastery and transcendence via technology, without registering the extent to which technological transformation has altered the social parameters within which individual identity is embedded.

REFERENCES

Banham, R. 1969. *The architecture of the well-tempered environment*. London: Architectural Press.

Benjamin, W. 1999. Reply to Oscar Schmitz. In *Selected writings. Volume 2, 1927-1934*;
eds M. W. Jennings, H. Eiland and G. Smith; trans. R. Livingstone et al. Cambridge, MA and London: Belknap Press.

Boyer, C. 1996. *Cybercities: Visual perception in the age of electronic communication*. New York: Princeton Architectural Press.

Buck-Morss, S. 1991. *The dialectics of seeing: Walter Benjamin and the Arcades Project*. Cambridge, MA: MIT Press.

Bukatman, S. 1989. The cybernetic (city) state: Terminal space becomes phenomenal. *Journal of the Fantastic in the Arts* 2 (summer): 48.

————— 1993. *Terminal identity: The virtual subject in postmodern science fiction*. Durham, NC: Duke University Press.

————— 1994. Gibson's typewriter. In *Flame wars: The discourse of cyberculture*, ed. M. Dery. Durham, NC and London: Duke University Press.

Davis, M. 1993. Apocalypse soon. *Artforum* December.

Eisenstein S. 1963. *The film sense*. Trans. J. Leyda. London: Faber & Faber.

Gibson, W. 1988. *Mona Lisa overdrive*. London: Victor Gollanz.

————— 1995. *Neuromancer*. First published 1983. London: HarperCollins.

Haraway, D. 1985. A manifesto for cyborgs: Science, technology and socialist feminism in the 1980s. *Socialist Review* 80:65-107.

Harbou, T. von 1927. *Metropolis*. London: The Reader's Library.

Heim, M. 1991. The erotic ontology of cyberspace. In *Cyberspace: First steps*, Benedikt, M. (ed.) London: MIT Press.

Jameson, F. 1991. *Postmodernism, or the cultural logic of late capitalism*. London: Verso.

Kellner, D. 1995. Mapping the present from the future: From Baudrillard to cyberpunk.
In *Media culture: Cultural studies, identity, and politics between the modern and the postmodern*. London: Routledge.

Koolhaas, R. 1994. *Delirious New York: A retroactive manifesto for Manhattan*. Rotterdam: 010 Publishers.

Lacoue-Labarthes, P. 1990. *Heidegger, art and politics*. Trans. C. Turner. Oxford: Blackwell.

Leary, T. 1989. High tech, high life. *Mondo 2000*, no. 7 (fall): 59.

McCaffrey, L. (ed.) 1991. *Storming the reality studio*. Durham: Duke University Press.

Mehlman, J. 1993. *Walter Benjamin for children: An essay on his radio years*. Chicago: University of Chicago Press.

Penley, C. and A. Ross. 1991. Cyborgs at large: An interview with Donna Haraway. *Social Text* 25/26:20.

Rheingold, H. 1994. *The virtual community: Finding connections in a computerised world*. London: Secker and Warburg.

Stone, A.R. (Sandy). 1991. Will the real body please stand up? Boundary stories about virtual cultures.
In *Cyberspace: First steps*, Benedikt, M. (ed.) London: MIT Press.

Vidler, A. 1990. The building in pain: The body and architecture in post-modern culture. *AA Files* 19 (spring): 3-10.

Virilio, P. 1986. The overexposed city. Trans. A. Hustvedt. *Zone* 1/2 1986:20.

————— 1993. Marginal groups. *Daidalos* 50:75.

————— 1997. *Open sky*. Trans. J. Rose. London and New York: Verso.

179

SECTION THREE

VISIBLE UNREALITIES:
ARTISTS' STATEMENTS

"Art – always – requires visible unrealities"
JORGE LUIS BORGES

Annemarie Jonson and Alessio Cavallaro

One of the premises of this book is that representation is constitutive of the contemporary moment we shorthand as cyberculture. Human beings proliferate virtualities – we always have done, as the historical dimension of the book documents – which in turn weave themselves into the fabric of the real. To this way of thinking, art is autochthonous to the book's terrain. Better, art is cyberculture's "natural" home. Perhaps unsurprisingly then, the artists gathered here probe many of this volume's foundational themes, offering both striking theoretical insights and compelling praxes that render the former immediate. Their critical interventions parallel *and* counterpoint those of the essayists, occasioning within these pages the lively buzz of a meta-conversation about what it means to be human now.

For example, while Erik Davis argues persuasively that aspects of cyberculture such as VR and AI may reinstate the Cartesian split, Stelarc takes a very different tack. For this artist, technologies of virtualisation such as the Internet present radically new possibilities for "projecting" embodied subjectivity into cyberspace. Putting himself literally on (the) line, Stelarc wires his body to the Web and surrenders it to capricious manipulation by remote, unseen agents, effectively inverting the techno-utopian paradigm of the disincarnate mind free-ranging in cyberspace. His version of telematic man is nothing so much as a meat-marionette animated by multiple, networked puppetmaster-avatars with whom it reciprocally interacts. A latter-day Netkenstein, Stelarc's prosthetic combinatorics of body and technology speaks to what Catherine Waldby, in her essay on Shelley's monstrous creation, eloquently neologizes as the human subject's open-ended "technogenesis". This technological process of becoming "sends naturalized modes of human embodiment into disarray".

Objectors might point to the metaphysical tenor of (a networked array of) incorporeal will(s) acting upon Stelarc's automaton-like body, the same trope that Davis locates at the epicenter of now discredited top-down approaches to AI. But Stelarc's deranged post-Cartesian theater explodes the one-to-one mind/body correspondence of classical subjectivity. Here, his work intersects with Umland and Wessel's fascinating account of recent neurophysiological research which shows that motor commands from the brain to the body in some inscrutable sense precede conscious awareness of an intention to act, a weird disjunction intrinsic to consciousness. This phenomenon has a pathological counterpart in so-called Alien Hand Syndrome. A lesion interrupts the brain's hemispheric connections, allowing the left hand to act unpredictably and independently of the person's will, as if it had its own volition (as was noted in the introduction to section one). In Philip K. Dick's memorable gloss, quoted by the writers, "whereas we have one body, [there is a possibility] we have two minds …".

Or in Stelarc's case, numberless more. Stelarc goes live with McLuhan's supraindividual "technological brain", a hydra-like entity indebted, as Donald Theall shows in his paper, to Teilhard de Chardin's "ultrahuman," a term which refers to the "noosphere" or "thinking layer" which supervenes on the material world by virtue of the interconnection of multiple minds in electronic networks. But, as Stelarc himself frames them, the "Net-body actions," like all of his other works, reverse the impulse to transcendence that is our metaphysical heritage. He relentlessly hyperbolizes brute carnality, giving the finger (and leg, and arm, and stomach …) to cyberculture's high-rotation advertorial pitching a "liberating" disinvestment of mind from flesh. For Stelarc, the virtual is always already the carnal.

Tropes of "bodymind" (pace Davis) which draw attention to these terms' interpenetration recur as leitmotifs in the work of many of the artists represented here, just as they exercise the essayists. Both Simon Penny and Char Davies offer acute critical rejoinders to virtuality's promissory inducements to disembodied life. Forged in the traditions of performance, sculpture and installation work, Penny's VR work *Traces* puts the body center-stage, forgoing the trad VR headgear and harness to afford the user "kinesthetically intuitive, unimpeded, dynamic, full-body interaction" with a real-time volumetric model of the user's body, which is "danced" into existence as he/she moves in the virtual space. Penny wants to evoke the embodied "being in the world" which he believes is neglected by conventional computational systems, a preoccupation which also motivates Char Davies in her immersive VR works *Osmose* and *Ephémère*.

Like Penny, Davies seeks to "reaffirm the role of the subjectively *lived* body" *in* the virtual realm. She adduces a chiasmatic reciprocity between the virtual and the actual in which the former manifests her richly-felt "sensory response to a real place," the tracts of virgin land which inspire her work. Justine Cooper, likewise, transliterates lived experience across the virtual-actual interface using high-end medical imaging technologies such as MRI to expose the body's recondite innards "in 5mm transverse slices." Cooper's bloodless eviscerations in *RAPT I* and *II*, while seemingly anodyne, still pack a potent punch, resembling grisly snapshots from a vivisector's high-tech photoarchive. If, via digital media, "living flesh is translated into malleable data," then these data, as Cooper claims, make merely "conceptual" inner spaces powerfully "experienceable." They transmit intact the visceral shock that we are, finally, meat. Davies, Penny and Cooper, together with Stelarc, respond provocatively to a central question put by Scott McQuire in his essay on William Gibson's *Neuromancer*, the oft-quoted (though, McQuire says, misinterpreted) *ur*-text of legion somatophobic cyberpunks: "How do we reconcile the body to the uncertain spaces in which it now dwells without giving in to the lure of technological transcendence or nostalgia for pre-technological existence?"

The categorical impossibility of human existence prior to technological accoutrement induces what McQuire aptly diagnoses as "crises of reference." Where do we find the real in a culture of

183

simulacra? None of the artists here appeals to an Edenic nature that escapes entanglement with technologies of production and representation. None opposes nature and technology. But some, like Davies and Jon McCormack, insist on a therapeutic dose of ecological realpolitik. Davies' VR practice is of a piece with her custodianship of a fecund swathe of the Quebec wild, the "palpably other" organic model for her virtual environments. This is an emblem of the endangered natural world to which, says Davies, "techné" – technology used as an "instrument of domination" – now catastrophically "lays waste," as Heidegger warned. Umland and Wessel echo this theme: the extreme complexity and hyperconnectivity that characterizes the human-nature-technology interface in postmodernity renders us, and the biosphere, vulnerable to massive and unpredictable systems failure.

McCormack, for his part, synthesizes teeming ecologies of digital "organisms" whose interactions mimic those of naturally-evolved vivisystems according to the principles of artificial life. At the same time, he cautions that the "beauty to be" of increasingly sophisticated ALife simulations like his own may be a poor substitute for what the poet Gerard Manley Hopkins called the "beauty been" of the devastated biota. While McCormack and Davies celebrate the aesthetic and affective kick of digital "poiesis," the "poetic bringing forth or revealing" of mimetic extensions of nature in and through technology, they also keep a vigilant weather-eye on terra firma. Their mantra might be Donna Haraway's injunction, taken up by McQuire, to take "pleasure in the blurring of boundaries" in technoculture but attend to the "responsibility in their construction."

Haraway, as is well known, indelibly inscribed the "cyborg" in the cybercultural lexicon, a legacy which is appraised in Zoë Sofoulis's essay and treated in others such as Waldby's. This ideographic figure for "many states of the human-technology interface" (Sofoulis) is the tacit companion of most, if not all, of the artists gathered here (pre-eminently perhaps Stelarc), assaying as they do "heterogenous assemblages" of humans and machines. And, in Patricia Piccinini's case, human/nonhuman hybrids *made* machine.

Piccinini's mordant gallery of biotech burlesques buff with the cheesy, high-gloss veneer of consumerist shtick is the mutant progeny of late capitalist technoculture in which cyborg hybridity *is* ontology. In *Protein Lattice*, a supermodel cradles the iconic "ear-mouse," a sorry, hairless lab rat cum in-vivo culture medium sporting an outsize prosthetic human ear on its back. A mid-1990s instance of interspecies "tissue engineering," the mouse supplied the circulatory system for a three-dimensional auricle-shaped framework of synthetic protein seeded with cartilage cells taken from the ear's intended human recipient.[1] Like Haraway's cyborg, a "monstrous chimera" whose identity is "permanently partial," the mouse inhabits a world "ambiguously natural and crafted" in which origin is a mirage and the "leaky and unstable boundaries" between human, animal and machine perfuse blood. Where, then, does mouse end and human begin? asks Piccinini. It's a trick question: there *is* no "mouse" and there *is* no "human," only what Waldby has styled as an "organic machine," an ambivalent figure fashioned for human ends, like the abject protagonist of Shelley's canonical text, in the "workshop of filthy creation."

VNS Matrix is the cyberfeminist spawn of that infamous bodyshop. Inspired by Haraway's Manifesto, VNS sought to "explore the artistic … potential of technologies and information age metaphors for women," at the same time "taking feminist activism into the virtual worlds" (Sofoulis). Francesca da Rimini, a former member of the group, continues the activist tradition in Net-based interventions which locate the political and social concerns of "post-media" feminism in a complex technological field in which militarism, masculine narratives of domination and

---

1   Information taken from Patricia Piccinini's artist's statement for *Protein Lattice* published on http//:www.patriciapiccinini.net

control, the flux of global capital, and first-world technophilia are inextricable. During their heyday as VNS, da Rimini and her co-conspirators, Julianne Pierce, Josephine Starrs and Virginia Barratt, were equally politically motivated, taking in-your-face cyberfeminism up to "big daddy mainframe" by insisting that "the clitoris is a direct line to the matrix." Wearing their feminine abjection loud and proud, these aggressive, hi-tech guerrilla-bricoleurs wanted to jam the masculinist machine, infiltrating it with grrrl-slime as they cobbled together profane techno-visceral contraptions from the iconography of the female body and the computer.

In putting the body back in the machine, it seems the grrrls were onto something. This has to do not only with the framing conundrum of this book – what it means to be human as technologies unravel and re-make us – but what it might mean to *be* cyborg. How do we get from here to HAL? a self-conscious entity not of our species, but one who nevertheless petitions persuasively that "I am."

It seems we may have to keep the meat, of one sort or another, or at least preserve the profound interconnection of self and substance. In *The Brain Project*, his work-in-progress which maps the field of consciousness studies, Stephen Jones makes the point that dualist conceptions of mind as an ineffable extra-material entity have given way to "physicalism" or "identism," locating cognition and self-consciousness squarely back in the grey matter. To this way of thinking, the brain is a sophisticated "meat computer," one of almost unthinkable organizational complexity, from which consciousness emerges.

The consequences of this conception of the brain/mind relation for AI are far-reaching. Attempts to synthesize consciousness by pre-programming vast quantities of data into serial processing computers have been ignominious failures. Traditional AI, which privileged software and data – contemporary versions of bodiless mind – produced chess grandmasters like Deep Blue but nothing that emulates subjectivity as we know it. Jones claims a true AI, one which approaches human consciousness, would need to reduplicate the brain's "massively parallel," hyperconnected "heterarchical" "neural architecture," the organized complexity that underpins the head-spinningly prodigious computational processing power of the human "wetware." The AI would need, moreover, to be in dynamic interplay with the world, as we are, and it would adapt heuristically to new inputs: it would learn. Appositely, Elizabeth Wilson notes in her essay that Alan Turing, a giant in the pre-history of computer intelligence, supposed that such a synthetic entity – one which interacts and learns, as we do – may require sense organs (ie., embodiment), and initially may be child-like.

Re-assessing Turing's legacy, Wilson shows that promising contemporary robotics research begins from the assumption that human-like intelligence is "situated, embodied and interactive." These same principles are implicit in Jones' prescription for AI. Furthermore, this research takes the "developmental" approach to machine intelligence that Turing outlined (the alternative to the top-down programming tradition of classical AI which he also mooted and for which he is most renowned). Such a computer would be designed and programmed to interact iteratively with human and nonhuman others, as an infant does. The realm of affect – including the phenomenon of surprise – would be a key quality of the robot's engagement with the world. Importantly, this intelligent machine should be able to act unpredictably; to throw up novel outputs; to surprise its creator. A domestic robotic companion which displayed something akin to human consciousness might cause us to observe: " 'My machine' (instead of 'my little boy'), 'said such a funny thing this morning' " (Wilson quoting an account of Turing).

Jones considers just this capacity to generate novelty, to innovate, and to act flexibly (ie., in a non-stereotypical, non-"programmed" way), the mark *par excellence* of consciousness. This, he says, is the quality which a genuinely conscious synthetic entity would need to evince: "The cardinal characteristic of human consciousness is that is generative, capable of producing novel linguistic constructions, new ideas, original works of art. A conscious machine must be able to extemporize about the neighbours or, better, confabulate about itself." The machine that Jones envisages, in other words, resembles the linguistically inventive little entity that Turing tentatively imagined some 50 years ago, a little known aspect of his work that Wilson brings to light.

These ruminations on subjectivity, infancy and embodiment merely sketch the highly-nuanced debates around consciousness and its virtual mimeses. More to the point, Jones' work, like all of AI, is indebted to the advent more than 50 years ago of Wienerian cybernetics and the sciences of complexity. Projects to simulate organic life and its cognitive properties collectively hang on a principal cybernetic assumption – that organisms and machines are complex, highly-structured, interactive systems whose dynamic principles of organization are isomorphic. As Evelyn Fox Keller writes in this volume, the notion of organization "unifies the animate and the inanimate." And with the epochal erasure of the distinction human/machine comes, inevitably, cyberculture, and its apotheotic figures, the cyborg, AI and ALife. Keller continues: "Just as the early cyberneticians envisioned, in this new age of complexity, life is not limited to biological organisms. Indeed for some, life is not limited to physico-chemical materiality. Artificial Life … can exist as pure simulation … What now is an organism? … It might be green or grey, carbon- or silicon-based, real or virtual."

Two artists explicitly address artificial life here. As was noted above, Jon McCormack contrives an artificial biosphere from "digital genes." These computer algorithms describe synthetic cultivars whose digital world is accessible to us through an elaborate interactive interface. The "lifeforms" populating *Turbulence* carry vestigial traces of organic form, inflected by the artist's speculative take on "life as it could be" to phantasmagorical ends.

To advocates of artificial life, the recombination of algorithms in the system corresponds to the intermixing of genes in nature, making such simulations independent, non-biological "instantiations" of life itself. The legitimacy of such claims is a moot point. McCormack's own more modest ambitions for *Turbulence* pertain, rather, to the relationship of its artificial world to the wild. He frames *Turbulence* as both a thought experiment on future life and an exemplum of *ars memoria*, "a natural history museum." A thing, in other words, to remember the world by. In this sense, whether or not its organisms are literally alive, the work has what we might call a rich "cognitive life," John Sutton's elegant term, developed in his essay in this volume, for the vivacious prosthetic consciousness of sorts that inheres in things of our own making (and in natural and animate things). In artifacts like computers, sketchpads and the Internet (but also in "bowls, building and knots," "pets and friends") we externalize, supplement, and embody cognitive processes such as memory. Reciprocally, our own minds are shaped and modified in intimate, "mutually modulatory" and "dynamic" "couplings" with things. In Sutton's evocative re-framing of the sciences of mind, cognitive life is "smeared across" "brain, body and world," enlivening our "natural, social and technological environments."

Thought in these terms, our mnemonic and cognitive technologies, like our pets and friends, or indeed the aberrant body part in Alien Hand Syndrome, might have lives or minds of their own. Troy Innocent introduces his ALife work with just this momentous question: "where do life and intelligence reside?" Eschewing contentions over the literal "aliveness" or otherwise of digital organisms, he turns his attention to the "unique internal logic" of interactive virtual space. Innocent notes that while we are skilled at "shifting perceptually between the real and the hyperreal," and at reading digital simulacra as "integral to reality itself," virtual worlds are also coherent alternative universes whose modus operandi stands, in some important ways, outside the operational logics of "real" space.

In the interactive three-dimensional computer game *Aretfact*, Innocent wants to examine the features native to virtuality. So first, he jimmies open the cracks between the real and the virtual. Innocent accentuates the sound and visual glitches (the eponymous artefacts) which are the "unintended side-effects of the algorithms used to construct virtual worlds," bringing the occult machinery of representation clunking into view. The digital world's seamless illusion of gestalt and narrative integrity is fractured, as it is in Neo's pixelated passage in *The Matrix* from that illusory space to devastating reality (discussed in detail in Davis' essay). This de-familiarizing perceptual re-orientation – in the normal course of our interface with computers, an uncontrollable and unpredictable aberration – exposes virtuality's conceit: there's life in simulation, but it's merely verisimilar. If the fire goes out, the shadow play in Plato's cave disappears. Mimetic representation, a "palimpsest" of the real, always fall short of its object, as McKenzie Wark suggests in his essay on Ray Bradbury's futuristic short story of 1952, *The Veldt*.

But there's another aspect of representation that outdoes mimesis. Innocent draws our attention here not to the (flawed) capacity of digital technologies to reproduce the appearance of the real, but to the formal properties which distinguish computational artefacts, as expressed in the game logic of the artwork. He argues that interactive technologies of virtualization afford a certain semiotic flexibility which renders representation slippery, generative and always in process. Things, in digital game play, morph fluidly into other things, making of virtual space a kind of machine for (things) becoming (something else), as any avid gamester would intuitively know.

As Wark shows, *The Veldt*'s holographic nursery is such a machine. A post-televisual divertissement in the Hadley family's "HappyLife" home, this in-house entertainment center simulates distant and exotic climes with striking veracity – and its effects are lethal. The Hadley children tinker with the program so that the wildlife populating the veldt devour their unsuspecting parents. Fluent in the elaborate formal logic of game play, the junior Hadleys are able to "exploit the machine's capacity to produce worlds," permuting mimetic causality so that the real itself is transformed to "conform to the structure of a game." Representation here exceeds the real, becoming what Wark calls the "too real," a "working model for another world" which breaks violently and tangibly into reality.

In 2002, Innocent's interpretation of the same kind of machine is less dystopian, but no less intellectually compelling than Bradbury's. For him, the ludic, interactive logic of digital technology also conjures into being other spaces, with inner lives all their own. The formal properties specific to computational media "animate digital space, lending it a peculiar kind of … denatured intelligence which is reflective of the 'nature' of the medium itself." Unlike Bradbury, writing presciently from the cusp of cyberculture, Innocent is sanguine about our encounters with this medium.

After all, aren't we now, 50 years on from *The Veldt*, the Hadley children all grown up? By now, isn't it the "nature of the medium itself" to be (our second) nature? The Platonic cave and the game parlour, TV and Web and playstation, cyborg, digital organism and organic machine, and all the artful devices and contraptions in between that accompany what we call, for convenience, cyberculture; don't they, and won't they in the future, spin us into being alongside the "less than real" and the "too real"? As the artists here show, these states are becoming two names for hearth and home: the domesticated gallery of visible unrealities.

RAPT

Justine Cooper

By virtue of digital technologies, electronic art aligns itself with the Einsteinian precept that space and time are "forms of intuition" – mutable, conceptual entities. As science has untethered itself from the conceit of a knowable material world, electronic art forms, equally, have dissolved the confining one-way link between representation and the object. Art is no longer embedded in the object, but evolves into an information processing event. Physical objects and the spaces that they occupy are transformed by the liquid topography of information.

My goal is for technology to open to experience the relationships of real and virtual, physical and immaterial spaces. I am interested in the ways information media dynamically interrelate science and art, the "inside" and the "outside," the architectonics of new spaces and old, collapsing the boundaries between the conceptual and the experienceable.

In my work, I have used Magnetic Resonance Imaging, a medical visualization technology, to probe and reconfigure the inner space of the body. In *RAPT I* (1998) MRI scans of my full body were made in a series of 365 transverse slices, captured every 5mm. The data slices are computer animated and rendered as video. The body is unmade and remade, dissected and reconstituted, in real time and in several orientations. In *RAPT II* (1998), a room-sized installation, 76 slices taken at regular intervals through the body were printed onto one-meter-square sheets of architectural film and hung sequentially with gaps between them. A dialogue is set up between the three-dimensional animated body on the flat screen of the video, and the two-dimensional slices occupying three-dimensional space in the installation.

The fluid transformations of the body in these works echo the materiality lost at the moment of its digital imaging, and its rematerialization as information. Just as the body is recodified through medical technology, so its internal spaces and brute physicality are remapped and made accessible in these works. Living flesh is translated into malleable data.

http://justinecooper.com

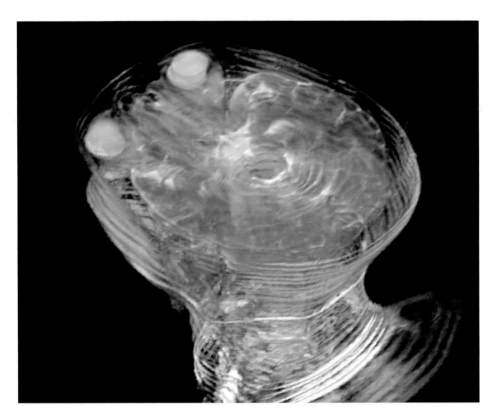

JUSTINE COOPER *Rapt* 1998

SOFTLY FROM THE RUINS...

Francesca da Rimini

efemera> she breathes the uncertainty of her time

packets of soft recognition

adrift in the network resembling the body called flesh

plastique> decay of overused images that barely elicit a pang of recognition

~~All post media direct action cells must pursue the instabilities in technologies even before they become metaphors.~~

efemera> this is a cry for new memory systems to build in

our dead must come out of the night and the earth

let them dress in the garb of war

so their voice may be heard in the empire of silence

Long-range precision strike. Exquisite precise location of targets.

plastique> a precision that is more much more than a mere casualty

Their VR helmets can't see the failure of Reality before the new fundamentalism of the telematic – they believe the lights they see from the midnight bombs they drop are coming from something that still exists: nation, justice, and democracy. These are now nothing more than the last signs of dead cultural stars.

We have revitalized our missile defense research and testing program, free of the constraints of the Anti-Ballistic Missile Treaty.

invisible artillery follows nurse with wound

endlessly uncoiling a spectacle of irretrievable situations

intolerable signs

ruined, all ruined

ErosloNE> in gathering we move closer to the feeds we are constantly circuiting but unable to access alone

~~Post media cells must fight the future with gestures that have no name in the present.~~

liquid_nation> veiled little sisters counter-surveilling Big Daddy

~~Post media cells must travel among strings of inventions that fall outside of the logomass, placing the impossible and the unexpected as our counter dialectics.~~

The 2002 Unified Command Plan prepares us for the future by assigning every area of the globe to a combatant commander's area of responsibility

come, she said

destroy, she said

~~Post media cells must create situations for mutation that can interrupt and reroute the protocols of acceleration, improvement and obsolescence that late capital is bound by. So that rational history will be broken and remade by the tiny hands of the intergalactic niños of the fifth world.~~

Now that is uncivilized behavior.

Texts sourced from: liquid_nation, Ricardo Dominguez, Diane Ludin, Julieta Aranda, United States Department of Defense.

http://sysx.org/gashgirl/Space Command

FRANCESCA DA RIMINI *Weirdmo1* 2001

ESPACES ENTRELACÉS :
VR AS POIESIS

Char Davies

Heidegger warned of technology's tendency to function as an instrument of domination laying waste to the natural world. His words resonate as I write on my laptop in the middle of a forest by the edge of the Pacific in British Columbia among sweet cedar, hemlock, fir and fern. Though encircled by a sea of noise, crisscrossed by communication transmissions, and tainted by Vancouver smog, this small pocket of ancient growth is palpably *other*.

Such fecund places are receding as tangible territory and as mythic elements of symbolic desire and dread. We scarcely notice the threads that ground our mortal bodies to the living earth, as our attention is diverted to hi-tech wonders, and the replacement of living nonhuman others with things of our own making. VR may well further the destructive trajectory which Heidegger signposted, distract from earthly crises by substituting the virtual for the real, and prove to be the nemesis of nature.

But Heidegger wrote also of an earlier techné, called "poiesis" by the Greeks, associated with a poetic bringing forth or revealing. Encouraged by his words, I have explored an alternative VR – a means of "bringing forth" subjective experiences of the natural world. My work is grounded in caring for 400 acres of semi-wild land in southern Québec, a landscape whose natural features and unseen animal presences populate the virtual spaces of *Osmose* and *Ephémère*. These works reveal a reciprocal link, *un espace entrelacé*, by which virtual landscape manifests my own responses to a real place.

In *Osmose* and *Ephémère*, virtual embodiment is kinesthetically explored through full-body immersion and interaction. I use participants' breath as interface (enabling them to "float"), and semi-transparency to evoke cognitive ambiguity, reaffirming the role of the subjectively *lived* body within the virtual realm's evocation of nature. And, while *Osmose* is an exploratory step, *Ephémère* goes further in reaffirming a poetic and mythic need for nature's otherness, returning attention to our fleeting existences as embodied beings embedded in a living and sensuous world.

http://www.immersence.com

CHAR DAVIES *Osmose* 1995

MAPPING THE DIGITAL REALM

Troy Innocent

Where do life and intelligence reside? Until recently, our only examples of these phenomena have been biological. But digital technologies invite us to consider what role life may play in electronic space. In the field of artificial life, complex real-world systems such as ecologies can be represented as a set of algorithmic rules which describe simulated lifeforms. Over time, an adaptive, dynamic computer-bound version of the given system emerges in accordance with the rule-governed behavior of the entities.

Iconica's synthetic world is populated by plastic knowbots which communicate amongst themselves and with the audience via an iconic language system peculiar to their universe. The work evolves in response to both audience interaction – users can create, construct and manipulate entities – and evolutionary principles.

On one level, *Iconica* is a stylized model or simulacrum of reality. The ubiquity of various simulacra in media culture (movies, computer games and so on) makes us adept at shifting perceptually between the real and the hyperreal, extrapolating from our own world to synthesized worlds, or accepting simulation as integral to reality itself.

But *Iconica* also actualizes a coherent *alternative* space, a project which I develop further in *Artefact*, an immersive three-dimensional computer game. *Artefact* investigates the particular "language" of computer-mediated space as it is expressed in games, its unique internal logic and formal properties.

Artefact, for example, foregrounds the shift between the real and the virtual by accentuating "artefacts" or errors, the sound glitches and aberrant visual patterns which are the unintended side-effects of the algorithms used to construct virtual worlds. It explores, also, what Joseph Goguen has called "semiotic morphism," a "systematic translation between sign systems" in which signified messages can be mapped onto various signifiers, multiplying instances of semiosis. This term captures the shape-shifting plasticity of relationships between sound, image, text, and users which characterizes computer-simulated worlds. Through these fluid, synesthetic interactions, meanings and identities remain constantly negotiable, processually emerging and transforming in real time. These formal properties "animate" digital space, lending it its own peculiar alternative logic – a denatured intelligence which is reflective of the "nature" of the medium itself.

http://www.iconica.org

TROY INNOCENT *Iconica* 1998

SELF PORTRAITS FROM THE INSIDE

Stephen Jones*

Consciousness, being private and experiential, appears to be something quite other than the physical world of molecules, machines and electromagnetics. It consists, at least, in the properties of our phenomenological encounters, qualia like the redness of a rose, the smell of freshly brewed coffee, weaving that ineffable sense that I am. It equips us with both a stable, temporally consistent centre of action and the ingenuity and nimbleness to innovate and adapt moment-by-moment in dynamic interplay with the world. It faultlessly computes the dizzying syntactic and semantic complexity of language. But how does this subjectivity, one's sense of a coherent inner life, arise, when to all appearances, one only has a physical system with which to experience it?

The news for those who want to preserve the mystery around consciousness is not good. The Cartesian, dualist view of the mind/body problem – the insistence for example that consciousness exists independently of the physical world – has been superseded by the identist or physicalist view which dominates contemporary philosophy, cognitive science and neurobiology. For physicalism the mind is the brain. I – *res cogitans* – am nothing more or less than my brain's physical process; the first-person experience of its electrochemical activity. Professor Ramsay in Woolf's *To the Lighthouse* put it with brutal concision: "The mind, Sir, is meat." After the epoch-making advent of the "electronic brain" in the 40s and 50s, Marvin Minsky lends the physicalist metaphor a more contemporary idiom: "The mind is a meat computer."

But we're not talking a three-pound wetware version of ENIAC here. Consensus is that the neural computer is a terrifyingly complex beast; an elaborately structured, multiply bifurcating, highly connectionist, massively distributed, multilevel, heterarchical, cybernetic, parallel information processing network of which consciousness is an emergent property; a dynamic system skeined in with the world, adaptive to new inputs, and, crucially, able to generate novel outputs.

What are the implications of this view of mind for AI? Can a computer be conscious? The cardinal characteristic of human consciousness is that it is generative, capable of producing novel linguistic constructions, new ideas, original works of art. An intelligent machine must be able to pass a math exam, but a conscious machine must also be able to extemporize about the neighbors or, better, confabulate about itself. Or perhaps compose *Ulysses*. But as Daniel Hillis has pointed out, the machines we make now lack the complexity to simulate even the cogitations of a resting fly. Until we're able to emulate the neural architecture that underpins consciousness, an entertaining robotic interlocutor, let alone a silicon Joyce, will remain strictly vapourware. The circuitous and exhilarating route to these possibilities is what *The Brain Project* maps.

*edited by Annemarie Jonson, based on personal communication with Stephen Jones and his collected works published in *The Brain Project* http://www.culture.com.au

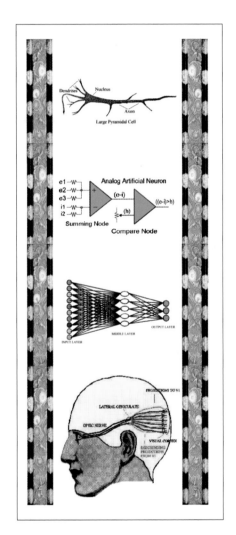

STEPHEN JONES *Neural* 1998

TURBULENCE :
AN INTERACTIVE MUSEUM OF UNNATURAL HISTORY

Jon McCormack

In *Binsey Poplars*, Gerard Manley Hopkins laments the felling of a row of trees to make way for a housing estate: "Aftercomers cannot guess the beauty been." The poem evinces what has been called "the beginning of a sustained note of mourning" (Hamilton-Paterson 1993) about the loss of natural landscapes. For humans, landscape, nature and the wild are more than just a physical resource to be consumed: they are an aesthetic necessity. While sentiments about a balance of nature that existed prior to our intervention are misguided, since we are as much a part of nature as anything else, nature *is* changed and diminished because of us, and we are emotionally the worse for it.

Media theorist Gene Youngblood claims that VR will allow us to create on the same scale as we have destroyed. But if the entities and spaces that we are now beginning to synthesize are indications of the possible aesthetic substitutes for what we are losing, what does this portend? Surely a virtual nature will have to be the aesthetic equal of its organic counterpart.

Turbulence is an evolutionary landscape made possible by technology – a digital "poiesis." It explores artificial life, a field which abstracts the essential principles of life from its biological substrate and applies them in the digital medium, simulating in the computer natural forms and processes. *Turbulence*'s menagerie of synthesized lifeforms is created by purpose-designed software and made discernible by graphic visualisation. The computer becomes the world within which a teeming macrocosm evolves according to simple algorithmic rules (digital DNA), and in which digital genes' survival depends on the "aesthetic fitness" of the artificial organisms they define – my perception of their visceral quality of "naturedness." The work is a kind of futuristic natural history museum, a living archive which subsists only in the abstract "pluriverse" of cyberspace. It is both a celebration and a lament: a lament for things now gone, a celebration of the beauty to come.

Hopkins coined the neologism "inscape" for the "individually distinctive" form which constitutes the rich "oneness" of the "natural object." Ultimately, *Turbulence* asks whether our synthetic constructions can ever have the inner oneness of naturally evolved spaces and entities. Won't new evolutionary landscapes be prone to mirror human limitations? Will the beauty to be ever equal the beauty been?

http://www.csse.monash.edu.au/-jonmc

REFERENCES

Hamilton-Paterson, James, 1993. *Seven truths: The sea and its thresholds*, London: Vintage.

Section Three

JON M^cCORMACK *Turbulence* 1994

TRACES *

Simon Penny

One may argue that computer science endorses Cartesian dualism and implements it in the hardware and software duality, and that conventional interfaces to digital systems filter out the complex performed and embodied intelligences which constitute our engagement with the world. My own recent work in this field has been motivated by a critical position with respect to the main currents of thought in computer science (and particularly in artificial intelligence and VR). This critique of the Cartesianism implicit in new technologies is informed by traditions and sensibilities of sculpture, performance and installation, which emphasize the immediate, dynamic, sensorial and embodied nature of engagement with the artwork. While criticality and intervention had existed in my previous works at the level of 'content', in my works of the 1990s the ostensible subject matter, that which was depicted, was largely irrelevant, and critical intervention occurs on a formal level. The choice of what phenomena to aim computational processes at, and how those processes would be conformed, became central to my project.

Traces consists of a custom infra-red multi-camera machine-vision system which constructs a volumetric model of the user's body in real time. This vision system functions as the sensor front-end to a suite of custom computer graphic operations which are displayed stereographically in an immersive environment called a CAVE. Unencumbered by wiring harness, pointing devices or headgear, nor any sort of textual, iconic or symbolic gesture interface, the user 'dances virtual sculpture into existence'. *Traces* affords kinesthetically intuitive, unimpeded, dynamic full-body interaction with increasingly autonomous computational entities, opening a new and rich conceptual space for design of behaviors in virtual environments. This transpires to be engaging, and encourages expansive bodily movement. After even a short period of engagement, users emerged breathless and sweaty.

Unlike most VR environments, *Traces* contains no 'virtual world' and requires no 'navigation'. Rather, the situated, freely-moving body is the engine of creation in the VR experience. *Traces* comes further towards the world humans inhabit, and 'understands' dimensions of human being-in-the-world seldom engaged by computational systems.

* *Traces* was realized 1998-9 with the assistance of Phoebe Sengers, Andre Bernhardt, Jeff Smith and Jamie Schulte at Carnegie Mellon University, Pittsburgh PA, and the GMD, Sankt Augustin. Germany. It premiered at Ars Electronica 1999. For further information, see *Traces*: Embodied immersive interaction with semi-autonomous avatars. Convergence, Luton UK, Vol 7 No 2, (summer) 2001; and www.ace.uci.edu/penny

SIMON PENNY *Traces* 1999

Visible Unrealities : Artists' Statements

LIFE IN THE MEDIA LANDSCAPE

Patricia Piccinini

I am interested in the politics and experiences of contemporary life. I am not especially interested in technology per se. However, it is impossible to deal with contemporary issues without reference to technology. I am interested in how media culture both reflects and shapes our lives, and how technologies impact on lived experience. I am interested in the myths, truths and "nature" of life in the contemporary media landscape.

An aspect of this, which I explore in my work, is the shifting status of "the natural." Nature is now the stratum through which the cutting edge of technology most regularly slices. The opposition of nature to technology is, to me, increasingly irrelevant. I have an ambivalent attitude towards both technology and nature. They are both forces that are too massive to be either good or bad. I attempt to chart their effects and create a place to reflect on them. I am interested in working along fault lines and exposing self-contradictory moments, and moments of compromised enjoyment.

I am interested in bringing the dark forces to light without denying the allure that I find there. This is because wonder is as important to me as politics. Something that makes art valuable is that it can create a new thing or experience that exists outside of the rules.

In *Protein Lattice*, I explored the commodification of life that is implicit in biotechnology. What struck me about the mouse with a human ear growing on it was not the giant technological leap it represents but the empathy I felt towards the mouse and the sense that reality was far weirder than science fiction or art. The work's form is like some strange advertising campaign that challenges the viewer to disbelieve its reality.

The significance of the juxtaposition of the attractive model and the grotesque mouse is their similarity, not their difference. To me they are both natural (organic) and artificial (constructed, retouched); both beautiful and empty, valued only for the intellectual property that they represent.

http://www.patriciapiccinini.net

PATRICIA PICCININI *Protein Lattice* 1997

Visible Unrealities : Artists' Statements

AVATARS AND THE INTERNET :
ALTERNATE, INTIMATE AND INVOLUNTARY EXPERIENCES

Stelarc

Contrary to the popular view that the Internet offers fulfilment of the metaphysical desire for disembodiment, I believe it presents powerful strategies for projecting body presence into cyberspace. But authenticity of this "Net-body" does not rely on the coherence of its individuality. Rather, it is emerges from the multiplicity of its collaborating agents. What is important is not this body's identity, but its connectivity; not its physical location, but its interface. Thought about in this way, the Internet doesn't presage the disappearance of the body and the dissolution of the self, it *multiplies* the body and self, generating new collective physical couplings in cyberspace.

Thus, just as the Net provides interactive ways of linking and retrieving information, it now allows unexpected ways of interfacing and even inverting the relationship between remote agents – avatars – and the body. It becomes both an external nervous system conjoining any number of bodies and machines, and a space inhabited by intelligent entities that reciprocally inform humans. The body is possessed – a body not (only) for itself, but for its avatars.

In my previous actions the (my) body has performed with: technology attached, the prosthetic "Third Hand" actuated by EMG signals; technology inserted, the "Stomach Sculpture" work, a self-illuminating, sound-emitting, opening/closing, extending and retracting camera-equipped endoscopic mechanism operating in the stomach cavity; and with the Internet connected, whereby the (my) body has been accessed and activated by remote agents. This (my) body has, in other words, been augmented, invaded and finally becomes a host; for technology, but also for intimately interfaced others acting on it at a distance.

Must a body continuously affirm its emotional, social and biological status quo? Imagine a body reconfigured, enmeshed in electronic circuitry with a fluid awareness that dims and intensifies as agents are connected and disconnected in an electronic space of distributed intelligence. A body that is inhabited, stirred, startled and caressed by remote promptings of other bodies in other places.

http://www.stelarc.vs.com.au

S T E L A R C *Split Body* 1998

VNS MATRIX

Josephine Starrs, Francesca da Rimini and Julianne Pierce (Virginia Barratt until 1996)

VNS's impetus was to decipher the narratives of domination and control which surround high technology using the various media at our disposal (video, sound, games, computer animation, photography). An activist collective as much as an art group – a sort Guerilla Girls of the Web – we incorporated feminist language and the female body into technology, subverting it for our own means and purposes, while exploring constructions identity and sexuality in cyberspace.

We wanted to debunk the masculinist myths which alienate women from technologies and their cultural products, hijack the technological tools of domination and control, introduce ruptures into cyberculture's rigid systematics, infect the technomachine with feminist thought, and divert technologies from their inherent purpose of linear, top-down mastery.

This was accomplished in part by aggressively (girl)monstering language in the *Cyberfeminist Manifesto for the 21st Century*, widely disseminated via the Web:

> we are the modern cunt positive anti reason
> unbounded unleashed unforgiving
> we see art with our cunt we make art with our cunt
> we believe in jouissance madness holiness and poetry
> we are the virus of the new world disorder
> rupturing the symbolic from within
> saboteurs of big daddy mainframe
> the clitoris is a direct line to the matrix
> VNS MATRIX
> terminators of the moral code
> mercenaries of slime
> go down on the altar of abjection
> probing the visceral temple we speak in tongues
> infiltrating disrupting disseminating
> corrupting the discourse
> we are the future cunt

http://sysx.org/vns

V N S   M A T R I X   *DNA Sluts* (from *All New Gen interactive*) 1994

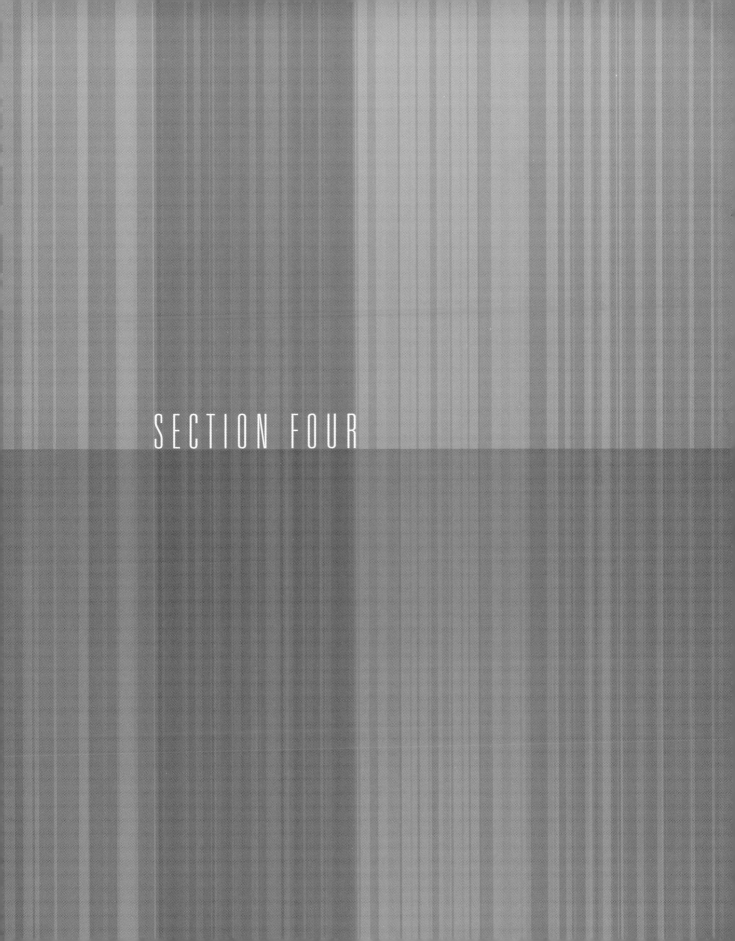

SECTION FOUR

FUTUROPOLIS:
POSTMILLENNIAL SPECULATIONS

"A hand-me-down dress from who knows where, to all tomorrow's parties"
THE VELVET UNDERGROUND AND NICO

Darren Tofts and Annemarie Jonson

How has the technological future been imagined? If the year 2001 was the beacon signaling its arrival (an expectation in part installed by Stanley Kubrick's eponymous film of 1968), how has it measured up to those imaginings? In a 1997 text devoted to assessing the accuracy of *2001*'s portrayal of the future of space travel and computer technology, Arthur C. Clarke admitted how difficult it was "to foresee the future!" (Clarke 1997, xvi).[1] Unwittingly or otherwise, he echoed, shifting tense, Marshall McLuhan's aphorism that we in fact see the future through a rearview mirror, noting that "the hindsight of specialists is bound to be more accurate than the foresight of a writer and filmmaker – especially in a field as rapidly changing as computer science" (Clarke 1997, xvi). As will become evident in these discussions, there is something oddly, even uncannily anachronistic about the very notion of futuristic speculation, since its visions are always, and of necessity, irresistibly measured against an eventual point that is designated, in an imminent present, as the future. The future is always history.

The talismanic year 2001 was the future. But so too was 1984 ("the year in which it happens"). Earlier in the 20th century Virginia Woolf, with beautifully imprecise precision, noted its arrival "[on] or about December 1910." These are just a few of the incalculable historical moments of epiphany when the present caught up with and became the image or relic of a projected future. How have philosophers and artists thought about the future of our involvement with technology and what it portends for human life? The essays in this section draw on a range of historical and contemporary sources of futuristic speculation and examine some of the issues to do with how the future – and in particular the informatic future – has been forecast, predicted or imagined.

---

1   The fact that this text was published in 1997, rather than held over until 2001 (in a commemorative sense the most obvious date), was to mark HAL's "birthday," January 12, 1997. It is also worth noting that in the film HAL "became operative" in 1992; Clarke changed the date to 1997 in the novelization of the film to make HAL seem more state of the art.

The idea of the future as utopia, as the best possible outcome of technological progress, is a common theme explored in these essays. In "Internet Dreaming: A Utopia for All Seasons," Margaret Wertheim traces the lineage of utopian thinking back to Thomas More's 16th century picaresque narrative, the text that gave the word utopia to the English language and futurist thinking. In More's *Utopia* (1516) Wertheim finds a vision of the kind of egalitarian society imagined and promised in the early days of the Internet. Wertheim argues, though, that by the time of Francis Bacon's *New Atlantis* (1627), a dramatic transformation had diminished the optimism of utopian thinking, revealing the decline of a "catholic communist dream" to a profit-driven elitist nightmare. In detailing this shift from an ideal utilitarianism to a utopia for one, Wertheim draws an analogy with cyber-utopian discourses surrounding the Internet. A decadent utopia, the keepers of the Internet (20th century counterparts to Bacon's "technology kings") have broken their original promise – and with more alarming speed than their 17th century forebears – of general equity and universal access to the information economy. Saddened by the increasing divide between the information rich and information poor, Wertheim is bewildered by the speed with which this has happened and concludes that it is in the nature of utopias to sour into decline.

The imaginary traveler's tale of a journey to an unknown land has been a distinctive trope in utopian narratives. Wertheim's portrait of More's character Raphael as a stranger in a strange land endows him with the qualities of a witness. It is this detached perspective that Bruce Mazlish identifies in Samuel Butler's 1872 novel *Erewhon*, a vision of a world turned upside down – "erewhon" spelled backwards is nowhere, or utopia – in which machines have become part of human evolution. In "Butler's Brainstorm" Mazlish argues that Butler had anticipated a posthuman future that we would now consider quite possible. In an age of prosthetic limbs, biotechnology and artificial intelligence, Butler's ruminations on machines occupying an advanced stage in human evolution do sound very familiar. Moreover, in terms of popular representations and acceptance of the reality of technologically augmented bodies, Butler's vision of human life being transformed by machines has proven to be quite accurate. From the point of view of cultural theory, Butler's discourse on the integration of humans and machines is very contemporary, an unstated intimation of the cyborg, "an entirely new class of life" – what Deleuze and Guattari have called the "machinic phylum." This breakdown of any qualitative or ontological distinction between the organic and the inorganic (discussed in previous essays by Waldby, Keller and Sofoulis in Section One) is, for Mazlish, an evolutionary issue of "human selection," an active choice in the next stage of human becoming.

The idea of utopia as an "other place," or uncharted territory of unpredictable human change, has become commonplace in discussions of cyberspace. Utopia is also of the future; it is, of necessity, a vision of tense – a time yet to come. As John Potts elaborates in his discussion of Futurism, "Nowhereseville: Utopia is No-Place," F. T. Marinetti was, like his utopian predecessors, a prophet of becoming through technological change. Progress, as evidenced in *Erewhon*, is indelibly linked to the fusion of humans and machines. Addicted to machines, explosive speed and caffeine, Marinetti was intoxicated with the dream of a 21st century "Futurist City" forged deliriously out of the doctrine of progress. As faith in the ideal of progress never even made it to the end of the 20th century, the desire for techno-utopia was resuscitated, Potts argues, in cyber-culture and the iconic, immaterial "nowhereseville" of cyberspace. Information technology is to 21st century utopia what electricity was to Marinetti's, an equally transforming and rhapsodic "technology of the immaterial traveling at the speed of light." In tracking the parallels between Futurism's accelerated idolatry of technological progress and the more recent cyber-utopian zeal for virtual reality and the mobility of place, space and identity, Potts concludes that contemporary techno-utopian desire is merely the "inflection of a much older ambition." In other words, a utopia well past its use-by date and wiped clean with the veneer of silicon.

When utopias decline their visions of the future appear either outmoded or positively undesirable. Russell Blackford, in "Stranger Than You Think: Arthur C. Clarke's *Profiles of the Future*," situates the great visionary – who had proposed telecommunications satellites as early as 1945 – as a kind of prophet of the cyber age. In *Profiles of the Future: An Inquiry into the Limits of the Possible* (1962) Clarke laid the foundations of what we now call cyberculture, a prophetic vision against which, Blackford argues, "almost all the current thinking on transhuman/posthuman themes can be seen as footnotes to Clarke's speculations." But in *Profiles of the Future* Blackford also sees a dystopian vision of the limits of human nature in the dread associated with technological evolution – the terrifying fear that "the machine is going to take over." Such a vision – given its ultimate expression in *2001: A Space Odyssey* – reveals the crucial and inevitable differential between expectation and the limits of the possible in futuristic speculation. The future, once it arrives, may indeed be stranger than you had thought possible, but it may also be nothing remotely like what you were expecting, or had imagined. The contradictory key to *Profiles of the Future*, Blackford argues, is not technological speculation, but its profound difficulty and limitation.

The difficulty of predicting the shape of things to come, to us H. G. Wells' famous dictum, attests to a perception of the future as a time and space of inevitable and unpredictable change. Seeking a more optimistic and tenable strategy for coping with the "accelerative thrust" of rapidly changing times – what Alvin Toffler famously called "future shock" – Richard Slaughter argues that the very notion of cyberculture needs to be recontextualized. In "From Future Shock to Social Foresight: Re-contextualizing Cyberculture," Slaughter asserts that the utopian impulse has lost its relevance for futurism and that, contrary to the general tenor of the essays in this book as a whole, "the future is not about massive computing, AI, and a vastly expanded Internet." Also contrary to the posthumanism discussed elsewhere in this volume, Slaughter posits a decidedly humanist perspective of the inner constitution of social and cultural significance, in which technology adapts to, rather than determines or transforms, human and social values. In other words, he argues for a symbolic domain of inquiry that eschews the over-identification of technology with the future, so commonplace in discussions of technoscience and cyberculture. Slaughter offers instead a series of mental models for thinking about the future in different ways, models for exploring a "wider stream of human responses to transformed futures." Consistent with Blackford's observations on the unforeseeable and the unpredictable, Slaughter's strategies for social futurism are responses to change that we may not – particularly in the context of cybercultural and informatic discourses – recognize as ways of thinking about possible futures.

Even within a reconfigured model of thinking about the future, there is nonetheless the problem of becoming. This is the event horizon that bedevils futurism. Is there a limit beyond which it is impossible to predict or even imagine the future, what we might become? And concealed within this *ne plus ultra* is the more specific recognition of the inadequacy of frameworks of speculative thinking (as Slaughter suggests), as well as the limitations of prediction in the face of relentless technological change. Damien Broderick, in "Racing toward the Spike," examines this problem by drawing on mathematician Vernor Vinge's concept of a technological singularity or "spike;" "a horizon of ever-swifter change we cannot yet see past." Looking into the future, Vinge sees a black hole, a point of phenomenal change that we can't see into, let alone understand. This moment is qualitatively different from futurisms of the past, since it involves a sense of place in which the model of physical reality fails. Echoing Samuel Butler, Vinge imagines the evolution of "superhuman intelligences" and a state of posthuman life about which we cannot possibly imagine anything. Avowing the posthumanism and emphasis on externalities – such as computational power, networks, telepresence – that Slaughter seeks to move beyond, Broderick sees in Vinge's portrait of computational complexity a technological impasse that will effectively make it impossible to predict what the future might be like. At the point of Vinge's singularity, change, becoming, so central to utopian thinking, have run out of control. Broderick's scenarios for the future of unimaginable technological complexity are thus predicated on the assumption that, as with all utopian explorers and visionaries, you can only try to imagine the unimaginable up to a point.

Given that the future is history, is always already history, it seems fitting to offer a coda to the future and to utopian thinking. And exquisitely appropriate, given that "that year," 2001, the year that was supposed to be the future, is now behind us. We couldn't help but view the year 2001 in terms of the film *2001*, though the excitement associated with its celebrated re-release spoke of a nostalgia for the 60s space-pop chic of its vision of the future, rather than the arrival of a new dawn of man. Yet what a vision it offered – ape becomes man becomes starchild. This expansive, incomprehensible evolution is condensed into one of the film's most famous montages, a 12 second sequence during which we bear silent witness, like so many utopian travelers before us, to the quantum leap from the prehensile flinging of bleached bones to the graceful magnitude of computer driven interstellar travel.

The space age promised escape velocity, a new domain of human experience. Mark Dery, in "Memories of the Future: Excavating the Jet Age at the TWA Terminal," finds himself drawn to Eero Saarinen's TWA Terminal at New York's John F. Kennedy airport to re-discover the traces of the golden age of transcendence, the liberation from gravity and the human occupation of space. What he finds, instead, are "the ghosts of modernism's obsolete tomorrows." The TWA terminal is for Dery a symbol of things to come that never came. His reflective journey into futures past is a *ricorso* to the 20th century's most enduring and signatory image of the technological future, the age of jet propulsion and flight. Representing a different accelerative thrust from Toffler, the TWA terminal is for Dery a "Jet-Age cathedral, consecrated to speed and – unintentionally – to the double-edged sword of technological utopianism, which staked its faith in a future whose chrome was never supposed to flake."

It's fitting, as the last essay in this book, that Dery's essay is framed around a terminal, and a vestigial, faded one at that. The terminal is an ambivalent image suggestive of closure and transition, an interstitial space between different states – terrestrial/airborne, human/cyborg. An apt index of the becoming informatic, the terminal, as a spatial thing and a temporal concept, is, as Scott Bukatman has argued, suggestive of the end of one regime of the self and the emergence of a new one, the subjectivity of networked terminal culture (Bukatman 1993, 9). Dery's excavation of the ruins of the TWA terminal reinforces assertions made by other critics (including Bukatman) that the information age, with its totemic cyberspace, revives the lost technological aspirations of the space age. The intimate symbiosis between pilot and aircraft, astronaut and rocket, is inflected with the intimacies of human interaction with informatic technologies that have been the object of attention of *Prefiguring Cyberculture*. The idea of flight is an apt figure for the transhuman, being a transitive condition of unnatural speed, on the way to somewhere, to something else, as well as a vital link in what Dery calls the cyborgian chain. But what are the consequences of the diminution of the human element in this chain, when flight becomes a posthuman trajectory of simulated flight attendants and cybernetically controlled aircraft? We have been familiar with the concept of the "auto-pilot" for a long time now, but are we ready for the agency that may come with an artificial intelligence beyond automaticity, as the aircraft itself becomes a form of life?

Open the pod bay doors, HAL.

REFERENCES

Bukatman, S. 1993. *Terminal identity: The virtual subject in postmodern science fiction*. Durham, NC: Duke University Press.

Clarke, A. C. 1997. Foreword to *HAL's legacy: 2001's computer as dream and reality*, ed. David Stork. Cambridge, MA: MIT Press.

Section Four

INTERNET DREAMING :
A UTOPIA FOR ALL SEASONS

Margaret Wertheim

Utopia ain't what it used to be. Particularly not in its cyber incarnation. Not so long ago cyberutopians were proffering a vision of a new age in which the Internet would enable everyone to become equal participants in the knowledge society. Age, race, and gender would supposedly be transcended as cyberspace leveled the playing field for all citizens. In the imminent era heralded we were all to become active producers, no longer mere spectators or passive consumers. Whether it was starting up an online business, hosting a website, running a newsgroup, putting together an online newsletter, opening a virtual store, or creating an online fantasy world, brothers and sisters together we'd be doing it for ourselves. The age of the Big Guys was over, it was said. But as the new millennium dawned it was clear that a very different scenario was unfolding.

Just days into the much vaunted year 2K the tone was set when AOL fused with Time Warner. In the *new* new Internet era what would matter most, raved a slew of awestruck commentators, was not individual ingenuity and creativity but brand name identity. Since the instigation of the World Wide Web in 1993 the Internet had been through a Cambrian explosion as millions of fledgling sites spawned from the digital mud. But when so many people were talking, who were you going to listen to? Judging by the insane valuation of AOL in early 2000, Wall Street agreed with AOL chairman Steve Case – only those with the Really Big megaphones were going to register on our radars. Time, CNN, Sports Illustrated, Turner Classic Movies – these were names with solid peerage that would stand out among the gazillions of dot.coms clamoring for our attention.

Hence the spectacle of Superbowl 2000. I don't mean the match itself, but the commercials which peppered the show, many produced specifically for this singular event. In 1999 the going price for a 60 second Superbowl spot was a million dollars. Industry watchers predicted that in 2000 the rate would climb to two million, but as the year's most heavily watched telespectacle approached prices soared to almost three million. The reason for this hike: demand from Internet companies. In the *new* new Internet era when priority number one was the perception of brand name leadership, the only game in town was to become the new Amazon.

Throughout 2000 the U.S. media was obsessed with the stock prices and market valuations of dot.com companies. What, if any, was the actual output of all this frenzy, few seemed to care; all that mattered (apparently) was the size of the IPOs. Judging by the media coverage, the most important outcome of the "wired" revolution was the elevation to mind-numbing wealth of a small elite who had leveraged their companies into the biggest bubble the stock market had ever seen. Unlike the geeks and hackers who had created the Internet, these newly minted millionaires showed little interest in the commonweal. As cyber-commentator Paulina Borsook has documented, the culture of Silicon Valley is profoundly libertarian. To be sure, we are still being told the entire society will benefit from the innovations of these cutting edge individuals, but it is *they* (aglow with the dual aura of the technosavvy and the vastly rich), not society at large, who now constitute the focus of mainstream Internet dreaming.

The burst of the dot.com bubble in the first half of 2001 has done little to change this trend. The NASDAQ may be down, but dot.dreams now seem irrevocably tied to Wall Street. No doubt the Net is no longer a focus for so many get-rich-quick schemes, but it remains a focus for get-rich schemes. The DIY, almost communistic visions that used to be the backbone of Internet dreaming have been superseded in what, I suggest, must be interpreted as a reassertion of the status quo.

What I want to explore in this essay is the parallels between this shift in Internet dreaming and an evolution that occurred in utopian thinking from the 16th through 17th centuries. In Thomas More's seminal *Utopia* (1516), the text from which the term derives, we also find a radical communistic vision of an egalitarian society. A critical feature of More's Utopia was that citizens produced their own goods; everyone was actively and productively engaged. A century later Francis Bacon's *New Atlantis* (1627), the second most famous modern utopia, posited a very different vision. Where More's tale focused on the lives of ordinary men and women, Bacon's dealt mostly with an elite cadre of scientist-priests who alone produced and controlled the technologies that constitute the foundation of his utopian world. These technology kings, like their Internet counterparts today, lived like kings, presiding over vast state-of-the-art facilities and going among their subjects attended by pages, bedecked in jewels.

My plan here is to trace the evolution of utopian thinking from Thomas More to Francis Bacon and to parallel this with the changing strands of Internet dreaming over the past decade. More's Utopia will form the basis for the first stage of our analysis; in the second stage we will turn our attention to the hermetic magic-inspired utopias of Tommaso Campanella and Giordano Bruno; and in third stage Bacon's New Atlantis will form the heart of the discussion.

A CATHOLIC COMMUNIST'S DREAM

One cannot but marvel at the modernity of Thomas More's *Utopia*. We know the world he writes of; in so many ways it is the communist society of the former Soviet Union. Whether or not this is your idea of earthly paradise, it is astonishing how much of communist idealism More prefigured.

Beating Marx to the punch by three centuries, More recognized that the key to an egalitarian society was the fair division of labor. Early on in *Utopia* the enigmatic Raphael – who has traveled to the island of Utopia and lived there for many years – rails against European societies in which hordes of noblemen and their idle retainers "live like drones on the labor of other people," i.e. the hard-working peasants. According to Raphael these upper-class layabouts not only "bleed"

217

the peasants dry; they drain society of vitality (More 1965, 44). In Utopia, by contrast, everyone works. Chief among their jobs is farming, which "they all do, irrespective of sex" (More 1965, 75). Aside from farming, men and women alike learn a trade, such as carpentry, stone masonry, spinning or weaving.

With everyone working industriously (except for a small number of scholars, and the government officials who oversee the smooth running of society) goods and services can be produced efficiently, so that no one need work above six hours a day, this being "more than enough to produce plenty of everything that's needed for a comfortable life" (More 1965, 76). This means there is plenty of time off for edifying activities such as practicing musical instruments and playing one of Utopia's two chess-like games. Most importantly, Utopia's citizens have lots of time for further education and improving their minds: public lectures on a wide variety of topics are freely available and well attended.

Crucially for More, Utopia's citizens are self-reliant: they make everything they need themselves. Each community produces its own food, giving any excess freely to neighboring communities; when new houses and public buildings are needed, or old ones have to be repaired, carpenters and stone masons all pitch in to help; periodically there are communal bouts of road repair. Utopians have no tailors or dressmakers – all clothes are made at home and everyone wears pretty much the same thing.

In Utopia, all goods and services are administered by a benign state which ensures that everyone has what they need. When a man requires tools or equipment he simply goes to a store and is given it free – Utopians have no need of money, its only purpose in their society is to hire outside mercenaries in the rare case of their nation being attacked. While Utopians have all they need, More stresses that they have nothing superfluous; there is no excess consumption or frivolous luxury. This restraint is epitomized by their dress, for, as Raphael explains, "the fashion never changes" (More 1965, 75); both men and women wear a sort of leather overall designed for maximum movement and comfort. Here again we see a precursor to modern communism, More's ubiquitous unisex outfits preempting the famous Chinese Mao suit.

And like Mao, Thomas More was propelled by a concern for social justice. *Utopia* begins with a scathing indictment of the British practice of hanging people caught stealing food. Whether or not More meant Utopia as a model for a real society – a point around which there has been considerable scholarly debate – he certainly intended the book as a rebuke to contemporary European societies, which he rightly saw as appallingly unjust; the nobles squandered while the peasants starved. The model of society he offered instead was one in which, to use Thomas Jefferson's famous phrase, "all men were created equal." Although women are ultimately subservient to their husbands in Utopia, they are full citizens to be treated with dignity and respect; More even allows that they can be priests, an extraordinary suggestion for a Catholic of the early 16th century.

In Utopia no one is left out of the social contract: the mentally deficient are cared for, the ugly and deformed are loved, the elderly are paid every respect, there is toleration of all religions. Meals are held communally, with the young and old interspersed together around the table.

More's vision of Utopia was rooted in his Catholicism and owes a great deal to the monastic way of life. It is worth noting that as a young man More considered becoming a monk and throughout his adult life wore a hair shirt under his clothes which "tamed the flesh till the blood was seen in his clothes" (More 1965, 14). Most of us today would not want to live in his Utopia – life there is as regimented as a monastery; you can't travel in your own country without a special permit; premarital sex is punished by compulsory celibacy for life, adultery by slavery; there is no drinking or gambling or frivolity of any kind; idleness is not an option – yet one cannot but admire the boldness and essential fairness of More's vision. Moreover, the active engagement of all Utopia's citizens and their self-sufficient productivity are remarkable.

The idea of an active, engaged, productive citizenry has also been a key feature of Internet dreaming from the Net's earliest days. One of its primary appeals has always been the promise of a more utopian community. At a time of widespread social breakdown in the western world, the Net, with its vast global web, beckons us all with the hope of inclusion in a wider and more supportive social whole. Howard Rheingold, one of the founders of the WELL, a seminal online community based out of San Francisco, has written about the utopian potential of the Internet. In his landmark study of online culture, *The Virtual Community* (1993), Rheingold recalled the prediction of legendary Internet pioneer J. C. Licklider that "life will be happier for the online individual because the people with whom one interacts most strongly will be selected more by commonality of interests and goals than by accidents of proximity." Speaking of his WELL colleagues, Rheingold noted that "my friends and I sometimes believe we are part of the future that Licklider dreamed about, and we often can attest to the truth of his prediction" (Rheingold 1994, 24).

According to Rheingold, the Internet could help us to (re)create the ethos of a better era: "Perhaps cyberspace is one of the informal places where people can rebuild the aspects of community that were lost when the malt shop became a mall" (Rheingold 1994, 26). Like More, Rheingold recognized the threat of consumerism to community health and he dreamed of a society free from this scourge. His ideal society also was one in which people worked together to fashion a greater common good, and in *The Virtual Community* he enthused about the ways in which the Internet was enabling groups of people "to rediscover the power of cooperation" (Rheingold 1994, 110). According to Rheingold, online communities enable everyone to be involved "in an ongoing process of group problem solving" (Rheingold 1994, 110). As in *Utopia*, with its communal house building, Rheingold's cybercitizens pull together to realize one another's needs. Here too, Rheingold tells us, all citizens will be fully appreciated: "People whose physical handicaps make it difficult to form new friendships find that virtual communities treat them as they always wanted to be treated — as thinkers and transmitters of ideas and feeling beings" (Rheingold 1994, 26).

MIT's William Mitchell is another who has championed the Internet's utopian potential. In his book *City of Bits*, Mitchell drew a parallel between the online world and the agora of ancient Athens (1995, 8). The center of the original democracy, the agora was the place where Athenian citizens met to discuss ideas for the common good and in this non-hierarchical place everyone was equal. So, too, Mitchell saw the Internet giving back a level playing field to all citizens.

Along with More, MIT Media Lab director Nicholas Negroponte also worried about inequality between the rich and poor, and in his 1995 book *Being Digital* he proposed that the Internet would provide a foundation for a more egalitarian world order:

> Today, when 20 percent of the world consumes 80 percent of its resources, when a quarter of us have an acceptable standard of living and three-quarters don't, how can this divide possibly come together? While the politicians struggle with the baggage of history, a new generation is emerging from the digital landscape free of many of the old prejudices. These kids are released from the limitation of geographic proximity as the sole basis of friendship, collaboration, play, and neighborhood. Digital technology can be a natural force drawing people into greater world harmony (Negroponte 1995, 230).

One of the recurring tropes of Internet dreaming has indeed been equality. Early Internet enthusiasts proposed that the Net would dissolve the very boundaries of gender, race and age, elevating everyone equally to a disembodied digital stream. In the physical world, the nearest one can get to effacement of physical difference is a unisex uniform code of dress, hence the monkish habit and More's adoption of this sartorial strategy in Utopia; but online you are physically invisible. While this doesn't mean, as some enthusiasts have claimed, that race and gender become meaningless or moot, on the Internet we are shielded from the relentless bodily scrutiny that has become a hallmark of life in America today – and in much of the rest of the western world as well. More, the closet monk, would have applauded.

Above all, More wanted a society in which all citizens were actively engaged in their communities. High-tech entrepreneur Esther Dyson echoed this desire in her 1997 book *Release 2.0: A Design for Living in the Digital Age*. Packed with utopian rhetoric, Dyson's book posited that the Internet would revitalize the very fabric of our communities:

> The Net offers us a chance to take charge of our own lives and to redefine our role as citizens of local communities and of a global society. It also hands us the responsibility to govern ourselves, to think for ourselves, to educate our children, to do business honestly, and to work with fellow citizens to design rules we want to live by (Dyson 1997, 2).

According to Dyson, "our common task is to do a better job with the Net than we have done so far in the physical world." Dyson is a libertarian and there are none of *Utopia*'s communist overtones in *Release 2.0* – many cyber-utopians have in fact a strong libertarian streak, and in that sense have always been much less radical than Thomas More. But as in *Utopia*, they have proffered a vision in which revitalization of community and active engagement of all citizens are key themes. One might well ask how seriously Negroponte or Dyson's rhetoric should be taken; how far, in practice, would they be prepared to work towards the future they envision? It is hard to know; yet however sincere the messengers, the messages have been clear – while certainly less than Utopian, these visions have nonetheless been utopian.

MAN AS MAGUS

The 16th century was a time of immense social upheaval: the reformation created an unprecedented level of religious and political tension across Europe, while the collapse of feudalism had left a vacuum that drew forth all sorts of proposals for new social models. Thomas More was ahead of his time, but he was emblematic of the century to come – from England to Italy people began to dream of a new and "better" society.

One of the chief currents that fed this dream was the stream of hermetic magic, which Europeans had begun to recover in the late 15th century. The hermetic magic texts, what came to be known as the *Corpus Hermeticum*, were actually written in the second and third centuries AD, but Renaissance scholars believed they contained the most ancient of all knowledge, revealed directly by god to the Egyptian sage Hermes Trismegistus. Viewed by his champions as a precursor to Christ, Hermes was said to have taught both Moses and Plato, and hence in his wisdom supposedly lay a reconciliation between the ancient Greek philosophers and Christian theology. Along with its theological import, the *Corpus Hermeticum* also contained an encyclopedic knowledge of practical magic by which a skilled initiate, or magus, could manipulate the forces of nature and the affairs of man. At its core hermetic magic was celestial magic – the world of man was seen to be embedded in a web of cosmic influences which the magus must learn how to draw down through the manipulation of herbs, crystals, talismans and symbols. Today the word "magic" is seen as the antithesis of "science," but 16th century hermeticists believed they were recovering the true science of nature through which they would ultimately effect all sorts of marvelous technologies.

At the same time, the hermetic project was a political project, one with a strong utopian cast. Christianity had been corrupted, renaissance hermeticists believed; the only way back to the true path was through the wisdom of Hermes. Among the most extreme of such reformers were the Dominican friar Tommaso Campanella, whose attempts to establish a new capital for worldwide religious and political reform in Calabria in southern Italy earned him 25 years in prison; and Giordano Bruno, whose plans for a religious overhaul the Catholic Church found so threatening they burnt him at the stake.

In his 1584 book *The Ash Wednesday Supper (Cena de la Cerneri)*, Bruno developed an extraordinary utopian vision, outlining the messianic role he himself would play in ushering in this bold new age. According to Bruno, ancient hermetic wisdom would serve as a means for overcoming the present age of darkness, for by this wisdom men become "judicious in contemplation, unique in divination, miraculous in magic ... irreproachable in morality, godlike in theology, heroic in every way" (Goselin and Lerner 1977, 96). As historian Stephen McKnight has noted, Bruno's texts make clear that "the efforts of the magus are essential in overcoming human ignorance and in installing a perfect social order" (1992, 107).

Campanella also dreamed of a perfect social order, and like More he imagined an ideal city in which such a society was realized. Written while in prison in the opening decades of the 17th century, Campanella's *The City of the Sun (Citta del Sole)* depicts a spectacular seven-tiered metropolis built in concentric layers ascending a hill. The Renaissance historian Frances Yates has described this City as "a complete reflection of the world as governed by the laws of natural magic in dependence on the stars" (1964, 369). Presiding over Campanella's City were an elite group of magi (headed by a priest-king) who understood how to manipulate these magical laws. Under the magi's benign governance, Yates writes:

[T]he people of the City lived in brotherly love, having all things in common ... They encouraged scientific invention, all inventions being used in the service of the community to improve general well-being. They were healthy and well skilled in medicine. And they were virtuous ... [T]here was no robbery, murder, incest, adultery, no malignity or malevolence of any kind (1964, 369).

As in More's *Utopia*, everything is provided by the state and all property is communally owned – including, Campanella tells us, wives, they being the property of their husbands (Claeys and Sargent 1999, 112).

Hermes Trismegistus also had envisioned an idealized city, Adocentyn, in which "all human physical, emotional and spiritual needs were met through the efforts of the magus." In the bubbling social cauldron of the late Renaissance, "this utopian element reemerged most fully in the work of Bruno and Campanella" (McKnight 1992, 110). As in More's Utopia, a communist spirit continues to inform Campanella's visionary society, but a fundamental shift had taken place between the writing of *Utopia* and *The City of the Sun*. In the former vision, all citizens are equally engaged in the functioning of the new society, in the latter, the community is now reliant on an elite group to create the perfect society for them. In short, active participation has given way to a technological oligarchy.

Examining Internet dreaming during the 1990s we can chart the emergence of a similar trend. While the older forms of cyberutopianism continued to be expressed, it is elements of this second strand that would soon come to prevail. Chief among the characteristics of this new strand of Internet dreaming was an increasing focus on an elite group with specialized knowledge and power, the men (and a few women) who were pioneering Internet technologies and putting in place the tools that would supposedly revolutionize life as we know it.

Foremost promoter of this emerging elite was none other than *Wired* magazine, the San Francisco based glitter zine founded in 1993 expressly to proselytize the "wired" age. Brewing together a heady mix of leading-edge technology, speculative science, new media, philosophical musing, digital art and reports from the front lines of electronic commerce, *Wired* conjured up a dazzling future. In the premier issue Stuart Brand – founder of the *Whole Earth Review* and its electronic spinoff, the WELL – encapsulated the magazine's zeitgeist with a piece entitled "Creating Creating." Here Brand declared: "The cutting edge of new media is the cutting edge of human cognition, which is the cutting edge of what it means to be human" (1993). There was little doubt where we should come to "meet" these cutting edge individuals.

In a 1995 dialogue in *Harper's* magazine, *Wired*'s executive editor Kevin Kelly looked to the future and saw a limitless expansion of our horizons. "The flourishing of digital communication will enable more options, more possibilities, more diversity, more room, more frontiers," Kelly burbled (1995, 44). With its sizzling electro-acid graphics and avant-garde typefaces, *Wired* sought to establish itself as the apotheosis of fin-de-millennium techno-chic. In effusive articles the magazine promoted the technologies and visions of the Internet vanguard and plastered their faces across its covers in a style that referenced the iconic celebrity portraits of Andy Warhol. Nicholas Negroponte is one who received the coveted cover treatment, as also have Microsoft's Bill Gates (twice), Amazon's Jeff Bezos, Internet sociologist Sherry Turkle, Net-futurist George Gilder, and cyberpunk writers William Gibson and Neal Stephenson. Negroponte, one of the first Internet gurus, was given a regular column – his book *Being Digital* is largely a reworking of his *Wired* pieces. While the magazine stressed that every one of us would benefit from the wired "revolution," the clear message was that the people creating it were a rather special lot – more forward thinking, more savvy, more daring than the average Joe.

What made it possible for such people to become stars at all was the fact that in 1993 the Internet suddenly acquired a user-friendly interface – the World Wide Web. Almost overnight the Net was transformed: by making the Internet easily accessible, the Web enabled its metamorphosis into a public playground. Until then the Internet and its tools had largely been developed by anonymous computer scientists and hackers, many of whom contributed their time and energy freely. A paradigmatic example was Tim Berners-Lee, a computer scientist employed at CERN (Europe's leading particle accelerator facility) who created the original World Wide Web interface and developed the protocols that made it work. Berners-Lee created the Web to enable scientists in different countries to easily exchange information and research online. Like most Internet pioneers, he never thought he'd become rich or famous; he gave his invention freely to the world and has not made money from it. Nor has he received anything like the media attention lavished on the dot.com billionaires whose rise to fame and fortune the Web made possible.

The emergence of the dot.com barons (the digital age equivalent of the old oil and railroad barons) signaled a new era for the Internet characterized by a vast expansion of its possibilities. No longer just a tool of academics, hackers, and bulletin-board geeks, it could now play a role on the main social stage. *Wired*'s timing was no accident.

Far more than passive observers, *Wired*'s writers and editors played a central role in creating the perception that a glorious new age was nigh. Of course it was in the interests of Kelly and others to promote such a view; the more exciting the coming revolution appeared to be, the more magazines they stood to sell. But this was not just a cynical exercise: Kelly seemed genuinely to regard himself and his peers as the ushers of a new era. In a 1997 *Wired* feature he ecstatically declared: "We are now engaged in a grand scheme to augment, amplify, enhance and extend the relationships and connections between all beings and all objects" (Kelly 1997, 39). He was not alone in such sentiments. John Perry Barlow, founder of the Electronic Frontier Foundation and occasional *Wired* contributor, put it this way: "… with the development of the Internet and the increasing pervasiveness of communication between computers, we are in the middle of the most transforming technological event since the capture of fire" (Barlow 1995, 36).

It was not uncommon to find a religious dimension mixed into this revolutionary rhetoric. In the *Harper's* magazine dialogue (1995) previously quoted, Kelly revealed that he had "experienced soul-data through silicon. You'll be surprised at the amount of soul-data we'll have in this new space," he continued. Kelly in fact is an evangelical Christian, though that was not widely publicized in the *Wired* universe, and at a Bay area conference I once saw him give a talk in which he presented a "hacker theology." The marriage of religious and utopian themes in the discourse about cyberspace is also to be seen in the seminal book of essays *Cyberspace: First Steps* (1991). In his introductory remarks editor Michael Benedikt compared cyberspace with the heavenly Christian city of the New Jerusalem, itself an early utopia (18). Outlining this theme, Benedikt tells us that "the impetus toward the Heavenly City remains. It is to be respected; indeed it can usefully flourish – in cyberspace" (18). In my own book *The Pearly Gates of Cyberspace* I have explored at length the religious undercurrents of Internet dreaming and I won't repeat that work here; suffice it to say that just as renaissance hermeticists wove their utopian ideas around a foundation of science, technology and religion, so we encounter a similar mix in the rhetoric of many Internet champions (Wertheim 1999).

RETURN TO ORTHODOXY

During the late Renaissance, the mixing of scientific, religious and utopian thinking reached its apotheosis in the work of Francis Bacon. Around the time that Campanella was released from prison and started promoting his utopian city to the French monarchy (it is from Campanella's *City of the Sun* that Louis XIV apparently derived his famous moniker), Bacon began to pen his own portrait of an ideal society. Presented in the *New Atlantis*, this vision is at once the epitome of Renaissance utopianism and also a new departure.

In classic utopian fashion, Bacon located his neo-Atlantis on an island in the southern hemisphere. Citizens there display all the usual utopian traits: they are virtuous, chaste, hard working, healthy, pious and so on. Everything they need is liberally provided, and in this case they are also deeply Christian – having been granted their own independent revelation of God and a copy of the Hebrew scriptures to boot. But there is nothing communistic about this society; instead Bacon heralds a return to orthodoxy. In the New Atlantis women are completely subservient to their husbands, having no voice of their own, patriarchy reigns in its strictest mode and everyone knows their place in the social order. Most of the narrator's account of the ordinary citizens' lives is devoted to showing how willingly they conform to the status quo.

The real heart of Bacon's society, and the focus of his book, are not the citizens, but an elite group of 36 scientist-priests, the "Fathers" of Solomon's House. The word scientist was not coined until 1834, but Bacon's Fathers are scientists in the full modern sense; indeed the New Atlantis is credited with being an inspiration for the establishment of the Royal Society, one of the first and still perhaps the most prestigious of all scientific societies.

The chief glory of the New Atlantis, Solomon's House, is the fount of the nation's prosperity and the wellspring of the people's well-being. It is Bacon's visionary scientific research center, wherein the Fathers devote their lives to pursuing "the knowledge of causes and secret motions of things; and the enlarging of the bounds of human empire to the effecting of all things possible" (1992, 27). In their vast laboratories, they are able, for example, to imitate the conditions of mines and produce wondrous new artificial metals. Likewise they produce artificial gems, crystals and other exotic materials. Atop gigantic towers the Fathers monitor the weather, enabling them to predict tempests and floods. Earthquakes, too, they can predict. They have enormous machines "for multiplying and enforcing winds," and they can simulate atmospheric phenomena such as rain, hail and snow.

They have wondrous optical devices giving them the "means of seeing objects in far off places" (telescopes), and powerful instruments "to see small and minute bodies perfectly" (microscopes). The aural sense also they can enhance, by conveying sound over long distances through "tubes and pipes." They also "multiply" light to produce intense, tightly focused beams, a description which to modern eyes seems very much like a laser. They have made flying machines and submarines, and mechanical devices that imitate the motions of birds and fishes.

Most remarkably, Bacon foresaw the science of genetic engineering. In a passage of historical significance one of the Fathers explains that he and his colleagues are able to tailor plants and animals to their own design. "By art likewise we make them greater or taller than their kind is, and contrariwise dwarf them and stay their growth; we make them more fruitful and bearing than their kind is, and contrariwise barren and not generative. Also we make them differ in color, shape, and activity, many ways" (Bacon 1992, 35-36).

The scale of the Fathers' laboratories is mind boggling – for their mining experiments they have excavated underground caves several miles deep and their atmospheric towers are several miles high; fleets of attendants assist them, and their range of equipment seems inexhaustible. Any scientist today would be aching with envy. When any Father goes among the people, he is carried in state in a cedar chariot adorned with gold, sapphires and emeralds, attended by 50 pages dressed in white satin; before the chariot walk more attendants carrying a pastoral staff, the symbol of a bishop (29).

To be sure, the people of the New Atlantis are happy and healthy, but how far we have moved from the communistic, self-sufficient citizens of More's Utopia. In Bacon's visionary society the "people" seem almost superfluous, their lives mere background, or wallpaper, to the Fathers and their laboratories. Bacon is almost ecstatic when writing about these men (they are all men) and their facilities. Unlike More, who seemed to care deeply about the lives of ordinary Utopians, one senses that for Bacon the only value of the citizens resides in their willingness to revere and support the Fathers – it is their lives, their wisdom, their virtue that is the true measure of this utopia.

So too, in the mainstream of Internet dreaming today, the "people" are becoming little more than wallpaper, as the focus of attention shifts to the high-profile CEOs. Like Bacon's scientific priesthood, these men (again, almost all of them are men) have resources beyond their wildest dreams, raising hundreds of millions of dollars from their splashy IPOs. Just as Bacon's populace turn out in droves to catch a public glimpse of the Fathers, so the new Internet "fathers" are watched by millions on CNN, on the cover of *Time* and in the pages of the *New York Times*. Every major newspaper, magazine, television and radio news program in America now has a section devoted to the Internet and its high-flying executives. As I suggested at the start of this essay, we witness here a return to orthodoxy. Far from overthrowing the old world order, the new order has simply allied itself with the old. Silicon Valley has not replaced Wall Street, it has merged with it.

A similar convergence occurred in the 17th century. Despite the widespread view today that Bacon and his fellow scientists were breakaway radicals, most of the founders of the "scientific revolution" wanted a science that would support the traditional regime of church and state. However radical they may have been epistemologically, most of these men were committed to the old social hierarchy. First in the Catholic countries of Italy and France, then later in Protestant Holland and England, the new men of science specifically sought a world picture that would serve as a natural model for a strict social order with king and pope at the helm.

The regressive social elements of Bacon's New Atlantis were indeed a reaction to the radically reformist visions that had been percolating throughout the 16th century. In particular, Bacon and his fellow scientists wanted to articulate an alternative to the heretical world-views of the hermetic magi. Despite the fact many of these scientists were themselves influenced by hermetic currents (including Bacon, and most famously, Isaac Newton), they viewed the magi as pariahs. With the hermeticists' religious unorthodoxies and their proposals for sweeping social change, they represented a genuine threat to the old world order, which is precisely why Campanella spent so much time in jail and Bruno lost his life. Against this threat, Bacon and his fellow scientists aligned themselves with the old powers of Church and State. In that sense the "scientific revolution" represented a new triumph for the orthodox powers. While it is true that Baconian science brought change to late Renaissance society, it was not the kind of foundational social change that Thomas More had envisaged a century before.

The present state of affairs vis-à-vis the Internet should be seen, I suggest, as a Baconian rather than a Morean triumph. Although there remain pockets of old-style utopian vision – notably among the hacker-programmers of the open-source movement, epitomized by the developers of the Linux operating system and the Bytes-For-All movement based in India and Pakistan – the mainstream of Internet visionaries today are dreaming not of a better world for the "people," but of how quickly they can launch their next IPOs.

As a coda to this piece it might be noted that this convergence of Internet technology and big business follows the current trend of science itself. Over the past decade the entwining of scientific research with business interests has reached unprecedented heights. Nowhere is this more evident than in the "new science" of the new millennium: genetic engineering. Along with the inflation of the dot.com bubble the year 2000 saw the sequencing of the human genome and in its wake a market frenzy over biotech stocks. Even Bacon, though he bedecked his Fathers in satin and jewels, still saw scientific research in essentially humanitarian terms. Today, however, it is increasingly viewed in financial terms: Tell us not what your research will do for the world, but by how much it will increase our bottom line. That Internet dreaming has now followed this pattern should not perhaps surprise us. In an age when the market has replaced the temple as the epicenter of our social landscape, when the "right" of multinational corporations to trade "freely" takes precedence over the well-being of entire populations and when material wealth has been invested with near-reverential status, it was, no doubt, naïve to imagine that cyberspace could provide a more "pure" foundation for our dreams.

REFERENCES

Bacon, F. 1992. *New Atlantis*. Kessigner Publishing Company.

Barlow, J. P. 1995. Quoted in What are we doing on-line? *Harper's* (August).

Benedikt, M. 1991. *Cyberspace: First steps*. Cambridge, MA: MIT Press.

Borsook, P. 2000. *Cyberselfish: A critical romp through the terribly libertarian culture of high tech*. New York: Public Affairs.

Brand, S. 1993. Creating creating. *Wired*, 1.01 (March).

Claeys. G. and Sargent, L.T. (eds.). 1999. *The* Utopia *reader*. New York: New York University Press.

Corpus Hermeticum. See the Gnostic Society Library, www.gnosis.org/library/hermet.htm. Accessed January 26, 2001.

Dyson, E. 1997. *Release 2.0: A design for living in the digital age*. New York: Broadway Books.

Goselin, E.A. and L. S. Lerner (trans. and eds.). 1977. *Giordano Bruno, The Ash Wednesday supper*. Hamden, CN: Archon Books.

Kelly, K. 1995. Quoted in What are we doing on-line? *Harper's* (August).

————— 1997. New rules for the new economy. *Wired*, 5.09 (September).

McKnight, S.A. 1992. Science, The Prisca Theologia, and modern epochal consciousness.
     In *Science, pseudo-science and utopianism in early modern thought*. Columbia, MO: University of Missouri Press.

Mitchell, W. 1995. *City of bits: Space, place and the infobahn*. Cambridge, MA: MIT Press.

More, T. 1965. *Utopia*. Trans. P. Turner. London: Penguin Books.

Negroponte, N. 1996. *Being digital*. New York: Vintage Books.

Rheingold, H. 1994. *The virtual community: Homesteading on the electronic frontier*. New York: HarperPerennial.

Wertheim, M. 1999. *The pearly gates of cyberspace: A history of space from Dante to the Internet*. New York: W.W. Norton.

Yates, F. 1964. *Giordano Bruno and the hermetic tradition*. Chicago: University of Chicago Press.

BUTLER'S BRAINSTORM

Bruce Mazlish

Samuel Butler was a curious, rather bittersweet, and certainly eccentric Victorian. He thought of himself primarily as a painter, was a middling musician, and achieved fame as a novelist. His upbringing, in an Evangelical Protestant family, scarred him for life, producing in him divided, or schizoid, characteristics, and, I would argue, turned him to scientific discourse in an effort to deal with his family and religious stigmata. As a result, he wrote a number of tracts and books on religious and scientific subjects, specifically on evolution, which are now neglected. His novels, *Erewhon* (1872) and *The Way of All Flesh* (1903), however, remain classics. And in *Erewhon*, Butler gave classic form to the view that machines might represent an advanced stage of evolution. In the context of an emerging cyberculture, in which the divide between humans and machines seems to be diminishing, Butler's work assumes a renewed importance.

In 1859, the same year as *The Origin of Species*, after attendance at Cambridge University, Butler emigrated to New Zealand to raise sheep. Darwin's book had a powerful effect on him, just at a time when Butler was wrestling with religious doubts. Admittedly ignorant of any actual science, as he confessed, "not as knowing anything whatsoever of natural history" (Butler, 1923a, 186), Butler nonetheless took it upon himself to write a dialogue, explaining and supposedly espousing the Darwinian theories. I say "supposedly," for although Butler so represented it later to Darwin, the dialogue is a satirical, ambiguous piece, whose final position is by no means clear. Indeed, the fictitious Darwinian protagonist claims belief in both Christianity and Darwin, though admitting that they appear irreconcilable (the dialogue, in fact, makes no effort at reconciliation).

The dialogue, published in *The Press*, a new New Zealand publication, occasioned a reply from an author subsequently identified by Butler as the Bishop of Wellington. In the exchange that followed, the Bishop caught Butler up short on his knowledge of natural history, challenging him to show that Darwin was anything more than a rehash of theories brought forth in the previous century by his grandfather, Erasmus Darwin, Priestley, and Lord Monboddo. Butler seems to have had the worst of the exchange, indulged in self-satire, and then wisely concluded that the truth of Darwin's theory of evolution "can be decided only among naturalists themselves" (Butler 1923a, 201). It was not advice he would later follow.

The surprising thing is that, somehow, the dialogue fell into Charles Darwin's hands and caused him to write to the young author. Was it that Darwin was desperate for allies and welcomed recognition from as far away as New Zealand, and even by an author who hardly represented his doctrines in a knowledgeable fashion? In any case, a correspondence sprang up between Butler and Darwin, and Butler, having published, in 1865, a pamphlet ("The Evidence for the Resurrection of Jesus Christ …") sent a copy to Darwin. The conflation of evolutionary and religious enthusiasm on Butler's part is clear. Darwin expressed polite interest, urging Butler to write "a work descriptive of a colonist's life in New Zealand" (Butler 1923a, 187). In one sense, unexpected, *Erewhon* was Butler's response to Darwin's suggestion.

Before that, however, Butler, under the pseudonym Cellarius, wrote an essay, "Darwin Among the Machines," which appeared in the *The Press* on June 13, 1863, and dramatically announced the theme of the machine as an advanced evolutionary species. Here, the purely satirical, scientific side of Butler, without religious concerns, appears. By an extraordinary leap of intuition, even if ambiguous, Butler foresees a possible future. Aware that his probe may humble human pride, Butler asks:

> If we revert to the earliest primordial types of mechanical life, to the lever, the wedge, the inclined plane, the screw, and the pulley … and if we then examine the machinery of the Great Eastern, we find ourselves almost awestruck at the vast development of the mechanical world, at the gigantic strides with which it has advanced in comparison with the slow progress of the animal and vegetable kingdom. We shall find it impossible to refrain from asking ourselves what the end of this mighty movement is to be? In what direction is it tending? What will be its upshot? (Butler 1923a, 208).

At a quick stroke, Butler draws a map of his proposed inquiry, admitting his own limitations to undertake it:

> We regret deeply that our knowledge both of natural history and of machinery is too small to enable us to undertake the gigantic task of classifying machines into the genera and sub-genera, species, varieties and sub-varieties, and so forth, of tracing the connecting links between machines of widely different characters, of pointing out how subservience to the use of man has played that part among machines which natural selection has performed in the animal and vegetable kingdoms, of pointing out rudimentary organs which exist in some few machines, feebly developed and perfectly useless, yet serving to mark descent from some ancestral type which has either perished or been modified into some new phase of mechanical existence (Butler 1923a, 209).[1]

The suggested classification of machines reminds us of Babbage's taxonomy; only Butler is proposing to "animate" his schema, just as Darwin did with Linnaeus' static system. Further, in

---

1 George W. Stocking Jr. tells us that Colonel Lane Fox (better known as Pitt Rivers), "classifying the 'various products of human industry' into 'genera, species, and varieties … discussed their development explicitly in terms of Spencerian evolutionary psychology and the processes of natural selection" (Stocking 1987, 180). Lane Fox was writing in 1874, suggesting either that he had read Butler or that the idea was in the air. The 1851 Crystal Palace Exhibition, with its arrangement of exhibits along a line of progress for both people and machines, lends support to the "in the air" hypothesis.

place of natural selection, Butler is urging human selection as the means by which machine "breeding," just as in animal breeding, takes place. It is an audacious analogy.

The rest of the short essay spells out a few details. As in animal evolution, vertebrata of great size, for example dinosaurs, gave way to smaller animals, so we may expect miniaturization (my word) among the machines. "The day may come when clocks … may by entirely superseded by the universal use of watches, in which case clocks will become extinct like the earlier saurians" (Butler 1923a, 210). Although Babbage is not mentioned (and apparently not thought about), Butler foresees machines acquiring "by all sorts of ingenious contrivances that self-regulating, self-acting power which will be to them what intellect has been to the human race" (Butler 1923a, 210).

It is this evolutionary development that raises the prospect of the machine becoming man's superior ("We shall find ourselves the inferior race," is how Butler puts it). We will have to "feed" them and in doing so become unto them as a kind of "slave." True, just as we take care of our domestic animals, the machines will take care of us; after all, we are useful to them. Put simply, machines will have dominion over man.

In our infatuation with machines, Butler continues, we even "desire … to see a fertile union between two steam engines" (Butler 1923a, 212) and instruct our machinery in "begetting" other machinery. In short, we are creating the means of our own inferiority and enslavement. Aghast at this vision, Butler has his presumed author, Cellarius, exclaim, "war to the death should be instantly proclaimed against" every machine (Butler 1923a, 212). With this trumpet call, the author ends his letter.

Two years later, Butler returned to his subject in another essay, "Lucubratio Ebria," published in *The Press* (July 29, 1865). The tone is now positive. Machines are, in fact, merely extra-corporaneous limbs added to the members of man's own body: a lever is an extension of his arm, microscopes and telescopes of his eyes, and so forth. Body changes, i.e., extensions, entailed mind changes as well, and man, Butler suggests, comes to be his civilization.

"It is a mistake," therefore, Butler says, satirizing his earlier letter, "to take the view adopted by a previous correspondent of this paper, to consider the machines as identities, to animalize them and to anticipate their final triumph over mankind." The machine simply makes for a difference in degree from man to man – the new limbs are preserved by natural selection, descend with modifications into society, and thus mark the difference between our ancestors and ourselves. What is more, within an advanced society itself, the command over these new limbs, for example, the seven-leagued foot we call a railroad train, is a matter of class. "He alone," declares Butler, "possesses the full complement of limbs who stands at the summit of opulence." A Rothschild is a "most astonishing" organism and "may be reckoned by his horse-power, by the number of foot-pounds which he has money enough to set in motion." As Butler concludes, "Who, then, will deny that a man whose will represents the motive power of a thousand horses is a being very different from the one who is equivalent but to the power of a single one?" (Butler 1923b, 217, 219).

Adam Smith, in his labor theory of value, had spoken of opulence as meaning command over labor; Butler has now mechanized the concept. He is aware that what separates the races of man is their level of culture, or civilization, which means their command over "extra-corporaneous limbs." In a spirit that Benjamin Franklin and Karl Marx, both of whom called man "the tool-making animal," would approve, Butler labeled man a "vertebrate machinate mammal" (Butler 1923b, 215). In "Darwin Among the Machines," the "limbs" are feared as taking on a life of their own; here they are benignly subordinated to man.

Yet, ambiguity and ambivalence still press down on Butler. In another essay, "The Mechanical Creation," written after "Lucubratio Ebria" but published shortly before it (in *The Reasoner*, July 1, 1865), his divided feelings re-appear. Now we are told that, though Butler does not wish to "throw ridicule on Darwin's magnificent work," further thinking suggests the possibility of "an eventual development of mechanical life, far superior to, and widely differing from, any yet known ..." (Butler 1923c, 231-232).

The process is Darwinian, but the end result is the replacement of the organic by the machinate. Was this a satiric *reductio ad absurdum* of Darwin's work? Why not conceive "of a life which in another ten or twenty million years shall be to us as we to the vegetable?" In "Lucubratio," Butler had looked at machines as mere extra-corporaneous limbs, hardly threatening to us; now he returns to the prospect of their supplanting us. A spade is simply an extension of our forearm. We must look at a steam engine to see the evolutionary direction. I quote Butler at length, for it is his working out of the analogy that is persuasive:

> It [the steam engine] eats its own food for itself; it consumes it by inhaling the very air which we ourselves breathe; it rejects what it cannot digest as man rejects it; it has a very considerable power of self-regulation and adaptability to contingency. It cannot be said to be conscious. It is employed in the manufacture of machinery, and though steam engines are as the angels in heaven, with respect to matrimony, yet in their reproduction of machinery we seem to catch a glimpse of the extraordinary vicarious arrangement whereby it is not impossible that the reproductive system of the mechanical world will be always carried on. It must be borne in mind that we are not thinking so much of what the steam engine is at present, as of what it may become. The steam engine of to-day is to the mechanical prodigies which are to come as the spade to the steam engine, as the ovum to the human being (Butler 1923c, 232, 233).

The agency of this mechanical evolution is "natural selection and the struggle for existence," in the form of man's exercise of self-interest – shades of Adam Smith! Men, pursuing a competitive advantage, invent new machines, which then eliminate or exterminate the older, inferior ones (Butler speaks of guns actually fighting one another for survival). In the process, the machines "evolve," threatening eventually to become man's superior.

Can men halt this process? In "Darwin among the Machines," Butler called for just such a halt. Here, in "The Mechanical Creation," he tells us that "man is committed hopelessly to the machines" (Butler 1923c, 236). So dependent are we now that, even if we wished, we cannot live without them, and one hears the voice of H. G. Wells in his story of the time machine. The dependency, however, will be mutual: the machines are not likely to want us as a delicacy on their table, and we shall be as serviceable to them as they are to us.

In three short articles, Butler explored, even if in mixed-up fashion and with tongue in cheek, the prospect of extending Darwin's theory of evolution to machines. On one side, he has taken up Ambroise Pare's hand as a prosthetic device and extended it to the idea of extra-corporaneous limbs, developed in an evolutionary manner, in the service of man's survival. On another side, Butler has taken up the Frankenstein-like fears of the monster as machine, now evolving, and threatening our own survival, or at least our dominion over species.

These are powerful premonitions that are particularly apposite in an age of cybernetic systems, advanced biotechnologies and the human genome project. What did Darwin think of these offspring of *The Origin of Species*? We have no hint of his reception of Butler's extension of evolutionary theory; as we shall see, the relation between Darwin and Butler was to take another turn.

Butler's seminal joining of mechanism and evolution, alas, must be thought of without the benefit of the great biologist's judgement.

In fact, Butler's three pieces, two of them in an obscure New Zealand journal, awakened little attention. Only when included, slightly revised, in his novel *Erewhon* (1872), did Butler's provocative ideas reach the general public. And even in this form, they were treated as rather foolish, satiric notions, *Erewhon* as a whole being either ignored or criticized by contemporaries, and enjoying only a poor sale. Yet it was in this version, rather than the shorter, more self-contained essays, that Butler's views on the evolution of machines came to enjoy a posthumous regard.

Does *Erewhon* add anything to Butler's earlier conception or argument? Let us first look at the book as a whole, serving as it does for the context of "The Book of Machines." *Erewhon* spelled backwards (with the *wh* left as is) is "Nowhere," or utopia. It is cast in the form of an imaginary explorer's tale: the hero wanders over a mountain range in New Zealand, into the land of the Erewhonians. The opening chapters, with their John Buchan-like tone, and air of excitement and adventure, lead us on into the book.

Once in *Erewhon*, however, we find ourselves engaged in a more serious exploration; we enter into culture of sorts. Butler's purpose is like that of Descartes in his *Discourse*, or Montesquieu in his Persian Letters: to examine and satirize his own culture by comparison with a fictitious one abroad. It was Descartes who remarked that customs are obviously uncertain, being different on one side of a mountain from the other. Over the mountain, Butler finds everything different from his own western, that is, English, customs. To make his point, Butler inverts everything he can. First, names: for example, the girl the hero first encounters is Yram (Mary), and we have already noted the spelling of *Erewhon*. Next, customs: in *Erewhon*, physical illness is thought criminal and immoral, but crimes such as embezzlement are considered to be a mental disease, to be cured by "straiteners"; unreasoning and Colleges of Unreason are preferred to reasoning and Colleges of Reason; the young inflict corporal punishment on the old, for otherwise the latter would be incorrigible (in fact, Butler as a boy had been repeatedly flogged by his parents);[2] progress is a bad word; and so forth.

It is in this context, of a world turned upside down, so to speak, that what Butler has to say on machines and their evolution is presented. The subject surfaces when the Erewhonians exhibit displeasure, rather than awe, at the sight of the hero's watch. The concern takes center stage when the hero is shown through a museum, in which, in addition to stuffed birds and animals, there are rooms filled with broken machinery. Here he is told the following history: that some 400-500 years ago, the state of mechanical knowledge being so advanced and threatening, a learned professor wrote an extraordinary book, calling for the extinction of all machinery not in use for over 271 years. In the massive outbreak of Luddism that the book inspired, machinists and anti-machinists engaged in a violent civil war for many years, in which half the population perished. Finally, the antis won, and, with unparalleled severity, eliminated every trace of opposition. The Industrial Revolution had been inverted.

Aside from this dramatic suggestion, it is the arguments of the professor, in his "The Book of the Machines" (purported to be translated by the hero), that are of interest. Mainly, they are a restatement of Butler's earlier essays, as I have outlined them. However, a few points are developed significantly further, or in more powerful form, and worth quoting at some length. Thus, on the subject of consciousness, Butler has his professor say (Butler 1954, 161-171):

---

2 See Jones 1986, vol. 1, 20.

'There is no security' to quote his own words 'against the ultimate development of mechanical consciousness, in the fact of machines possessing little consciousness now. A mollusc has not much consciousness. Reflect upon the extraordinary advance which machines have made during the last few hundred years, and note how slowly the animal and vegetable kingdoms are advancing. The more highly organized machines are creatures not so much of yesterday as of the last five minutes, so to speak, in comparison with past time. Assume for the sake of argument that conscious beings have existed for some twenty million years: see what strides machines have made in the last thousand!'

Then, continuing on the subject of consciousness, but adding a further thought, the author adds:

But who can say that the vapour-engine has not a kind of consciousness? Where does consciousness begin, and where end? Who can draw the line? Is not machinery linked with animal life in an infinite variety of ways? The shell of a hen's egg is made of a delicate white ware and is a machine as much as an egg-cup is; the shell is a device for holding the egg as much as the egg-cup for holding the shell: both are phases of the same function; the hen makes the shell in her inside, but it is pure pottery. She makes her nest outside of herself for convenience' sake, but the nest is not more of a machine than the egg-shell is. A 'machine' is only a 'device.'

There is no limit, however, to the development of these "devices," with their consciousness. Answering the argument that the machine, even when more fully developed, is merely man's servant, the writer contends:

But the servant glides by imperceptible approaches into the master; and we have come to such a pass that, even now, man must suffer terribly on ceasing to benefit the machines … Man's very soul is due to the machines; it is a machine-made thing; he thinks as he thinks, and feels as he feels, through the work that machines have wrought upon him, and their existence is quite as much a sine qua non for his, as his for theirs. This fact precludes us from proposing the complete annihilation of machinery, but surely it indicates that we should destroy as many of them as we can dispense with, lest they should tyrannize over us even more completely.

And, finally, another dealt with the latent sexual threat:

It is said by some with whom I have conversed upon this subject, that the machines can never be developed into animate or quasi-animate existences, inasmuch as they have no reproductive systems, nor seem ever likely to possess one. If this be taken to mean that they cannot marry, and that we are never likely to see a fertile union between two vapour-engines with the young ones playing about the door of the shed, however greatly we might desire to do so, I will readily grant it. But the objection is not a very profound one. No one expects that all the features of the now existing organizations will be absolutely repeated in an entirely new class of life. The reproductive system of animals differs widely from that of plants, but both are reproductive systems. Has nature exhausted her phases of this power?

Inspired by fears such as these, which sound like our present realities, the Erewhonians rise up and destroy almost all their machines.

In a letter to Charles Darwin, Butler referred to "the obviously absurd theory" that he had developed in the chapter upon Machines (Jones 1986, vol. 1, 156). Butler is too quick to dismiss his own theory; it deserves a full hearing. I suspect, in fact, that Butler was satirizing his own satire, in a deprecating manner we have encountered in his earlier exchange with the Bishop of Wellington.

Most of *Erewhon* is not about machines. It is about what Butler in *The Way of All Flesh* referred to as the "question of the day ... marriage and the family system" (Butler 1950, 531). To this question of questions for Butler we must add religion: the hero of *Erewhon* thinks of converting the Erewhonians to Christianity, and spends much time implicitly satirizing Christianity by comparisons to the strange religious customs and beliefs of the local inhabitants. Similarly, the hero satirizes Victorian marriage and the family system by the same device of inversion. The satire on machinery and progress fits into the general picture. It can, however, be taken out of this frame, and left to stand on its own, as it does in the earlier essays, as a profound statement about man, machines, and evolution.

The Way of All Flesh (published posthumously, in 1903) seems far removed from *Erewhon* and its "The Book of Machines." It is an autobiographical novel, a sort of *Bildungsroman*, tracing the development of Butler's hero, Ernest Pontifex. In fact, as we see from Butler's own statement, it was also a device for him to promulgate his general views. As he wrote to his friend Miss Savage, who had urged him to write a novel, "But I am very doubtful about a novel at all; I know I should regard it as I did *Erewhon*, i.e., as a mere peg on which to hang anything that I had in mind to say" (Jones 1986, vol. 1, 160). Charles Darwin also sensed what Butler was about when he wrote to him shortly after Miss Savage, "What has struck me much in your book (*The Fair Haven*, Butler's religious tract of 1873) is your dramatic power that is to [say] the way in which you earnestly and thoroughly assume the character and think the thoughts of the man you pretend to be. Hence I conclude that you could write a really good novel" (Jones 1986, vol. 1, 187).

Butler was obsessed with the question of heredity. It is, fundamentally, what *The Way of All Flesh* is about. It is why Butler was so concerned with religion – were the sins of his religious forebears to be visited on him? – and so ready to be seized upon by Darwin. While there is, surprisingly, nothing about the industrial revolution or its machines, so omnipresent in 19th century England, the book is immersed in the controversies of religion and science of its time. Outwardly a novel, and a very successful, important, and interesting one, it is also in many ways "a mere peg," transmuted though it is into a whole tree.

I limit myself here, however, to the peg. In the figure of Ernest, Butler recounts the battle between religion and science waged by so many caught "Between Two Worlds," as Matthew Arnold puts it of his generation. At Cambridge, in 1859, Ernest reads Darwin's *The Origin of Species* and "adopted evolution as an article of faith" (Butler 1950, 500). In fact, though he never repudiates evolution, Ernest, or rather Butler, comes to a Lamarckian version of it, which in the end allows for a return to religion of a sort.

The path is through Butler's obsession with the subject of heredity. A characteristic reflection is the following:

Accidents which happen to a man before he is born, in the persons of his ancestors, will, if he remembers them at all, leave an indelible impression on him; they will have moulded his character so that, do what he will, it is hardly possible for him to escape their consequences. If a man is to enter into the Kingdom of Heaven, he must do so, not only as a little child, but as a little embryo, or rather as a little zoosperm – and not only this, but as one that has come of zoosperms which have entered into the Kingdom of Heaven before him for many generations (Butler 1950, 300).

Butler's message is conveyed by vague hint and suggestion. Distinguishing between our reflex actions and our reflections, he says:

Man, forsooth, prides himself on his consciousness! We boast that we differ from the winds and waves and falling stones and plants, which grow they know not why, and from the wandering creatures which go up and down after their prey, as we are pleased to say, without the help of reason. We know so well what we are doing ourselves and why we do it, do we not? I fancy that there is some truth in the view which is being put forward nowadays, that it is our less conscious thoughts and our less conscious actions which mainly mould our lives and the lives of those who spring from us (53-54).

What does Butler mean here? Is he anticipating Freud's theory of the unconscious? I think not. Butler's unconscious, I believe, represents man's inherited memory. And the latter is Lamarckian, in the sense that an ancestor's thought can descend to his offspring, as if carried in the germ cells. "As I watched him," Butler says of his hero, "I fancied that so supreme a moment of difficulty and danger might leave him with an increase of moral and physical power which might even descend in some measure to his offspring" (380).

It is difficult in *The Way of All Flesh* to trace either the origin or descent of Butler's ideas on evolution, though they are there, animating the book. At one point, we are told, in fact, that the novel was meant to be illustrative of the theory of *Life and Habit*, his book of 1874. In another strange passage, written in 1898, he offers a further puzzling hint when he says:

If in my books, from *Erewhon* [1872] to *Luck or Cunning?* [1887] there is something behind the written words which the reader can feel but not grasp and I fancy that this must be so it is due, I believe, to the sense of wrong which was omnipresent with me, not only in regard to Pauli, the Darwins, and my father, but also in regard to my ever-present anxiety about money (Jones 1986, vol. 2, 39).

Whatever does Butler mean? To answer this question, I believe, we must turn back to his relations to Charles Darwin and Thomas Huxley.

The connection starts before Charles Darwin and Samuel Butler. Both men had grandparents who were leading figures in Shrewsbury. Dr. Samuel Butler, whose life his grandson was to write, was Bishop of Lichfield (1774-1839), who also served as Headmaster of Shrewsbury School, in which not only his own son Thomas, but also Robert Waring Darwin's sons Erasmus (named after his grandfather) and Charles were scholars. Thomas, our Samuel's father, was at Cambridge at the same time as Charles Darwin, and the two spent the summer of 1828 together, though never becoming intimates. It was a matter of some pride to young Samuel Butler that the great Charles Darwin had been a student under his grandfather, Samuel.

We can only guess at the mixed feelings that Butler brought to the relation with Darwin. At first, as we have seen, Butler found in *The Origin of Species* a liberating force, replacing his familial religion with a new "article of faith." He was full of praise for this new god, and declared himself one of his "many enthusiastic admirers" (Jones 1986, 99). As late as 1872, and the publication of *Erewhon*, he was still kneeling before Darwin, writing of the *Origin* as "a book for which I can never be sufficiently grateful, though I am well aware how utterly incapable I am of forming any opinion on a scientific subject which is worth a moment's consideration." About the same time, Butler sent Darwin some drawings of dogs made by a friend of his, which Darwin used in his book *The Expression of the Emotions in Man and Animals* (1872).

Such humbleness and desire to assist did not last. A first meeting with Darwin had gone off most pleasantly; at a second lunch about this time, Butler fancied an initial coldness on the older man's part, which he felt subsequently wore off. But he received no further invitations. For whatever reasons, Butler began to feel hostility to his erstwhile mentor. The explosion came seven or eight years later over a preface Darwin wrote to the life of his grandfather, Erasmus Darwin, by a German scholar, Dr. Krause, which seemed to cast aspersions on Butler. Without going into details over this sad and distorted matter, the episode gave occasion to Butler to release his accumulated grievances. He now referred to Darwin's theory as "puerile" (Jones 1986, 291). Echoing the Bishop of Wellington, he declared that Charles Darwin had basely filched the ideas of his grandfather, Erasmus, on evolution, and made them his own.

Charles Darwin, shocked and upset, asked, among others, his friend Huxley's advice about how to respond to "a clever and unscrupulous man like Dr. Butler" (Jones 1986, 326). Fortunately, Huxley counseled silence, and the episode blew over. But Butler had lost his faith. In 1890, he would write, "I have so often found Charles Darwin neglectful of every canon I have myself been taught to respect that his name carries no weight with me, and I am not disposed to attach much importance to him or to his work" (Jones 1986, vol. 2, 96). Ambivalence, however, still remained, and Butler never ceased to give Darwin credit for enunciating in bold terms the theory of evolution.

Why do I tell this story? It is because the personal relations are intermingled with a profound philosophical disagreement. In 1874, Butler began to strike off from Darwin, in his book *Life and Habit*. Seeking in later life to explain what he was writing there, he remembered that:

> The first passage in *Life and Habit* which I can date with certainty is the one on page 52 which runs as follows: 'It is one against legion when a man tries to differ from his own past selves. He must yield or die if he wants to differ widely, so as to lack natural instincts, such as hunger or thirst, and not to gratify them … His past selves are living in him at this moment with the accumulated life of centuries. 'Do this, this, this, which we, too, have done and found our profit in it,' cry the souls of his forefathers within him (Jones, I, 212-213).

Though his theory frightened him – "it is so far-reaching and subversive it oppresses me and I take panics that there cannot really be any solid truth in it" (Jones 1986, 242-243) – he persisted. He knew it was subversive of his previous adherence to Darwin and his theories. As he wrote to Darwin's son, Francis, "I confess that I do not like the thought either of his seeing it, or of your doing so; for it has resolved itself into a downright attack upon your father's view of evolution, and a defence of what I conceive to be Lamarck's" (Jones 1986, vol. 1, 257). The die was cast, and Butler admits that "the full antagonism between him (Charles Darwin) and Lamarck came for the first time before me; and I saw plainly that there was no possibility of compromise between the two views" (Jones 1986, vol. 1, 258).

Memory and teleology are the two key terms in Butler's Lamarckism. He summarized his theory as follows:

> Firstly, that there is a bona fide oneness of personality existing between parents and offspring up to the time that the offspring leaves the parent's body. Secondly, that in virtue of

this oneness of personality the offspring will remember what has happened to the parent so long as the two were united in one person, subject, of course, to the limitations common to all memory. Thirdly, that the memory so obtained will, like all other memory, lie dormant until the return of the associated ideas. Fourthly, that the structures and instincts which are due to the possession of this memory will, like every other power of manufacture or habit due to memory, come in the course of time to be developed and acted upon without self-consciousness (Jones 1986, vol. 1, 265).

Habit, then, is founded on memory; and reflex action, it appears, is merely unconscious memory.

Next, he tell us, in writing *Life and Habit*, he "cut out all support of natural selection and made it square with a teleological view for such, I take it, Lamarck's is, and only different from Paley's in so far as the design with Paley is from without, and with Lamarck from within" (Jones 1986, vol. 1, 258). What this means, apparently, is that for Butler a creature can will its modifications, and then pass them by inherited memory to its offspring. Butler came to believe in this Lamarckian position because he had already envisioned man's conscious manufacture of "extra-corporaneous limbs" (that is, machines) in such terms; now, he extended the idea back into evolution, and considered man's actual limbs and organs as consciously manufactured by him for his convenience (Jones 1986, 233).

Is this a ludicrous idea? Butler appealed to the work of Auguste Weismann for support, for that eminent scientist had said:

I am also far from asserting that the germ plasm which, as I hold, is transmitted as the basis of heredity from one generation to another, is absolutely unchangeable or totally uninfluenced by forces residing in the organism. I am also compelled to admit it as conceivable that organisms may exert a modifying influence upon the germ cells, and even that such a process is to a certain extent inevitable (quoted in Joad 1924, 31).

Did Butler have first-hand scientific knowledge to uphold his speculations, or was the appeal to Weismann merely to a convenient authority? He had already admitted that he was utterly deficient in such knowledge. Thus, it was as a matter of philosophical, and, I would argue, religious belief that Butler embraced Lamarckism.

It was Charles Darwin's son, Francis, to whom Butler had written, who carried on the debate. Responding to Butler's description of his intentions in *Life and Habit*, Francis Darwin asked:

Have you ever read Huxley's article or articles on 'Animal Automatism,' two or three years ago in *The Contemporary*? He tried to show that consciousness was something superadded to nervous mechanism, like the striking of a clock is added to the ordinary going parts. I mean that the consciousness as we know it has nothing to do with the act, which is a mere question of nerve-machinery (Jones 1986, vol. 1, 263).

Butler never replied, and we must assume he had not read Huxley, whom I insist is so seminal (Huxley's articles appeared in 1870 and 1874) to the climate of opinion that could lead Butler to write "Darwin Among the Machines" and *Erewhon*. In any case, Francis went on to identify Butler's position as the reverse of Huxley's: where the latter made consciousness passive, the former made it an active cause. At this point, Francis declared:

It seems to me that being personally identical with all the people you have been is a different thing from being identical with all those you are going to be. The inventor of the telephone may be a lineal descendant of, and therefore personally identical with, the man (Thales?) who rubbed a bit of amber and attracted a bit of pith; but no one supposes that he investigated amber and pith in the hopes of making a telephone. He couldn't have wishes so utterly beyond his private (own + ancestral) experience. Perhaps I have been taking your meaning too literally, but it looks to me as if you thought that a fish consciously desired lungs when its air-bladder began to turn into one (Jones 1986, vol. 1, 263-264).

Most of biological science in the 20th century has been anti-Lamarckian. True, Charles Darwin himself, in *The Descent of Man*, allowed more and more for it. A few thinkers here and there, like Nietzsche, and Butler's friend George Bernard Shaw, became declared adherents; and even Sigmund Freud, in *Totem and Taboo* (1913), insisted on its necessary truth. But most bona fide scientists, and most thinkers, rejected the teleology involved in Lamarckism and relegated it to the same category as Aristotelianism, with its object "desiring" to remain still. Will seemed as out of place in animate evolution as in inanimate physics.

Why, then, was Butler so adamant, though openly professing his ignorance of science (Weismann, as I have noted, being only a handy stick with which to beat Darwin)? The real issue was religious. Butler, in a most convoluted way, was returning to the beliefs of his clerical ancestors. His fundamental objection to Darwin was that he had emptied the universe of mind. Earlier evolutionists had seen design and intelligence in the world; Darwin saw only random variation. Butler, commenting on his book, *Luck or Cunning?* (1887), written to further the ideas of his *Life and Habit* and *Evolution Old and New*, declared fervently:

> Its very essence is to insist on the omnipresence of a mind and intelligence throughout the universe to which no name can be so fittingly applied as God. Orthodox the book is not, religious I do verily believe and hope it is; its whole scope is directed against the present mindless, mechanical, materialistic view of nature … (Jones 1986, vol. 2, 41).

How can I, or anyone, possibly reconcile this statement with my assertion that Butler seriously, and not simply satirically, foresaw and advocated machines as an evolutionary species? The line of explanation, I believe, lies in what Butler called his "modest Pantheism" (Jones 1986, 303). As he wrote in 1880, "I have finally made up my mind that there is no hard and fast line to be drawn, and that every molecule of matter is full of will and consciousness, and that the motion of the stars in space is voluntary and of design on their parts" (Jones 1986, vol. 1, 333). So, too, life and death are on an unbroken continuum, a new kind of salvation and immortality. Thus, Butler wrote, "we are all one animal, and reproduction and death are phases of the ordinary waste and repair which go on in our bodies daily" (Jones 1986, vol. 2, 445).

Darwin and Professor Huxley had become the devils, advocating discontinuity and death. Butler saw himself as redeeming the world for continuity and life, with will and intelligence animating the whole. Earlier, I had quoted him as sensing "something behind the written words which the reader can feel but not grasp" (Jones 1986, vol. 2, 39). In part, the something was a deep sense of resentment – his omnipresent "sense of wrong" linking the Darwins and his father; in part, it was a strange memory of an ancestral religion.

Whatever the unconscious "will" behind Butler's venturesome speculations and however oddly arrived at, the result is a shocking challenge to our ideas about the continuity or discontinuity existing between man and machines. The essence of that challenge is as follows: if one takes Darwin seriously and accepts the fact that man is in an evolutionary relation to the other animals; adds to this the idea, advocated by Huxley (building on Descartes), that the animals are animal-machines, and man therefore the same (allowing for whatever differences in degree exist); admits that man then creates machines, which are merely consciously contrived versions of the animal-machine (again only differing in degree); then one can conclude that we are on an evolutionary continuum, with machines as a new, and possibly advanced, species.

What this conclusion actually entails, in the detail of evolutionary science, is not clear. What is clear is that Butler's view cannot merely be laughed out of court: matter, possessing the potentiality for becoming life – the ameba emerging out of inorganic matter – can be considered as already alive: there is no real distinction between the organic and the inorganic. No matter how satirically worked up, for example, in "The Book of the Machines," Butler's is a serious perspective on the evolving world that has profound implications for our own cybercultural moment.

Now, in my view, Butler confuses the way in which cultural and physical evolution takes place. In the former, with man, intelligence and will can and do enter into the process of "selection." We consciously create machines, art, and civilization (just as we breed domestic animals).

In his discussion of these matters, Butler ostensibly ignored Huxley's views, while accepting Darwin's general theory of evolution (though not its means). In fact, without the view on automatons enunciated by his hated contemporary Huxley, and others like him, Butler could not have pushed on to his own speculations. In some moods, Butler might disown his prophecies about machines and evolution; but they remain like Frankenstein monsters casting their shadows before them. Himself a divided soul, Butler mirrors our own ambivalent, and ambiguous, feelings on the subject.

Butler's lucubrations on machines have seemingly disappeared as a serious subject for discussion. I believe, however, that, viewed in terms of the continuity among man, animals, and machines, and the inextricable links between these categories within cyberculture, a revival of sorts of this thesis is in order. A further confession: I do not like much Butler, the man. Nevertheless, unlikely vessel though Butler is, his intuitions are important and far-reaching; Darwin was right: it is as a novelist of ideas, satiric and paradoxical, that Butler contributes to the continuing discourse on man and machines. Indeed, his work has forecast some of the possible futures that still await us and we see developing in the ongoing integration of humans and machines.

REFERENCES

Butler, S. 1923a. Darwin among the machines. In *A first year in Canterbury settlement and other early essays*.
    Vol. 1 of Shrewsbury edition of the works of Samuel Butler, ed. H. F. Jones and A. T. Bartholomew. London: Jonathan Cape.
————— 1923b. Lucubratio Ebria. In *A first year in Canterbury settlement and other early essays*.
    Vol. 1 of Shrewsbury edition of the works of Samuel Butler, ed. H. F. Jones and A. T. Bartholomew. London: Jonathan Cape.
————— 1923c. The mechanical creation. In *A first year in Canterbury settlement and other early essays*.
    Vol. 1 of Shrewsbury edition of the works of Samuel Butler, ed. H. F. Jones and A. T. Bartholomew. London: Jonathan Cape.
————— 1950. *The way of all flesh*. Harmondsworth: New York. Modern Library.
————— 1954. *Erewhon*. Harmondsworth: Penguin.
Joad, C. E. M. 1924. *Samuel Butler*. London: Leonard Parsons.
Jones, H. F. 1986. *Samuel Butler, author of* Erewhon *(1835-1902): A memoir*. 2 vols. London: Macmillan.
Stocking, G. W. Jr. 1987. *Victorian anthropology*. New York: The Free Press.

NOWHERESEVILLE. UTOPIA IS NO-PLACE

John Potts

In 1915, the futurist leader Filippo Tommaso Marinetti penned a "futurist vision-hypothesis" on life in the 21st century. He rhapsodized, enviously, of those future men whose "flesh, forgetful of the germinating roughness of trees, forces itself to resemble the surrounding steel" (Marinetti 1972, 105). The industrial basis of futurism, its obsession with technology and speed, are the informing factors here, inspiring an early glimpse of the cyborg. Marinetti's next step, however, is less predictable. He allows himself a vision of Utopia in the futurist city: hunger and poverty have disappeared, "the bitter social question, annihilated," electrical energy inexhaustible and transmitted without wires. It is a vision, from within the industrialist age, of an information society, in which "Intelligence finally reigns everywhere … Every intelligence grows lucid, every instinct is brought to its greatest splendor, they clash with each other for a surplus of pleasure" (Marinetti 1972, 106). Decades later, utopian fantasy in the age of the Internet revisits this intoxicated dream.

The futurists were prophets of "becoming" through technological means. Utopia was to be made with machines, steel and architecture. Limitations were to be transcended by technological speed. Marinetti was a techno-utopian, energized by the ancient zeal for utopia, but filtering it through the modern industrial world. This essay traces the utopian impulse as it ran through the futurists in the early 20th century and follows that impulse through to its current manifestation in the zone known as cyberspace.

NO PLACE

The word "utopia" entered the English language in 1516, with the publication of Thomas More's *Utopia* (1965). Writing in Latin, More had recourse to Greek in naming his ideal state: *ou + topos* = not place, or no place. This linguistic construct has been recruited to name a contrary bundle of pursuits, from literary works to political programs. Religion, art, science and politics are all enmeshed in utopian desire. Utopias have been mythical paradises, the ideals of political philosophy, the vehicles of satire, the earnest goals of communities motivated by politics or religion. Utopias have been situated in the metaphysical world of myth, in a lost past, in a distant future, on islands, on other countries or planets, in imaginary zones, in computer networks.

Utopia is always displaced. It is not of this time, not of this earth, not of this territory. It belongs to the future, but it may be built of the virtues from an imagined past. Utopia is made of unique materials. It may take its shape from nature, but of a magical nature capable of transformations favorable to humanity. It may take its cue from technology, but a magical technology of marvel and delight, an alchemist's city. It is nature, or technology, without hazard or side effect. Utopia is not Alphaville, nor Omegaville. It is not of this order, it exists in a different space: it is Nowheresville.

The first religious utopias dreamed a perfect nature: the Garden of Eden before corruption by sin. Or a perfect afterworld, a celestial city. Or a golden age of the mythical past, in which everything was infinitely superior to the tawdry present. Fantasy has been integral to the generation of utopias, yet there is also a long tradition of scientific speculation. Francis Bacon's *New Atlantis* of 1627 was written with a Renaissance confidence in scientific method and its product, technology. The New Atlantis incorporated modern telecommunications and transport as constituent parts of the perfect society; this model of utopia was assayed many times in later centuries. While political and religious visionaries of the 19th century were doggedly attempting to wring utopia out of the present, fiction writers projected their utopian visions into a technological future guided by reason.

The 20th century inherited these twisting lineaments of utopia. They intersected with the doctrine of progress, with the exaltation of technology, with the fervor of modernist aesthetics. Futurism trumpeted its technological utopia: a city to be forged of dynamism. A city of energy, speed and movement; a city of perpetual renewal. Yet that city was never built. The closest it ever came to being realized was in the drawings of Sant'Elia – or, more closely still, in the torrid invocations of Marinetti's writings. His manifestos continually called the future into being: new forms of art, of cities, of politics. The first surge of futurism promised an industrialist utopia; like all utopias, it could not exist in the real world.

And now, in the postindustrial, postmodern era, where is utopia? Removed as we are from the futurist faith in the machine and in progress, skeptical and more inclined to look backwards than forwards, is there room for utopia? The answer is that utopian desire is stronger now than ever, because postindustrial technology has opened up a new dreaming-space. Utopia thrives in the contemporary immaterial sphere, the electronic no-place: cyberspace.

CITY OF BECOMING

We declare that the world's splendor has been enriched by a new beauty, the beauty of speed … Why should we look behind us … Time and Space died yesterday. We want to destroy the museums, the libraries … We shall sing of the great crowds in the excitement of labor, pleasure and rebellion … of the nocturnal vibration of arsenals and workshops beneath their violent electric moons … of broad-chested locomotives prancing on the rails … and of the gliding flight of aeroplanes … (Marinetti 1961, 124).

Futurism was founded on February 20, 1909, its first manifesto typed by Marinetti in, no doubt, a frenzy of speed and aggressive keystrokes. Presumably he fuelled his typing-assault with strong draughts of espresso: a high-octane beverage to drive the avant-garde of culture. He proclaimed himself "the caffeine of Europe" (quoted in Hughes 1991, 43), jolting drowsy Italy from its agrarian slumber. He and his fellow futurists were gorged on cultural caffeine, slapping the face of any European contaminated by tradition. The futurists exalted a feverish insomnia. In a delirium of speed, metal and violent force, they declared war on the past and all its relics.

The futurists asserted the dynamic over the static, the technological over the natural, the street over the academy. The sickly crescent moon of the 19th century symbolists is overpowered by the electric lamp, which "shrieks out the most heart-rending expressions of color" (Boccioni et al. 1961, 124); Science – "victorious Science" – was not to be resisted, but welcomed as the liberator of art, its "vivifying current" (Boccioni et al. 1961, 126). The X-ray, cinema, and high-speed photography revealed new aspects of movement and perception. Futurist art responded to these innovations by charting the force-lines of the object-world, the psychic architecture of the viewer.

Previous artists had represented technology, but the futurists went further; they sought to absorb the properties of the machine into their art. A new century called for a new art with a new vision. The futurists' utopian zeal would surge forward on the back of technology and its unlimited promise. Sant'Elia's futurist architecture existed only on paper, but it was an early blueprint for the grand visions of modernist architecture:

> We must invent and remake the Futurist city to be like a huge tumultuous shipyard, agile, mobile, dynamic in all its parts; and the Futurist house to be like a gigantic machine … (quoted in Taylor 1961, 106).

The dominance of the machine aesthetic, the sculpting of whole cities according to new principles, the shaping of social and cultural life by architectural form, the solving of social problems by technological means: these emerged as tenets of Soviet constructivism in its brief celebration of the artist as engineer, then of the International Style at the height of modernism. This was the utopian spirit firmly welded to the properties of technology. Yet even so, it was a slowed-down version of futurism. It was the dynamic become static, energy solidified into form. It partook of the futurist zeal, but it was not futurism. Futurism was about becoming; it could not take the step into being without disastrous compromise, as evident in Marinetti's alliance with Fascism:

> When we are forty, let others, younger and more valiant, throw us into the basket like useless manuscripts! (Marinetti 1961, 125).

Futurist writings are full of the thrill of becoming. Stasis is the enemy, even in architecture:

> To a finished house we prefer the framework of a house in construction whose girders are the color of danger … The frame of a house in construction symbolizes our burning passion for the coming-into-being of things (Marinetti 1972, 81-82).

All the futurist aesthetic devices – words in freedom, plastic dynamism, the art of noises – reveled in liberation from the old. The futurists were "mystics of action," celebrating the vigor of youth, dreading decay.

The key influences on futurist thought were Sorel (the politics of violence), Nietzsche (the overcoming transvaluation of values), and Bergson (a generalized flux and dynamism). Of the three, it is Bergson whose influence is the most significant. His ideas are most transparently applied in futurist photo-dynamism, obsessed as it was with the interplay of time, movement, and perception. But they are also more generally apparent in the futurist love of the impermanent, the idea of constant renewal. Dynamism was the essential element in anything futurist. Decades before Deleuze inscribed Nietzsche and Bergson into a philosophy of desire, Marinetti invoked them in his impossible politics. Bergson's philosophical engagement with the developments of modern science provided a great stimulus, as reflected in Marinetti's statement of 1913: "Futurism is grounded in the complete renewal of human sensibility brought about by the great discoveries of science" (Tisdall and Bozzolla 1993, 8).

Marinetti was the first poet of this 20th century brave new world and futurism was to be more than an art movement – it had to be a total way of life. Marinetti's was a poet's vision. Politically naïve, he elevated artists – "the vast proletariat of gifted men" – to the status of government, so that life itself would become a work of art. The futurist artists were the avant-garde of this life-as-art politic, showing the way for the masses to follow. There was futurist clothing and futurist sleeping – standing up, aggressively, briefly. Futurists incorporated diverse media – performance, visual art, music, film – into public events five decades before Fluxus. Indeed, the 1960s caught up with a version of futurism: happenings, free love, utopian counter-culture, "elastic liberty," staged media events, replaying the techniques devised by Marinetti in the pursuit of radical art and the "morality of danger." Even the futurists' "kill us at 40" assertion found an echo in the "hope I die before I get old" ethos of rock music; rock stars, like futurists, were not meant to become middle-aged.

Marinetti's ideas reached forward in other ways as well. He saw the mediatized global village and its psychic impact – "Man Multiplied by the Motor" – decades before McLuhan (Tisdall and Bozzolla 1993, 8). He saw profound consequences for the human engagement with technological speed. This could entail the notion of transcending the body itself in the quest for speed and sensation. In Marinetti's time, the engines that might drive such a transcendence were the motorcar, the airplane and the factory, precursors of a future in which flesh would come to "resemble the surrounding steel." But Marinetti's emphasis on the transforming powers of technological speed is his greatest legacy for the 21st century. Computer processing speed and the speed of connection now set the pace for the postindustrial era, an age intoxicated by the rapidity of information flow through cyberspace.

THE POLITICS OF UTOPIA

There remains the problem of Marinetti's politics, specifically his association with fascism. There is no doubt that futurism has been stained for decades by its fascist affiliation. Is it possible to disengage the spirit of becoming from the reactionary direction taken by the Italian futurists? What are the politics ensuing from the celebration of technological speed? Does utopia favor any political creed?

The political economy of Thomas More's *Utopia* is, at least on the surface, communist. His narrator Raphael insists that the Utopian economy of communal ownership provides the best social model: "… I'm quite convinced that you'll never get a fair distribution of goods, or a satisfactory organization of human life, until you abolish private property altogether …" (More 1965, 66). Generations of commentators have claimed More could not have meant this literally: it must surely be ironical, or metaphorical, or merely provocative. But Raphael supports his passionate defense of the Utopian way with the authority of Christ himself: "… Christ prescribed of His own disciples a communist way of life, which is still practiced today in all the most truly Christian communities" (More 1965, 118). Raphael (or More) could justifiably point to the disdain for private property evinced by Christ, likewise to the organization of the early Church on communist lines, which was preserved in More's time in the monasteries. Reference to this Christian basis for communal ownership could shelter the author from his political critics. It was also intended, undoubtedly, to reveal the stature of the Church in ironic contrast. The vast accumulated wealth of the Church – as institutional power – stood in lurid contradiction to the early Christian spirit and monastic lifestyle.

Yet whatever More's intentions, his *Utopia*, like Plato's *Republic*, provided the perfect society with a specific political structure: harmonious and classless, as distinct from Plato's authoritarian state. Radical political programs of succeeding centuries dreamed of overcoming the inequities of capitalism. While their longed-for end point may have resembled the communal bliss of More's vision, the reality of the alternative state, when it was finally achieved, came much closer to the rigidity of Plato's prescription. Marx and Engels explicitly disparaged utopians, yet Marxism is now generally considered to have had a utopian dimension, especially as its political ambitions – including the "withering of the state" – produced such contrary results in practice.

But in the years before communism became a mass industrial reality, politics and the utopian imagination fired each other in unpredictable ways. In the turbulent years of the early 20th century, competing ideologies smashed against each other, nowhere more virulently than in Italy. Anarcho-syndicalism, communism, fascism, nationalism, and others all jostled for political ascendancy – and it was not uncommon for individuals to switch allegiance from one to the other, as did Mussolini, the former revolutionary socialist.

In this overheated environment, what kind of utopia did Marinetti have in his feverish mind? What were the colors of its political flag? The most forceful impulse was anarchist. Tradition, respectability, conservatism – in short, bourgeois values – were the target. This was recognized by the Marxist theorist Gramsci, who praised the futurists as cultural revolutionaries. Writing in 1921, Gramsci celebrated their "destruction of spiritual hierarchies, prejudices, idols, traditions that have become rigid" (Tisdall and Bozzolla 1993, 201), as the necessary first stage of revolution:

> The Futurists … destroyed, destroyed, destroyed … they had a precise and clear conception that our era … had to have new forms of art, philosophy, customs, language; this was their clearly revolutionary concept, and an absolutely Marxist one (Tisdall and Bozzolla 1993, 201).

As several commentators have remarked (Perloff 1986, 30; Tisdall and Bozzolla 1993, 200), left-wing attitudes to futurism have been influenced not by Gramsci, but by Walter Benjamin's famous "Work of art in the age of mechanical reproduction" essay of 1936. Emphasizing Marinetti's fascist affiliation, Benjamin rejected the fascist aestheticization of politics, to which he opposed communism's politicizing of art. While this dichotomy made urgent sense in 1936, it makes little in a world that has seen the unyielding communist aesthetics of the former Soviet Union. Nevertheless, the critical tradition has been maintained by Paul Virilio, who has regularly yoked together Marinetti's aesthetics and the fascist war machine. Virilio has argued that the imagery of aerial assault and speed, drawn from the second wave of futurism after 1918, served as symbolic re-enforcement not just for Italian fascism, but for later technological military constructions, such as Ronald Reagan's Star Wars and the Gulf War. These and other endeavors rendered their high-tech military apparatus aesthetically dazzling, a feat to be admired.

No-one can dispute that the futurist glorification of machinery, aggression and war – "the world's sole hygiene" – was readily recruited for fascist purposes. But this was only part of the story. Russian constructivism was directly inspired by Italian futurism, as Marinetti was quick to point out. He celebrated both the Bolshevik Revolution and the agitprop art of its cultural wing:

I am delighted that the Russian Futurists are all Bolsheviks and that for a while Futurism was the official Russian art … This does honor to Lenin and cheers us like one of our own victories (Marinetti 1972, 153; Tisdall and Bozzolla 1993, 15).

While Marinetti's comments here are typically self-serving, they at least indicate the mixed political legacy of futurism, even in its leader's own time. Further remarks on the same theme illuminate Marinetti's highly distinctive political theory. He could recognize a commonality with radical movements elsewhere because "every nation had or still has, its own form of passéism to overturn." But all such movements – grouped together as "futurist" – must necessarily remain "autonomous": "We are not Bolsheviks because we have our own revolution to make" (Marinetti 1972, 153).

Marinetti's anarchism was defiantly individualist. The nation was considered an extension of the individual; hence anything which vitiated the nation's sovereign strength was to be condemned. Marinetti despised internationalism as weakening, Christian and effeminate. He considered socialism "cowardly" and communism doomed to failure because of its "bureaucratic cancer": "the fixed, restful uniformity promised by Communism." Any impediment to the individual will – personal or national – was to be condemned because "[h]umanity is marching toward anarchic individualism, the dream and vocation of every powerful nature" (Marinetti 1972, 148).

Many decades later, there is a contemporary echo of this rampant individualism, even "the dream … of every powerful nature," in the rhetoric of the most zealous cyberculture theorists. For the performance artist Stelarc, the cyborg is not the collectivist cyborg theorized by Donna Haraway; it is rather a lone, technologically enhanced hero of the future. The technomystics known as the Extropians, and many others who dream of future perfectibility by technological means, place a similar emphasis on the individual – not the community – as the locus of perfection.

In the name of individualism, Marinetti defended his coupling of patriotism and anarchism against all critics. The nation for him was vital, whereas the Church, or monarchy, or international cooperation, were enfeebling. Futurist democracy was to be above all vitalizing, as he declared in 1919: "We unite the idea of Fatherland with that of daring Progress and of anti-police revolutionary democracy" (Tisdall and Bozzolla 1993, 19-20). The sad truth is, of course, that such a revolutionary democracy was a long way from being realized. Instead, Marinetti's fervent and militant nationalism had enough in common with Mussolini's vision to bring the two men together, although their relationship was at times discordant. Even at his most sympathetic to Mussolini, Marinetti was a most paradoxical fascist. Idealist, anarchist, utopian, he looked incessantly to the future while fascism looked to the past; to ancient Rome in the Italian version, to the soil and an imagined folk past in the German. Hitler's utopian vision – anti-industrialist, green, overseeing the cultivation of grain in the Ukraine to provide spaghetti on a mass scale (Carey 1999, 425) – would surely have appalled Marinetti, who not only elevated the industrial far above the agrarian, but rejected pasta as too slow and backward a food for futurists. Fascist architecture was grand, monolithic, narcissistic with power, built to last; futurist architecture was meant to be pulled down soon after it was built.

Marinetti itemized the demands of a futurist political program in 1913. Here we glimpse political ideals not yet contaminated by the lure of power offered by fascism. The futurists called for abolition of the monarchy, expulsion of the Papacy, socialization of land and Church property, equal pay for women, free legal aid, devaluation of matrimony to prepare for free love, freedom of the press, and later, the razing of prisons (Tisdall and Bozzolla 1993, 203). This program in its entirety is radical at any time; in Italy of 1913 it could be nothing other than utopian.

Marinetti's writing was a hymn to progress, the engine of a technological society. "Progress ... is always right even when it is wrong," he wrote, "because it is movement, life, struggle, hope" (Marinetti 1972, 82). The doctrine of progress that was forged in the 18th century enlightenment was itself part of a utopian vision: reason was meant to create a profoundly just society. Reason would drive a continual social progress until perfectibility was reached. This was the enlightenment-utopian creed. But progress took a technocratic turn in the Victorian period and by the early 20th century was equated exclusively with technological progress and considered a goal in itself. This was the techno-utopian drive articulated by Marinetti, and suffused throughout much of modernism, most obviously in architecture. Urban design was not only considered a technological fix for social ills, but the purity of architectural design held a spiritual dimension which, as stated in the gospel according to Le Corbusier and Mies van der Rohe, could not help but uplift all who experienced it.

Modernism's faith in technological progress was aligned with the culture of industrial capitalism in the first six decades of the 20th century. The shock of the new in aesthetics ran parallel with the triumph of the new in production, and the thrill of the new in consumption. Industrial capitalism declared war on the past just as ferociously as had Marinetti in his first manifesto. Planned obsolescence was a production strategy: why drive last year's car when this year's is so much better? Advertising sold the virtues of the new more effectively than any manifesto. Coffee was replaced by instant coffee (surely even Marinetti, as the caffeine of Europe, may have had his doubts about that innovation); tea became tea bags; soup became instant soup; razors became electric; cotton and wool were replaced by synthetics; seasonal vegetables were usurped by frozen packaged replacements. In the name of progress, nature was improved upon by technological means. Each of those transformations constituted a demonstrable loss of quality, but that was ignored for the sake of convenience, speed and the certainty that these emblems of progress represented a constantly improving way of life. Technological progress presided over a shift in cultural values; quality of life was mapped over by the commodities of a modern lifestyle.

When did the faith in progress begin to decline? A list of key names and terms sketches a chronological map: Mutually Assured Destruction, Agent Orange, Harrisburg, Chernobyl, Bhopal, Exxon oil spill, Challenger, acid rain, global warming, ozone depletion.[1] The protest movement of the 1960s grew into the environmental concern of the 1970s; by the 1980s ecological sensitivity had become a mainstream issue. Anxiety about degradation of the natural environment was complemented by a more general technofear: technology became no longer the benign agent of progress but a potential harbinger of disaster. The sense of ceding human agency to technological systems, combined with the Frankensteinian fear of technology out of control, generated the dystopian vision of sci-fi films such as *Blade Runner, Terminator, Robocop, Strange Days* and *The Matrix*.

The linear march of progress has been challenged in many ways since the 1960s. Recycling emerged as an alternative to planned obsolescence; conservation in all its forms aims to preserve the past rather than supersede it. Quality of lifestyle came to be asserted, by many, over convenience. The International Slow Food Movement, founded in 1988, expresses in its manifesto defiantly anti-futurist principles:

---

1   A similar list is provided by Leo Marx (1995, 11).

We are enslaved by speed and have all succumbed to the same insidious virus: Fast Life, which disrupts our habits, pervades the privacy of our homes and forces us to eat Fast Foods … May suitable doses of guaranteed sensual pleasure and slow, long-lasting enjoyment preserve us from the contagion of the multitude who mistake frenzy for efficiency … In the name of productivity, Fast Life has changed our way of being and threatens our environment and our landscapes. So Slow Food is now the only truly progressive answer (Slow Food 1989).[2]

Known as "the Greenpeace of gastronomy," this association, founded in Turin, is constitutionally anti-Marinetti. In "The New Religion – Morality of Speed" of 1916, Marinetti revered speed as "naturally *pure*," slowness as "naturally *unclean*." Slowness is "evil" because of its nostalgia and its "pessimism about the unexplored" (Marinetti 1972, 95-96). By the end of the 20th century, technological speed had been uncoupled from progress, in many quarters. The Slow Food writings are in praise of slowness and rest; their anti-progress vigilance extends to a manifesto against biotechnology. Yet they too have a utopia, drawn from Rabelais' *Gargantua* (1534). In that work, the Theleme Abbey is inhabited by an order of monks and nuns, whose sole motto is "Do what you like." Needless to say, food and wine, slowly prepared, play a major role in this paradise.

The demise of progress ran parallel with the decay of modernism, as visions of technological utopias became ossified, brittle, finally to be exploded. Architectural historian Charles Jencks pronounced the death of modernism on July 15, 1972, at 3.32 p.m. (or thereabouts) when the "infamous Pruitt-Igoe Scheme" in St. Louis Missouri was given "the final coup de grace by dynamite" (Jencks 1987, 9). The modernist ideal of the home as a machine for living in, of a city as a rationally ordered paradise, produced instead alienation and resentment. Would-be inhabitants refused to live in such constructions; likewise a new generation of artists could no longer inhabit the territories mapped out by modernism. Postmodern artists, armed with scanners and samplers, turned to the recorded past as a vast databank to be accessed. The notion of an avant-garde was jettisoned. The shock of the new was giving way to the lure of the old. It may have seemed that Marinetti – avant-gardist, preacher of progress and industrialism – had been thrown into the basket like a useless manuscript. But the spirit of futurism has survived, running through new circuits, infusing new technological spaces.

RETURN TO NO PLACE

Is progress banished from our midst? Has utopia taken fright and vanished? Advertising doesn't think so: we are still presented with an endless array of products, each one superior to the last. Technological progress is now measured in terms of smaller, faster, greater capacity and flexibility; the industrial criteria of production have been replaced by postindustrial equivalents. Speed remains as a basic criterion: the touchstone of processing speed at the heart of information technology.

And if utopia continues in public discourse, it is to be found in the domain of computer networking. Politicians compete in their promise of funding for the Information Superhighway and access to information, whereby social ills will be remedied when every child has a laptop. This is the old modernist techno-utopia speaking through politicians: a technological fix for social problems. Indeed, in the American context, the term "futurist" has been reborn and applied to business people and enthusiasts agitating for a high-tech future. Across the world, developed and

---

2   The official Slow Food Manifesto was endorsed and approved by 20 countries in 1989.

developing nations are racing to connect their citizenry via fiber optics. Electronic networking is being asked to carry the hopes of brave new digital societies in a new millennium.

While the construction of this information age is being overseen by a congruence of government and corporate investment, artists and theorists have contributed to the utopian hope invested in the networked future. Much of the early digital zealotry emanated from the American west coast, where magazines like *Mondo 2000* and *Wired* celebrated the wondrous potential of digital technology. A technomystical aura was created around virtual reality, which was affirmed by its guru Jaron Lanier, as "a sort of talisman for Western Civilization, a way for people to get ecstatic and be with each other" (Barlow 1990, 45). The virtual "telecollaboration" of this technology would, it was envisaged, overcome the alienation and isolation within contemporary culture. This was a new utopian space, reflecting the immaterial nature of digital information itself. If reality could be simulated and manipulated at will, if new forms of consciousness could be revealed to the VR user, then this virtual reality came to represent, at least in theory, an idealized, alternative world.

The hype generated around VR was quickly transferred to the emergent system of computer networking. The World Wide Web came to represent a prototype for a future globalized digital network. Utopia in the 21st century resides in the transforming power of this information technology. The dream is of an electronic democracy on a global scale, characterized by dissolving hierarchies, offering universal access, untrammeled creativity, unlimited knowledge, liberating interactivity and empowering connectivity. The hope once invested in industrial technology has been displaced on to its postindustrial successor – a technology of the immaterial travelling at the speed of light.

The utopian dream of a universal language haunts a wired world based on computer language. In the 17th century, Leibniz proposed a logical language to be learned by everyone. The echo of this ambition can be heard in the claims made by contemporary techno-mystics that computers are the ultimate "abstract structure manipulators," whose creative use will "open the doors to new worlds" (Holtzman 1994, 252). Less fancifully, the Internet, as precursor of a future connectivity, accommodates the dream of a universal network, if not of a universal language.

Consider this statement from Pierre Levy, from his opening address to ISEA 94, Helsinki:

> If we were to take the route of the collective intelligence, we would gradually invent techniques, systems of signs … permitting us to think together, to concentrate our intellectual power, to multiply our imaginations and our experiences, to work out practical solutions for the complex problems affronting us in real time and on all levels (Levy 1994, 17).

Levy's vision is of "a new stage in the evolution of man," achieved through digital networking. Others take the evolutionary concept more literally, proposing a transformation of the body as well as of consciousness. Stelarc, a self-proclaimed pioneer of the human-as-cyborg, advocates technologizing the body as the next stage in our "postevolutionary development" (Potts 1993, 120). Max More, leader of the Extropians, sees in contemporary technology the potential to perfect the flawed organism that is the human being:

> We challenge the inevitability of aging and death, and we seek continuing enhancements to our intellectual abilities, our physical capacities, and our emotional development. We see humanity as a transitory stage in the evolutionary development of intelligence. We advocate using science to accelerate our move from human to a transhuman or posthuman condition (More 1999).

There is a clear echo of the futurist zeal here. As techno-evangelists, Stelarc and the Extropians share the futurist distaste for the past and nature, adding to Marinetti's program a literal interpretation of the will to supersede the body's limitations. The Extropian Principles include Perpetual Progress, Practical Optimism and Intelligent Technology. In this there is a direct line connecting the enlightenment trust in progress toward perfectibility, the futurist faith in the machine, and the Extropian techno-utopia. Max More's faith in technology – in genetic engineering, virtual reality, artificial intelligence, worldwide data networks, nanotechnology, and "neural-computer integration" – is indeed evangelical:

> When technology allows us to reconstitute ourselves physiologically, genetically, and neurologically, we who have become transhuman will be primed to transform ourselves into posthumans – persons of unprecedented physical, intellectual, and psychological capacity, self-programming, potentially immortal, unlimited individuals (quoted in Davis 1999, 120).

This drive connects with the 19th century roots of modernism: a distorted version of Darwinian evolution fused with technological progress. Those individuals intoxicated by "posthuman" fervor share a desire for transcendence with the cyber-utopians: both groups envisage a postcorporeal, postindustrial future. From the wreckage of industrial progress has arisen a dream of escaping physical limitations: a digital no-space beyond the restraints of body and locality.

Techno-utopian desire is the current inflection of a much older ambition. In its industrial phase, hope was invested in the material forms of architecture and the futurist energy of explosive speed. The postindustrial version venerates a greater speed but an immaterial no-place. Industrialism impinged on bodies, polluted them, stressed them, violated them. Postindustrial utopia is disembodied. John Perry Barlow, prophet and advocate of virtuality, celebrates this aspect of the datasphere:

> Cyberspace consists of transactions, relationships, and thought itself, arrayed like a standing wave in the web of our communications. Ours is a world that is both everywhere and nowhere, but it is not where bodies live … (quoted in Davis 1999, 110).

Taken to its most extreme point, cyberculture dreams of pure consciousness on the Net, in computer memory. Modern day sects like the Extropians look to computer networking to realize the religious longing for transcendence, by uploading consciousness directly into a computer (Davis 1999, 121). This is the hope for a distilled essence of consciousness – a soul – to be stored as immortal information. The gnostic goal of pure spirituality is here fused with the futurists' exultation in the transforming powers of technology. The result is a techno-religion. The religious impulse of utopia is intact – computer technology has become its instrument.

There are other underlying reasons for this development besides the disillusionment with industrial progress. A puritanical distaste for the body, in an age of heightened awareness of viruses and debilitating illnesses, fuels the flight from physicality into virtuality. The western metaphysic has been marked since Plato with a desire for absence of the body, an otherworldliness central to Christianity and assorted mysticisms. A mystical aura has enveloped many technological forms at their inception, from electricity to radio to VR. Cybermysticism inherited utopian characteristics from the 1960s counterculture; its proponents, such as Timothy Leary, infused the digital realm with notions of transformed consciousness and alternative community structure.

Yet cyberspace has specific properties which set it apart, making it a prime candidate for utopian desire. It is an ill-defined, immaterial space, like the utopias conjured by imaginations through history. It is set free from the limitations of bodies, the restrictions of the past. It holds out the hope for a new, better world; indeed it *is* a new and better world for those inside it. Cybercommunities make their own rules, their own guidelines. They take in something of real communities, but create virtual communities freed from the tyranny of geography. Released from body and place, minds may interact in the manner dreamed of by Marinetti: "Intelligence reigns everywhere … Every intelligence grows lucid, every instinct is brought to its greatest splendor …" (Marinetti 1972, 106).

In cyberculture, identities may change and merge with others. Virtual space is a transformational device, where the imagination is encouraged. This may entail escape from the social order, or the will to improve it – the two complementary drives of utopia. These drives may be fanciful, deluded, self-important. Or they may be much-needed dreaming-places in a rationalized society. Or they may represent the latest version of an essentially religious longing. Or they may be all these aspects of utopia.

REFERENCES

Barlow, J. 1990. Life in the DataCloud. Interview with Jaron Lanier. *Mondo 2000*.

Boccioni, Carra, Russolo, Balla, and Severini. 1961. Futurist painting: Technical manifesto. In J. C. Taylor, *Futurism*. New York: Museum of Modern Art.

Carey, J. (ed.). 1999. *The Faber book of utopias*. London: Faber & Faber.

Davis, E. 1999. *TechGnosis*. London: Serpent's Tail.

Holtzman, S. R. 1994. *Digital mantras*. Cambridge, MA: MIT Press.

Hughes, R. 1991. *The shock of the new*. London: Thames & Hudson.

Jencks, C. 1987. *The language of postmodern architecture*. London: Academy Editions.

Levy, P. 1994. Toward superlanguage. In *ISEA94* catalogue. Helsinki: University of Art and Design.

Marinetti, F. T. 1961. Initial manifesto of futurism. In J. C. Taylor. *Futurism*. New York: Museum of Modern Art.

————— 1972. *Selected writings*. Ed. R.W. Flint. New York: Farrar, Strauss and Giroux.

Marx, L. 1995. The idea of technology and postmodern pessimism. In *Technology, pessimism and postmodernism*, ed. Ezrahi, Mendelsohn and Segal. Amherst: University of Massachusetts Press.

More, M. 1999. *Extropian principles (version 3.0: A transhumanist declaration)*. The Extropy Institute. At http://www.extropy.org/ (accessed February 1, 2001).

More, T. 1965. *Utopia*. Trans. P. Turner. London: Penguin.

Perloff, M. 1986. *The futurist moment*. Chicago: University of Chicago Press.

Potts, J. 1993. Techno-bodies: Report on TISEA. *West* 6/7:120.

Slow Food. 1989. *Slow Food Manifesto*. At http://www.slowfood.com/ (accessed February 4, 2001).

Taylor, J. C. 1961. *Futurism*. New York: Museum of Modern Art.

Tisdall, C. and A. Bozzolla. 1993. *Futurism*. London: Thames & Hudson.

STRANGER THAN YOU THINK :
ARTHUR C. CLARKE'S PROFILES OF THE FUTURE

"The Age of Postbiological Man would reveal the human condition for what it actually is,
which is to say, a condition to be gotten out of"
REGIS (1990, 175)

Russell Blackford

I.

Wherever we live on planet Earth, science and industrialization have molded our environments
and the kinds of people we are. In the postindustrial west's urbanized, knowledge-oriented societies, change is in our bones, an everyday fact of life. Whether we welcome or resist it, we expect
change in all forms of technology and all facets of the social world, and we find it natural to imagine the decades and centuries ahead. We have learned to view ourselves, as if from a distance, as
merely creatures of a particular time. In a later era, we realize, what is known, believed, or done
may be altogether alien to us.

Our ability to distance ourselves from our own stratum of time and our merely "current"
knowledge and beliefs, attitudes and fashions, may inspire a sense of absurdity.[1] Then again, the
implications may seem more frightening than that, or perhaps more liberating. In any case, the
pastoral conception of time, in which there is nothing new under the sun, now seems untenable,
as do the supernaturalist eschatologies that satisfied our ancestors' craving for cosmic justice.
Likewise, there is no return to the self-congratulatory conceptions of progress that appeared
plausible to generations of white western conquerors. Knowing ourselves as the naturalistic
products of events in vast space and deep time, we rate all these notions as fanciful.

This is a transhuman or posthuman moment in the history of ideas. It has become reasonable
to postulate radical technologies that could alter the fundamentals of the human condition and
human nature. Our particular societies and our species itself seem ready for transformation or

---

1   This is a special, but compelling, case of Thomas Nagel's reinterpretation of existentialist absurdity. Nagel refers to "the collision
between the seriousness with which we take our lives and the perpetual possibility of regarding everything about which we are serious
as arbitrary, or open to doubt" (Nagel 1981, 153).

supersession. These ideas encounter resistance, but they are growing increasingly attractive and are squarely on the agenda for intellectual debate. Some educated westerners have already embraced them, particularly those involved in the high-tech virtual communities that comprise cyberculture.

Ideas like these have many sources, merging and strengthening over decades and centuries from the time of Copernicus, Galileo, and Descartes. But much of their content and current popularity is directly attributable to Arthur C. Clarke, one of the handful of great science fiction writers to emerge in the 20th century. In the 1950s and 60s, Clarke prophesied the cyber age and started to set its philosophical direction. Clarke's breakthrough into a posthuman vision of the future was his 1956 novel, *The City and the Stars* (Clarke 1993), which portrays a world separated from ours by a billion-year gulf of time. In this classic work, he imagined the possibilities of a future race of mankind, finding inspiration in what were then recent ideas in the field of information theory. Ed Regis describes Clarke's fundamental insight:

> If you got enough information out of the brain and reproduced it with enough accuracy elsewhere, then you'd have a way of re-creating that person's memories, their innermost thoughts, feelings and everything else. You'd be able to remake the person in a form other than his original flesh-and-blood physique. You might even be able to transfer an entire human mind into a computer. It was hard to know what this would *mean*, exactly, but at the very least it was clear that our concept of what it was to be a human being would be changed forever (Regis 1990, 150).

Here, in the mid 1950s, we have the beginnings of a posthuman, postbiological, conception of future intelligence, inspired by what were then recent ideas in the field of information theory whose development underpins modern computer technology.

In the most recent edition of his nonfiction work *Profiles of the Future*, Clarke quotes the whimsical slogan of the Science Fiction and Fantasy Writers of America (SFWA): "The Future Isn't What It Used to Be." That is true in ways more profound than are probably intended either by Clarke or the SFWA. He goes wrong only when he adds, rather vacuously, "But, of course, it never was …" (Clarke 1999, 211). The element of commercial slickness displayed here should not detract from Clarke's insight nor his importance. He is the major exponent of a future that is stranger than we can yet imagine, in which, as has happened in the past, the limits of the possible may turn out quite different from what we currently expect.

II.

Throughout his long career, dating back to the publication of his first short stories in 1946, Clarke has had few rivals for pre-eminence in the science fiction field. He may be the most famous living science fiction writer, having outlived other giants of his time, such as Isaac Asimov, Robert A. Heinlein, and John Wyndham. Of current writers, only William Gibson, the dazzling cyberpunk author of *Neuromancer*, has a comparable public profile.

Now in his 80s, based in Sri Lanka, and socially restricted by the debilitating effects of post-polio syndrome, Clarke seems almost a legendary figure, distant from his admirers but larger than life, often present at international events via the satellite transmissions that he predicted before anyone else. His classic stories and novels remain among the genre's finest, while the film *2001: A Space Odyssey*, on which he collaborated with director Stanley Kubrick, is an immensely significant and influential production that some commentators judge the most important work of science fiction to date in any medium.[2] Clarke is probably best known for his creative guidance of *2001* and for his detailed proposals for telecommunications satellites, elaborated as early as a 1945 article in the journal *Wireless World*.

In his huge retrospective collection of nonfiction pieces, *Greetings, Carbon-based Bipeds!*, Clarke describes *Profiles of the Future* as "what may be my most important work of nonfiction" (Clarke 1999b, 195). It was originally published as a series of essays in *Playboy* magazine, which were then collected in 1962. Revised editions appeared in 1973 and 1982, making the 1999 "Millennial Edition" (or "Millennium Edition" according to the front cover) the fourth version of the book. As its blurb states, the theme of the essays "was one of ultimate possibilities, rather than achievements to be expected in the near future." Indeed, the book is subtitled *An Inquiry into the Limits of the Possible.* For that reason, it has remained relevant and suggestive.

As a futurist, Clarke has made some unfortunate predictions. For example, chapter four of *Profiles of the Future* contains a lengthy discussion of Ground Effect Machines (hovercraft), for which he anticipated a great future, believing they would become a ubiquitous means of public and private transport. The Millennial Edition contains some amusing reflections by Clarke on the book's original text and, in one such passage, he describes his subsequent unhappy experiences with hovercraft. He calls the experience of reading chapter four, nearly 40 years later, "chastening – even embarrassing" (Clarke 1999a, 40).

In a thoughtful review of the Millennial Edition, Damien Broderick points out that the hardware and operational principles involved in modern satellite communications are very different from Clarke's original proposal. When Broderick compares Clarke's 1945 vision with the sophisticated reality half a century later, the original conception shows up as primitive and clunky:

> Today the world is skeined with computerised comsats in low-earth orbit worth many billions of dollars, linking the planet into a global culture of many mansions. Clarke's triumph of prophecy is matched in hindsight by its comic pratfalls: his 1945 invention conceived three mighty space platforms high in geostationary orbit, crewed by technicians handling switchboards (Broderick 2000, 11).

Despite such "pratfalls," Clarke has been unusually insightful in his reflections on the future. In his essays for *Playboy* he sought to explore the possibilities of scientific and technological innovation without projecting one consistent line of development. In his introduction to *Profiles of the Future* he states that "many of the ideas developed in this book will be mutually contradictory" (Clarke 1999a, 8). As an example, he argues that perfected means of transportation and communication would render each other unnecessary. He examines these in separate chapters with no implication that both would develop, as described, in any actual society. More dramatically, his prediction of the obsolescence of humanity – our inevitable supersession by machine intelligence – involves scenarios that are inconsistent with all the others in the book. In his final chapter, "The Long Twilight," he restates his strategy of inconsistent lines of speculation, stating that he is "unrepentant" about the mutually incompatible prospects he has examined, and highlighting the fundamental issue of whether there is a long-term future for humanity at all:

---

2 See, for example, Rabkin 1981 (117).

In attempting to explore rival and indeed contradictory possibilities, I have tried to go to the end of the line in each case; sometimes this has led to a sense of pride in Man's past and possible future achievements – sometimes to a conviction that we represent only a very early stage in the story of evolution, destined to pass away leaving little mark on the universe (Clarke 1999a, 207).

III.

Profiles of the Future is most frequently cited not for its speculations about particular technologies, but for Clarke's reflections on the requirements and difficulties of technological prophecy. He states that it is impossible to predict the future, so "[t]his book has a more realistic, yet at the same time more ambitious, aim. It does not try to describe *the* future, but to define the boundaries within which possible futures must lie" (Clarke 1999a, 5).

Clarke proposes his three celebrated laws about the feasibility and character of technological innovations and describes two kinds of cognitive failure that lead to ill-founded pessimism about what is technically achievable. As expressed in the Millennial Edition, the three laws are these (1-2):

1. When a distinguished but elderly scientist says that something is possible, (s)he is almost certainly right. When (s)he says it is impossible, (s)he is very probably wrong.

2. The only way of finding the limits of the possible is by going beyond them into the impossible …

3. Any sufficiently advanced technology is indistinguishable from magic.

In his foreword to this edition, Clarke offers modern word processing as an example of his third law:

If anyone had told me, in 1962, that one day there would be book-sized objects that would hold the contents of an entire library, I would have believed them. But if they had said that I could find any page – or even word – in an instant and display it in scores of different typefaces ranging from Albertus Extra Bold to Zürich Calligraphic, any font size from 8 to 72, I would have protested that no imaginable technology could perform such a feat. I can still remember seeing – and hearing! – Linotype machines slowly converting molten lead into front pages that required two strong men to lift them. Now, of course, Microsoft Word performs far greater miracles, every day, in millions of homes all over the world (Clarke 1999a, 2).

As this implies, modern word processing would seem magical to a visitor from only a few decades in the past. That is not so surprising when we reflect on how much computer power is expended to achieve essentially trivial, though convenient, effects. The uses to which increasingly powerful computers are put in the entertainment industry and the business world in general demonstrate another law, one which Clarke does not expressly enunciate, that people and institutions find unpredicted uses for new technologies. These then shape the technologies' further development and add to the minefield of difficulties that confronts would-be prophets.

The failures in our speculative thinking identified by Clarke are closely related to his first law, with its droll commentary on the thought processes of distinguished, elderly scientists. He refers to "Failures of Nerve and Failures of Imagination" (1999a, 9). For Clarke, a failure of nerve is the inability to see, or accept, a conclusion even when given all the relevant facts. He states that "[t]he most famous, and perhaps the most instructive, failures of nerve have occurred in the fields of aero- and astronautics." As an example of this, he cites a declaration by Simon Newcomb, a "great American astronomer," that heavier-than-air flight was impossible, even though the relevant aeronautical theory was already available (Clarke 1999a, 10). Newcomb's view, expressed at the turn of the century, was not idiosyncratic but reflected those of most scientists of the time. They were soon proved wrong by the Wright brothers' experiments with their lightweight biplanes.

Similarly, there was deep and widespread skepticism about rocketry in the early decades of the 20th century, with attempts to "prove" the technical impossibility of a rocket-powered device that could deliver a payload to the moon. Clarke discusses in detail why seemingly formidable arguments that bristled with technicalities and were propounded by seeming experts were nonetheless wrong, even based on what was known at the time. Goddard, Oberth, and Tsiolkovsky had already published accurate technical analyses of the engineering problems associated with rocketry, but these were ignored or dismissed.

As conceived by Clarke, failures of the imagination are not so blameworthy:

[A failure of imagination] arises when all the available facts are appreciated *and* marshalled correctly – but when the really vital facts are still undiscovered, and the possibility of their existence is not admitted (1999a, 19).

As an example, he cites a statement by August Comte in 1835 to the effect that it was impossible to determine the chemical composition of astronomical bodies. As Clarke puts it, the stars could never be more than "celestial reference points" (1999a, 20). Unfortunately, Comte did not know about the spectroscope, which revolutionized our ability to obtain detailed chemical data on the stars and opened the way for modern astronomical science.

Such failures of the imagination attract relatively little blame because no-one can be expected to make predictions based on laws or techniques that are not yet known. *Some* blame, however, attaches to Comte for making such a sweeping and confident prediction about what would never be possible, since it was at least conceivable that an empirical means would be found to determine the composition of stars. There was nothing unclear or inconsistent in a description of what would be involved.

In the absence of appropriate theory or instrumentation, it would have been foolhardy to predict that such analyses would be done in the future, but that is not the point. Such examples suggest that nothing should be ruled out as impossible merely because the required technology cannot be extrapolated from current physical and engineering principles. However far-fetched an idea might currently seem, if it is logically coherent and does not violate any fundamental, well-tested physical laws, it may yet eventuate, even if in an unexpected form.

Clarke lists technologies that would have been "not merely incredible, but *incomprehensible* to the keenest scientific brains" of 1900 (1999a, 23; Clarke's emphasis). He includes a range of atomic and electromagnetic phenomena, and various techniques and devices based upon them. These include X-rays, radio, television, nuclear energy, and carbon dating. By definition, we do not know what unknown physical laws and phenomena await discovery, so we cannot deduce the future technologies that will be based on them. When we try to predict phenomena that are incomprehensible within current scientific theories, our thoughts reach a kind of boundary, but we can at least restrain ourselves from the gratuitous prediction of negatives: that such and such a technology will *never* be developed.

Many of the chapters of *Profiles of the Future* discuss the technical limits of everyday technologies such as devices for transport and communication, analyzing the possibilities in a way that is daring but not fantastic. Clarke's discussion of the quest for speed is taken to its limit in chapter seven, about instantaneous travel, or teleportation. Even in the less visionary chapters in the earlier part of the book, he pushes up against the limits of what might be possible and begins raising issues as much philosophical as technical or scientific. He considers, for example, the possibility of a form of teleportation that involves the following steps: detailed scanning of the composition of whatever or whoever is to be teleported; the transmission of the information obtained; and reconstitution at the destination point. He worries about whether the scanned and teleported being would still be the same at the other end of the process, even if it were physically indiscernible to whatever arbitrary degree might be specified.

This kind of speculation leads into such problems as the identity of the multiple selves that could be generated by a technology with these components. Yet, there is nothing incoherent about the combination of devices and processes postulated; it is quite simple to imagine what is involved. Nor are any fundamental physical laws broken, though the process is many orders of magnitude beyond our current abilities to analyze or assemble truly complex objects such as biological organisms. For Clarke, that is not enough to rule out the development of such a technology. He writes, "I suspect that some day we will have to face the problems it raises" (1999a, 74). However, he also suggests that truly instantaneous travel will ultimately be developed and "may involve the very nature of space itself" (74).

By contrast, his discussion of time travel and related concepts (chapter 11) demonstrates the kinds of technology that he *does* rule out as impossible. Clarke considers travel into the past a genuine impossibility, not because the necessary technology cannot be described or because the scientific principles on which it would be based are currently unknown, but because he finds the concept logically incoherent. He falls back on standard arguments about temporal paradoxes, rather than using detailed scientific reasoning.

Similarly, he suggests in chapter 10 that humanity will never truly conquer space, which is so vast and complex as to defy us eternally. Though he is confident that we will traverse the cosmic gulfs between the stars in vehicles that travel near the speed of light, he adds that "the whole concept of interstellar administration or culture is seen to be an absurdity" (1999a, 109) when the distances involved are considered. Even if some method of human transportation could exceed the speed of light and instantaneous interstellar travel were developed, the sheer complexity of the universe, with its incomprehensible number of stars, would defeat detailed understanding. Writing of the prospect of interstellar colonization, Clarke insists that such colonies would lose contact with the home planet and forge their own, isolated, destinies:

> Though Earth will try to keep in touch with her children, in the end all the efforts of her archivists and historians will be defeated by time and distance, and the sheer bulk of the material (1999a, 111).

At the same time, he begins to postulate societies that have moved on from traditional conceptions of human nature and its flourishing, such as Aristotle elaborated two and a half millennia ago with an intuitive appeal that remains almost undiminished. Accounts such as Aristotle's have provided a foundation for both religious and secular ethical systems, but these are now undermined by challenges to "the human" as the deepest source of value.

257

The posthuman element in Clarke's thinking appears when he discusses space travel. Though he does not believe we can conquer the vastness of space, in the sense of administering a universal network of human societies, he does advocate space exploration and colonization. In chapter eight, entitled "Rocket to the Renaissance," he develops the argument that humanity needs a frontier in space for economic and "spiritual" reasons. He foresees a new heroic literature emerging from the challenges and dangers of space travel, and speculates about the effects on architecture, music, and dance if our species became established in extraterrestrial environments with very different physical conditions from Earth's. Never somber in his approach, he also imagines new developments in games and (appropriately for an essay originally published in *Playboy*) eroticism.

In the chapter's closing words, he displays two elements of his thought that are in no way inconsistent – rather, they are complementary – but which make him a complex thinker who sometimes slips through the conceptual nets of his critics. Clarke is an advocate of technological innovation, with a sound grasp of well-established science and real-world engineering techniques, but he is also a visionary, straining at the limits of the possible and speculating about the transformation or supersession of humanity. Both elements are present when he acknowledges a somewhat Aristotelian criticism by Lewis Mumford of the prospect of space colonization, then responds with a striking image that creates one of the many posthuman moments in Clarke's writings:

> But when [Mumford] wrote: 'No one can pretend ... that existence on a space satellite or on the barren face of the Moon would bear any resemblance to human life,' he may well be expressing a truth he had not intended. 'Existence on dry land,' the more conservative fish may have said to their amphibious relatives, a billion years ago, 'will bear no resemblance to piscatorial life. We will stay where we are.'

They did. They are still fish (1999a, 90).

V.

When Clarke discusses the future of communications, manufacturing, and the exploitation of natural resources, his practical engineering bent shows as markedly as the breadth and depth of his vision. He explains his pet interests in loving detail, among them the idea of an elevator into space, the grand engineering concept that is the heart of his prize-winning 1979 novel, *The Fountains of Paradise*. Then he turns to nuclear engineering, for the manufacture of elements that we need in greater supply, and to the implications of replicator technology, by which the composition of any material item might be analyzed, then reproduced by molecular level assembly. Unlike thinkers who have approached the idea more recently, such as Eric Drexler, Clarke sees replicator technology as lying far in the future after "many social revolutions" (Clarke 1999a, 150). He expresses an idealistic hope for such magical-seeming technology:

> [O]ur descendants, no longer cluttered up with possessions, will remember what many of us have forgotten – that the only things in the world that really matter are such imponderables as beauty and wisdom, laughter and love (150).

Clarke is a technological meliorist, an engineer of utopian dreams. He hopes and believes that the coming age of abundance will greatly alter the human condition for the better, rendering irrelevant the struggles for resources and power that have dominated human societies to date, and allowing our descendants to concentrate on truly worthwhile things such as "the search for knowledge and the creation of beauty" (1999a, 83). This attitude to technology is a recurrent theme in Clarke's writing, making him an object of suspicion for those readers who advocate ameliorative change through political activism, rather than technological advancement, and who regard the latter as a distraction or even a tool of oppression.

In *Profiles of the Future*, he acknowledges that he concentrates on changes brought about by technology, rather than by economic or political pressures, but defends this by suggesting that "science will dominate the future even more than it dominates the present" (1999a, 5). He expresses a view that economic and political struggle will diminish in importance:

> [T]he time will come when most of our present controversies on these matters will seem as trivial, or as meaningless, as the theological debates in which the keenest minds of the Middle Ages dissipated their energies. Politics and economics are concerned with power and wealth, neither of which should be the primary, still less the exclusive, concern of adult human beings (5).

Clarke's vision of a utopian future is not altogether unlike Marx's: both look forward to a society of material abundance in which the state has served its purpose and withered away. The difference is that Clarke expects to reach this outcome through a gradual and unguided process of technological development, rather than through political consciousness, class-based revolution, and an interim dictatorship of the revolutionary class. Tom Moylan, a Marxist literary critic, has used bitter language to attack Clarke's utopian dreaming in the passage just quoted:

> This is typical of the cynical, mystified language that expressed the end-of-ideology in the 1950s; by denying politics and economics a place in the world of grown up people and by fetishizing science, the workings of capitalism are covered up. Change is limited to efforts of idealistic technocrats making a "better world" for "everyone" (Moylan 1997, 156).

Well, Clarke's language is far from "cynical"; if anything, it shows a touching idealism, not only about the uses of technology, but also about the capacity of human beings to abandon struggle and competition when these cease to have the rational purpose of obtaining scarce resources. On the other hand, Clarke is not naïve about technological change. In his most impressive display of imagination and nerve, his prediction of a wired, computerized world (chapter 16), he displays remarkable shrewdness about the likely benefits and costs.

His discussion of the future of global communications paints a picture very like the reality emerging in technologically advanced societies at the start of the 21st century. He postulates the development and convergence of communications and information technologies, pointing out, for a start, how much could be accomplished in unifying the world with just two or three high-altitude geosynchronous satellites (that clunky approach with its comic pratfalls):

> [T]he great highway of the ether will be thrown open to the whole world, and all men will become neighbours – whether they like it or not. Any form of censorship, political or otherwise, would be impossible; to jam signals coming down from the heavens is almost as difficult as blocking the light of the stars. The Russians could do nothing to stop their people from seeing the American way of life; on the other hand, Madison Avenue agencies and censorship committees might be equally distressed – though for different reasons – at a nationwide switch to uninhibited telecasts from Montmartre (1999a, 172).

Clarke foresees the benefits and problems of global communications very clearly, discussing their positive effects, while also fearing that content will be leveled down and diversity stifled. He considers the possibility of personal radio transceivers; then, in wonderfully prescient passages, imagines a version of the Internet, with computer addresses analogous to telephone numbers, and files and records in space rented in the memories of computers that could be anywhere on Earth – all this connected by satellite links. He describes possibilities such as telemedicine and the instantaneous transmission of electronic correspondence, making airmail obsolete. As Clarke predicted, we live in such a world, increasingly linked and shaped by computerized systems of communication.

V I .

Clarke has a practical and ethical interest in the advancement of technology to create a world of material abundance in which beauty, curiosity, and heroism may flourish, free of the age-old struggles for resources and power. Beyond this, however, he has a long-range view in which humanity may be superseded by posthuman intelligences of some kind, or may encounter them in space, with a transformational impact on our species, as shown in *2001: A Space Odyssey*. Peter Nicholls writes that Clarke imagines humanity "reaching out like a child to an alien Universe which may treat us as a godlike father would, or may respond with cool indifference"(Nicholls 1993, 232).

Moylan identifies contrasting technological and "mystical" directions in Clarke's work (Moylan 1977, 152), but it is all of a piece. Clarke advocates the exploration and colonization of space, and he prophesies a world of connection and abundance, with powerful new means of communication and manufacture. But this is only a small, if immediately useful and impressive, part of a much broader view of the universe and the future. In some of his work, he goes so far as to speculate that it may be scientifically possible for mind to exist in a form that does not depend upon matter as we know it, though his presentation of this idea in *Profiles of the Future* has an amusing, self-consciously disingenuous, touch:

> Though intelligence can only arise from life, it may then discard it. Perhaps at a later stage, as the mystics have suggested, it may also discard matter, but this leads us in realms of speculations which an unimaginative person like myself would prefer to avoid (Clarke 1999a, 201).

By the time he wrote the essays that comprised *Profiles of the Future*, Clarke had already published a number of works that portrayed disembodied mental beings, including *Against the Fall of Night* (1948), its lengthier, re-written version, *The City and the Stars* (1956) and *Childhood's End* (1954), in which the children of the Earth ultimately rise from their bodies, under alien guidance and set out for the stars. *Childhood's End* displays what might be considered an un-rationalized "mystical" element in Clarke's work and commentators such as Moylan frequently speak of Clarke's contrasting technological and "mystical" directions (Moylan, 1977, 152).

Nonetheless, it is better to see Clarke's work, at least since 1956, as all of a piece. By the time he wrote *The City and the Stars*, he had absorbed any genuinely "mystical" tendency into his practice of pushing the scientific imagination to its logical limits. *The City and the Stars* portrays a disembodied mental being called Vanamonde, shown as the product of advanced scientific experiments far in Earth's future. However, Clarke does not suggest here that mind could exist independently of physical reality in a spooky Cartesian sense. Instead, he speculates that it might be able to be "imprinted" on some physical aspect of space unknown to current science. Interestingly, in *Beyond the Fall of Night*, his authorized sequel to *Against the Fall of Night*, Gregory Benford rationalizes the immaterial minds described in the book as beings composed of complex magnetic fields.[3]

---

3   I have consulted the Orbit paperback edition of *Against the fall of night*, first published 1948, collected in a single volume with Benford's sequel, *Beyond the fall of night*, first published 1991 (London: Orbit, 1992). For Benford's use of minds constructed from magnetic fields, see p. 261. Benford again uses the idea of such a mind in his more recent novel, *Eater* (2000).

Similarly, the *Space Odyssey*[4] series describes the alien intelligences that shaped the evolution of humanity as having passed through radically distinct forms of evolution, from biological life, to a cyborg form with biological brains in machine bodies, then machine intelligence:

First their brains, and then their thoughts alone, they transferred into shining new homes of metal and of plastic. In these, they roamed among the stars. They no longer built spaceships. They *were* spaceships (Clarke 1968, 187).

Ultimately, they transform themselves "into pure energy." Referring to their material machine-bodies, the author informs us: "the empty shells they had discarded twitched for a while in a mindless dance of death, then crumbled into rust" (188). This image reflects the conclusion to *Childhood's End*, but Clarke now gives it a physicalist rationalization.

It is not necessarily mystical to ponder the far future, as Clarke does, or to wonder what forms of life or consciousness will one day supersede humanity. That kind of contemplation may be of no direct social use, but it is consistent with the development of science ever since Copernicus and Galileo. Since we are embedded in and produced by events in the vastness of space and time, everything about us is contingent and impermanent. When we consider the issue coldly, without a touch of mysticism, it is scarcely thinkable that humanity will continue in its current form for the billions of years ahead. Contemplating such distant events is legitimate play for a scientific imagination such as Clarke's.

In *Profiles of the Future*, Clarke offers a choice of posthuman scenarios: between postbiological enhancement of the brain and body (chapter 17) and the supersession of humanity by intelligent machines (chapter 18). Though he sometimes writes of the latter as inevitable, he clearly prefers the former. In the earlier chapter, he states the compelling need for a "mechanical educator," a device or technique capable of manipulating the brain directly, imprinting information to enable us to cope with the ever-increasing explosion of knowledge. If this is not invented, he suggests, "then the line of evolution discussed in the next chapter [i.e., machine evolution] will soon pre-dominate, and the end of human culture will be in sight" (1999a, 182). Such a device, the "Braincap," figures prominently in *3001: The Final Odyssey*.

Chapter 18 is entitled "The Obsolescence of Man." Here, Clarke writes that "[b]iological evolution has given way to a far more rapid process – technological evolution. To put it bluntly and brutally, the machine is going to take over" (1999a, 194). Earlier in the book, Clarke has speculated that the experience of extraterrestrial environments might prompt our evolution into something more than human, as the movement of fish onto dry land led to amphibians, reptiles, and mammals. He sees the challenge of space as one spur to "the evolution of the mechanical" since there is so little of the universe where humans can live "without elaborate protection or mechanical aids" (201).

Clarke imagines "a brief Golden Age when men will glory in the power and range of their new partners" (1999a, 203). He discusses cyborgs and mind/machine symbiosis, then asks, "But how long will this partnership last?" (205). Though he does not envisage an active war between humans and machines, as described in generations of SF novels and movies, he does propose, seriously and plausibly, that humanity is approaching its supersession by mechanical intelligence. One way or another, Clarke envisages the transformation or replacement of humanity as we know it.

---

4   The editions of Clarke's *Space odyssey* series that I have consulted are as follows: *2001: A space odyssey* (London: Hutchinson, 1968); *2010: Odyssey two* (1982; rpt. London: Voyager-HarperCollins, 1997); *2061: Odyssey three* (1988; rpt. London: Voyager-HarperCollins, 1997); *3001: The final odyssey* (London: Voyager-HarperCollins, 1997).

Though Clarke's vision is a unified whole, its articulation is not beyond criticism and argument. In *Profiles of the Future*, he admits to a failure to discuss medical and biological developments in detail and states that there are many possibilities for innovation in this direction, including "intelligent animals to do jobs that could otherwise be performed only by very expensive and sophisticated robots" (1999a, 207). It is noteworthy that the contraceptive pill was invented in the early 1960s, as the first edition of Clarke's book was finding readers. The rapid development and distribution of powerful contraceptive technology provided an initial challenge to Aristotelian assumptions about the nature of "the human," and dramatically altered the western world's sexual mores in the process.

Yet, if this omission is a weakness in *Profiles of the Future*, specifically, it does not necessarily show a gap in Clarke's overall thinking. Some of his fictional works, such as *Rendezvous with Rama* (1973), depict intelligent "superchimps," or "simps," engineered to carry out repetitive or routine tasks aboard space vessels. Unfortunately, he does not explore the obvious issue of what moral obligations we would have towards such creatures. Perhaps more impressively, he imagined something like the 1960s sexual revolution in *Childhood's End*, several years before the pill was developed, refined and marketed.[5]

Others may accuse him of a failure of imagination, or even of nerve, in his assertion that we will never acculturate and administer the vastness of space. At one point, he concludes that "no man will ever turn homewards from Vega to greet again those he knew and loved on Earth" (1999a, 111). This, however, does not take into account the possibility of postbiological intelligences, whether transformed humans or superintelligent machines, discussed later in the book – beings that might be effectively immortal and possess enormously enhanced mental capabilities.

In chapter 15, he notes that there are fundamental limits to miniaturization but there is no limit at the other end of the scale as to how vast a living thing might be: "There may be intellects among the stars as vast as worlds, or suns … or solar systems" (1999a, 169). There may be creatures on a scale to be truly at home in the universe; humanity might evolve into such a race of creatures. If we go with Clarke to the end of the line, infinite possibilities may await us.

All that said, Clarke was and still is a giant in his field, a breakthrough writer and thinker. Other visionaries came after him and elaborated their visions in detailed and attractive ways, among them Eric Drexler, Hans Moravec, Ray Kurzweil, Frank Tipler and a host of science fiction writers, of whom Vernor Vinge may be the most original and Greg Egan the most rigorous. But when we return to Clarke's fiction and essays of the 1950s and 1960s, they remain lyrical, compelling and surprisingly modern and fresh. The directions taken by more recent, more self-consciously transhumanist thinkers show how Clarke led the way. Contemporary cyberculture and current thinking on the transhuman/posthuman often seem like footnotes to Clarke's remarkable speculations.

---

5 *Childhood's end* was first published in 1954. I have consulted the revised edition, which includes a new Foreword by the author and contains changes to the opening chapter to reflect geopolitical events since the first edition (London: Pan, 1990). See p. 60 for Clarke's speculation about a revolution in sexual attitudes.

REFERENCES

Broderick, D. 2000. Still space for Odyssey in 2000. Review supplement, *Weekend Australian* 3-4 June.

Clarke, A. C. 1968. *2001: A space odyssey*. London: Hutchinson.

——— 1993. *The city and the stars*. VGSF edition (first edition 1956). London: Victor Gollancz.

——— 1999a. *Profiles of the future: An inquiry into the limits of the possible*. Millennial Edition (first edition 1962). London: Gollancz.

——— 1999b. *Greetings, carbon-based bipeds!: A vision of the 20th century as it happened*. Ed. Ian T. Macauley. London: Voyager-HarperCollins.

Moylan, T. 1977. Ideological contradiction in Clarke's *The city and the stars*. *Science-Fiction Studies* 4:150-157.

Nagel, T. 1981. The Absurd. In *The meaning of life*, ed. E.D. Klemke. 151-161. New York: Oxford University Press.

Nicholls, P. 1993. Clarke, Arthur C(harles). In *The encyclopedia of science fiction*, 2nd edition, ed. J. Clute and P. Nicholls. 229-232. London: Orbit.

Rabkin, E. S. 1981. Clarke, Arthur C(Charles). In *Twentieth century science fiction writers*, ed. Curtis C. Smith. New York: Macmillan.

Regis, E. 1990. *Great mambo chicken and the transhuman condition: Science slightly over the edge*. London: Viking.

FROM FUTURE SHOCK TO SOCIAL FORESIGHT : RE-CONTEXTUALIZING CYBERCULTURE

"The most interesting puzzle in our times is that we so willingly sleepwalk through the process of reconstituting the conditions of human existence"
WINNER

Richard A. Slaughter

The notion of "future shock" attracted widespread attention in the early 1970s but never became intellectually respectable. What it did do was to help express widely felt concerns about the nature of "changing times." The context in which it arose was that of a rapidly transforming world. As late as the mid-19th century human life was framed by what appeared to be the vast and inexhaustible realm of nature. But by the mid-20th century this relationship had been inverted: humanity had expanded to occupy nearly every niche on the planet and nature was in retreat. Similarly, the products of high industrialism (such as those that were proudly displayed at the Great Exhibition at the Crystal Palace in London in 1851) had been discarded, transformed, miniaturized or transcended. The replacement of old-fashioned radio tubes by tiny and more durable solid state transistors became one symbol of this transformation. Another was the rise of the conservation movement, first in the U.S., then in other places too. For many people the revolutions and changes of the early 20th century overturned their sense of "normalcy," of a predictable and settled social order. Instability became the norm in many domains of social and economic life. Consequently, "the future" no longer appeared normal and natural. It increasingly looked more like some kind of artifact – a consequence of what people did or failed to do. It was this sense of continuing transformation, existential threat and the intuition that the future would be very, very different that Alvin Toffler expressed in *Future Shock*.

Cyberculture is a relatively recent development that explores some aspects of highly technologized near future worlds. This essay seeks to contextualize cyberculture in a wider stream of human responses to transformed futures. It begins with a critical overview of the future shock thesis and attempts to situate this within the mindset of American futurism in the 1970s. It suggests that Toffler's work was one source of ideas that led toward the development of futures studies as a substantive field of inquiry. The essay considers how the latter has evolved into an intellectually robust and pragmatically useful field of inquiry and action. It then looks briefly at the interconnections between futures studies, cyberculture and the "real" future.

SHAPE OF THINGS TO COME

"The shape of things to come" had been a preoccupation of many writers, artists, visionaries and social critics for several centuries. For example, it generated an extensive utopian literature that speculated on future societies. But in the 20th century the utopian impulse was buried under the collective experience of two world wars, the coming of the nuclear bomb and the environmental crisis. Utopia gave way to dystopia – darker visions of futures gone sour. The human race would be overwhelmed by its own procreative powers, by pollution or perhaps by "intelligent" machines that no longer supported its existence. Such collective fears gave rise to responses in at least three domains.

One was popular culture. Science fiction writers from John Brunner to William Gibson and Neal Stephenson produced a variety of highly credible dystopian future novels. Filmmakers explored the cinematic possibilities of dystopia in films such as *Blade Runner*, *Terminator II* and the highly successful *The Matrix*. The term "technofear" was coined to describe much of the work of this genre and suggested that the latter provided ways for the popular imagination to come to grips with some of the implications of rapidly advancing science and technology.

Another domain of response embraces a modern lineage of writers, who articulated concerns about the future(s) of humanity through polemical non-fictional writing. Into this group fall such names as H. G. Wells (who wrote fiction as well), Lewis Mumford, Rachel Carson, Herbert Marcuse, and Theodore Roszak.

Finally there were, and continue to be, a number of more formal disciplinary and institutional responses to a drastically altered civilizational outlook. They include the development of futures studies as a distinct discipline, the rise of a wide variety of explicitly futures-oriented nongovernmental organizations such as the World Future Society and the World Futures Studies Federation, the evolution of Institutions of Foresight (IOFs) and, much later, the emergence of cyberculture.

When *Future Shock* was first published in 1970 it became an instant bestseller. It drew together many of the threads of these challenges and transformations. It also proposed measures for dealing with them.

THE FUTURE SHOCK THESIS

Writing during the late 1960s Toffler summarized this thesis thus:

> In three short decades between now and the turn of the next millennium, millions of psychologically normal people will experience an abrupt collision with the future. Affluent, educated citizens of the world's richest and most technically advanced nations, they will fall victim to tomorrow's most menacing malady: the disease of change. Unable to keep up with the supercharged pace of change, brought to the edge of breakdown by incessant demands to adapt to novelty, many will plunge into future shock. For them the future will have arrived too soon (Cross 1974).

He argued that a new force had entered history – what he called "the accelerative thrust" – and further, that individuals, organizations, society, and the entire world were completely unprepared for dealing with it. This led to a "sharp break with previous experience." We were now living in times that were "no longer normal." At the physical level we were "tampering with the chemical and biological stability of the human race," while at the psychological level we were subjecting whole populations to various forms of over-stimulation via "sensory, cognitive, and decision stress." The main thrust of the argument was that both individuals and societies needed to learn how to adapt to and manage the sources of over-rapid change. In particular, this meant bringing technological innovation under some sort of collective control. The bulk of *Future Shock* is devoted to exploring these themes in different areas of human experience and culture.

The keys to the book, however, lie in the final section, which is devoted to what Toffler termed "Strategies for Survival." Here are four chapters on "coping with tomorrow," "education in the future tense," "taming technology," and "the strategy of social futurism." Here is where Toffler set out his best ideas for responding to the situation he had described. Under "coping" were grouped proposals for "personal stability zones," counseling, halfway houses, the creation of "enclaves of the past" and "enclaves of the future," and the deliberate reinvention of coping rituals.

Possibly the best section in the book is that on education. Here he advanced a powerful critique: "… what passes for education today, even in our 'best' schools and colleges, is a hopeless anachronism …" (360). He then added:

> [F]or all this rhetoric about the future, our schools face backwards towards a dying system, rather than forwards to an emerging new society. Their vast energies are applied to cranking out Industrial Men – people tooled for survival in a system that will be dead before they are (Toffler 1971, 360-361).

The thesis was then advanced that the prime objective of education should be to "… increase the individual's 'cope-ability' – the speed and economy with which he can adapt to continual change …" (Toffler 1971, 364). Central to this was "the habit of anticipation." Assumptions, projections, images of futures would need to become part and parcel of every individual's school experience. Learning contracts would be needed, along with mentors from the adult population. The student's "future-focused role image" (that is, his or her view of their future self) would be nourished along with these capabilities. A democratic "council for the future" was needed in every school. Science fiction was an appropriate form of literature to encourage these capacities.

Regarding technology, Toffler put forward the view that a "powerful strategy in the battle to prevent mass future shock … involves the conscious regulation of scientific advance" (Toffler 1971, 387). For Toffler, "the horrifying truth is that, so far as much technology is concerned, no one is in charge." Hence what was needed was "far more sophisticated criteria for choosing among technologies" (Toffler 1971, 391). The option of what was later to be called an "expert system" named OLIVER was canvassed. Perhaps this would help diminish the demands on people. Overall, serious efforts needed to be devoted to anticipating the consequences of technological developments. Referring to changes in sexual habits consequent upon the contraceptive pill he asserted that:

> We can no longer afford to let such secondary social effects just 'happen'. We must attempt to anticipate them in advance, estimating, to the degree possible, their nature, strength, and timing. Where these effects are likely to be seriously damaging we must also be prepared to block the new technology. It is as simple as that. Technology cannot be permitted to rampage through the society (Toffler 1971, 396).

The writer concluded that "a machinery for screening machines" was needed. This could be created by appointing a "technology ombudsman" as part of an "environmental screen" for protecting society from untoward effects.

The culmination of *Future Shock* is a long final chapter on "the strategy of social futurism." It begins with a rhetorical flourish – "... can one live in a society that is out of control?" – and then goes on to outline some of the social innovations needed to ameliorate change. There is an emphatic call for social indicators:

> A sensitive system of indicators geared to measuring the achievement of social and cultural goals, and integrated with economic indicators, is part of the technical equipment that any society needs before it can successfully reach the next stage of eco-technological development. It is an absolute pre-requisite for post-technocratic planning and change management (Toffler 1971, 413).

A Council of Social Advisers could be created to complement an existing Council of Economic Advisers. The "proliferation of organizations devoted to the study of the future" is noted and their long-term time horizons commented on with approval. "Scientific futurists" would work hand-in-hand with them to explore possible, probable, and preferable futures. In Toffler's view the utopian impulse could be "used as a tool rather than an escape" and used to stimulate the social imagination in pursuit of better futures. But this would need institutional support:

> Scientific futurist institutes must be spotted like nodes in a loose network throughout the entire governmental structure ... so that in every department, local or national, some staff devotes itself to scanning the probable long-term future in its assigned field (Toffler 1971, 423).

In addition:

> We need to train thousands of young people in the perspectives and techniques of scientific futurism, inviting them to share in the exciting venture of mapping probable futures (Toffler 1971, 423).

In what was, perhaps, an unconscious echo of Wells' notion of a "global brain" (Wells 1938, 1971), Toffler suggested that "as the globe is itself dotted with future-sensors, we might consider creating a great international institute, a world futures data bank" (Toffler 1971, 424). This, in turn, would support what Toffler termed "anticipatory democracy." The latter would set up "a continuing plebiscite on the future," simulations of various kinds and "social futures assemblies," all designed to encourage wide participation in social decision-making. Toward the end of the chapter Toffler summarized his position thus:

> This, then, is the ultimate objective of social futurism, not merely the transcendence of technocracy and the substitution of more humane, far-sighted, more democratic planning, but the subjugation of the process of evolution itself to conscious human guidance (Toffler 1971, 438-439).

He added:

> For this is the supreme instant, the turning point in history at which man either vanquishes the process of change or vanishes, at which, from being the unconscious puppet of evolution he becomes either its victim or its master (Toffler 1971, 439).

These ideas and proposals drew widespread attention because they attempted to craft a number of responses to issues and questions that concerned many people but which fell outside the usual problem detection, problem resolution systems of an industrial society. They also drew attention to defects in the way that society operated and considered a range of innovative responses. But most of all, perhaps, what people responded to was the notion that here were clues to a very different future. Here, indeed, was a whole new way of responding to change that people felt they could begin to grasp and possibly use.

FUTURE SHOCK 30 YEARS ON

Three decades later, the underpinnings of many of the ideas advanced in *Future Shock* remain problematic. There is no doubt, however, that the thesis focused many people's attention on futures-related concerns. These included: the difficulties of understanding and responding to complex processes of change; issues of human and environmental adaptation to unprecedented rates of change; the problem of subjecting ever more powerful technologies to some form of effective social control; and, overall, the problem of how to come to terms with the wide range of futures clearly implied by all of the above.

Like others before and since, Toffler rightly argued that these transformations in the conditions of human life were unprecedented in human history. His work aligned with that of countless other people in many countries to help stimulate a range of social responses. Among them were the development of futures studies, the application of futures approaches in education and the growth of future oriented NGOs.

As noted previously, the *Future Shock* thesis portrayed people as being "overwhelmed" by change to a point of widespread dysfunctionality that might prefigure widespread social breakdown. But "change" was seen as a wholly external force, rather than something that worked through specific social formations and through the structures and processes that maintain their interests. Such a diagnosis placed the onus for response rather heavily upon these decontextualized and "shocked" individuals. It overlooked the social entities that were (and remain) complicit in generating and sustaining "change." Overall, this was a disempowering approach that displaced autonomy from individuals and groups into poorly defined and shadowy social locations that could not be readily located or challenged.

Linked with this is the way that Toffler ascribed the prime responsibility for "rapid change" to "technology" – not to the agencies and powers that have the ability to define, focus, develop, market, and apply it. This analysis was mystificatory in effect, though not, I am sure, in intent. While Toffler sought to encourage "social futurism" and "anticipatory democracy," he did so in a way that completely overlooked the difficulties people face in (a) understanding and (b) attempting to intervene in their historical context.

In summary, the *Future Shock* thesis can be seen as an expression of a journalistic view of macro-change from a very particular viewpoint in space and time. It foregrounds the habits of perception that are characteristic of that time and attempts to universalize them. As noted, this framework certainly provided some useful suggestions for possible ways forward. But as an interpretive agenda it was unworkable in practice. Conspicuously lacking were ways of understanding, and coming to grips with, other dysfunctional imbalances in culture. "Change" is only one of them. Meaninglessness, lack of purpose, hyper-materialism, technological narcissism, and spiritual hunger are a few of the others that might be encompassed within a wider view. But *Future Shock* was silent upon them all.

Early American futurist work based on *Future Shock*-type analyses was nothing if not ambitious. It attempted to monitor global trends (some of which were and are poorly understood), act as a societal early warning system, explore and illuminate a bewildering range of possibilities and choices, influence public and private decision-making in a multitude of contexts, and disseminate its ideas and conclusions as widely as possible. Its practitioners wanted to help "create the future."

In view of the enormity of these self-imposed tasks there is a strong case for a low-key, self-effacing mode of discourse hedged around with qualifications of various kinds. But when Edward Cornish wrote of the "great future that we all know is possible," he articulated a deeply felt and widely shared American attitude that "if we can create believable dreams of a better future world, then we can build for that world, for we live in an age when a peaceful, prosperous, and happy world is a genuine possibility." This view reflected a sense of optimism and power that was, perhaps, central to the American experience (Cornish 1980, 15-19). Another, much reprinted, paper exposed the darker side of this sensibility by concluding that "the only possible conclusion is a call to action." Along with, "the task is clear. The task is huge … time is horribly short … today the whole human experiment may hang on the question of how fast we now press for the development of a science for survival" (Platt 1973, 16).

Statements of this nature sprang from the same sources of self-understanding, concern, and limitation as *Future Shock*. They attempted to express the ideals and the fears of much of humankind. But, sympathize though we may, they simply did not travel well, and it is important to understand why. In these, and countless other cases, it was not always clear how, or in exactly what sense, people could begin to exert control over events or act to prevent threatened crises. Regardless of whether the view expressed was optimistic or pessimistic, whether the task was to create utopia or merely to avoid dystopia, something was missing.

People who were deeply involved in particular ways of life, values, logics-in-use, traditions, and so on – people whose world-views differed in many substantial ways from those of futurists – were being asked to cooperate from a great social distance in a demanding series of more or less well-defined tasks that lacked historical precedent, or, so far as they were concerned, contemporary sanction. Thus, generalized "calls to action" were not an effective way to make progress. The implicit view of individuals and societies was an under-dimensioned one that glossed over more than was prudent of the substance of social life and social being.

Then and now most people realize that the future is inherently uncertain and conditional. Its relation to the present is complicated by a host of social, cultural, and ideological factors. The sense in which it may be "built" or "chosen" needs to be clarified in some detail by those with ideas about what it should be like. That this seldom happens is not really a comment upon individuals. It is founded on a universal dilemma. People who are necessarily embedded in their own historicity cannot readily aspire to the almost supernatural (or supra-historical) powers involved. As Radnitzky put it, "what is 'irrational' in human history is that men make their history but … do not know the history they make. They have not yet been able to make it with full consciousness" (Radnitzky 1972, 119). This is a dilemma facing anyone wishing to direct change. To avoid "future shock," to build a "science of survival" or to design a "peaceful, prosperous, and happy world" not only begs a number of very important questions, it also requires the development of more inclusive and enabling forms of consciousness and action.

Hence, during the 1970s and 1980s the presentation of particular futures ideas – and indeed, of the futures field more generally – was marred by exaggeration, by a rather naïve view of human capacities and by over-optimism about the potential for social change. Yet, despite these drawbacks, the future shock thesis also helped to stimulate a number of constructive social responses.

FUTURE SHOCK AS A STIMULUS TO SOCIAL INNOVATION

Toffler was dissatisfied with what he regarded as "technocratic" forms of decision making and social administration. PPBS (planning, programming, budgeting systems) and a president's council set up by Nixon fell a long way short. Rather, he called for a "revolution" in the way long-term social goals were formulated. What he wanted was a "continuing plebiscite on the future." To this end he proposed the creation of what he called "social futures assemblies" throughout America, coupled with a range of social simulation exercises in schools.

Yet Toffler's vivid social imagination exceeded his practical grasp of what would be needed to enable such innovations. To read *Future Shock* 30 years on is to be struck by the disjunction between the power of the vision and the poverty of means. The vision stimulated a number of attempts to set up such assemblies. For example, in Hawaii citizens were polled as to how they saw likely and desired futures. The results were summarized as scenarios in a newsletter and acted out on television. A television vote then followed. A book entitled *Anticipatory Democracy* provided a showcase for ideas and experiments of this kind (Bezold 1978). So there is no doubt that *Future Shock* stimulated the social imagination. But most of Toffler's ideas needed a lot more work before they could be put into practice.

Part of the explanation lies in Toffler's habit of privileging aspects of the outer empirical world (facts, trends, change processes) and overlooking the inner interpretive one (worldviews, paradigms, social interests). In subsequent years it became clear that to carry futures proposals from the realm of ideas into social action requires far more than a description of the organizational forms they might take. What Toffler, and indeed many futurists, overlooked is that *the futures domain is primarily a symbolic one*. To operate successfully within it requires a working familiarity with the language, concepts and frameworks that support future-oriented modes of inquiry and action. While Toffler's research had provided him with an elaborate futures vocabulary and a rich store of futures-related ideas and proposals, most of those reading his work were not able to translate his proposals into action because they could not cross this symbolic gulf. To move from ideas to action in fact requires progress though several "layers of capability" which had not yet been described at that time (see "From future shock to social foresight" later in this chapter). Thus, the main drawback of the future shock thesis was that it did not help people find their way into that domain and hence discover the deeper sources of understanding and insight that Toffler had himself overlooked.

Toffler was equally adamant about the need for technology assessment – and in principle he was right. In the chapter "Taming Technology" he put forward the notion of a "technology ombudsman," a public agency that would investigate complaints about irresponsible applications of various technologies. Closely related to this was the idea of an "environmental screen" that would assess the impacts of technologies before they were adopted. Companies would employ their own "consequence analysis staff" to carry out this kind of work. In both cases it is possible to see one of the starting points of the OTA (Office of Technology Assessment) that was established some years later (only to be axed by Reagan). Similarly, the environmental screen may be seen as a precursor of environmental impact statements, which later became common practice.

In these cases a generous interpretation of the role of *Future Shock* would see it as helping to popularize the need for such arrangements in a rapidly changing society.

On the other hand, since Toffler did not attempt a deeper analysis of the worldviews, presuppositions, ideologies and embedded interests that were driving (and continue to drive) the global system, he was in a weak position to call into question the apparent inevitability of technological advance or to propose means of dealing with it at a constitutive level. Hence his well-meaning suggestions were, in effect, outstripped by vastly more powerful forces.

Legend has it that in 1966 Toffler was involved in one of the first high school courses in futures studies. What is certainly the case is that a few years later he edited a wide-ranging book called *Learning for Tomorrow* in which he collected together articles by many future-oriented educators in the U.S. (Toffler 1974). Here were displayed some of the early formulations of theory, practice, and self understanding that later were incorporated into more durable approaches to futures education. While the book was by no means as successful as the earlier one, it achieved a significant readership in the U.S. and elsewhere.

Toffler's ideas about future-oriented education certainly provided a stimulus to this hitherto neglected area. But, over time, it became increasingly clear that the foundations of futures in education were shaky. A close look at American classrooms during the 1970s and 1980s made it clear that innovative futures work had been successful in practical terms. But a search for durable underpinnings was fruitless for one very simple reason: there were none. The pop-psychology approach taken by Toffler served to initiate, and perhaps to inspire up to a point. But it could not nourish and support. Thus during the time of Reagan and Thatcher futures education initiatives were perceived to be inessential and were widely discarded. It would be some years before a more durable foundation would be constructed and a new wave of future-oriented educational work taken up by other hands and minds elsewhere (Hicks and Slaughter 1998).

In summary, the *Future Shock* saga provided a particular sort of thesis about social change, economic development, the role of technology, and, overall, the ways that organizations and individuals might begin to come to grips with them. But it did so in ways that failed to enable the very category of human agency that it sought to assist. Toffler went on to other work on other projects: *The Third Wave, Powershift, War and Anti-War* and the diminutive but ambitious paperback *Creating a New Civilization* (Toffler, A. and H. 1994). Perhaps the chief outcome of all this activity was to establish Toffler, and as time went by his wife Heidi also, as highly "mediagenic" futurists who not only earned a handsome living with their speculations and proposals, but also were sought out and promoted by politicians such as Newt Gingrich, one-time leader of the U.S. House of Representatives. The path from public engagement and discipline-building to lucrative private consulting is regrettably, however, a common one. It helps to explain why futures studies has taken longer to advance than it might otherwise have done.

DEVELOPMENT OF SUBSTANTIVE FUTURES INQUIRY

Futures studies has been described as:

> [A] field of intellectual and political activity concerning all sectors of the psychological, social, economic, political, and cultural life, aiming at discovering and mastering the extensions of the complex chains of causalities, by means of conceptualizations, systematic reflections, experimentations, anticipations, and creative thinking. Futures studies therefore constitute a natural basis for subnational, national, and international, and both interdisciplinary and

transdisciplinary activities tending to become a new forum for the basis of political decision making (Masini and Samset 1975, 15).

Another definition was offered by Roy Amara in 1981. He saw it as an exploration of possible, probable, and preferable futures (Amara 1981, 25-29). However, by the 1990s it became more appropriate to consider it as an emerging "meta-discipline." "Meta-" because of the way it integrates material, data, ideas, tools, etc. from a wide variety of sources; and "discipline" because when done well it clearly supports disciplined inquiry into the constitution of human futures (Slaughter 1988, 372-385). By the end of the 1990s four main traditions, or paradigmatic ways of framing and approaching futures work, were defined. These were as follows:

- **The empirical/analytic tradition.** This is basically data-driven, positivistic, often corporate and hence identified most strongly with North American sources. The names of Herman Kahn and Julian Simon are often identified with this approach.

- **The critical/comparative tradition.** This is a more socially critical approach which recognizes different approaches to knowledge and its use, and different social interests. It takes a more comparative approach and is linked with this writer, Hazel Henderson and Sohail Inayatullah, among others.

- **The activist/participatory tradition.** This is very much about facilitation and activism. Hence it has links with some of the social movements that are close to futures studies, such as the peace, women's, and environmental movements. The approach is expressed most directly in workshop formats such as those created and implemented by Robert Jungk, Elise Boulding, Warren Zieglar and Joanna Macy.

- **The multicultural/global tradition.** This more recent approach springs from the emergence of futures studies, and its underlying concerns, from many non-western contexts. It has been supported by UNESCO and by the courses run in various countries by the World Futures Studies Federation. Those associated with this arena include Zia Sardar, Tony Stevenson and Sohail Inayatullah, as well as a growing number of non-western futurists.

Besides these four traditions, or paradigms, of futures work, there are also a number of substantive levels at which this work can take place (Slaughter 1993). These include the following:

- **Pop futurism.** This is trite, superficial work. It is media-friendly and often seen in weekend newspaper supplements and on brief television features. It is summed up by statements such as "how science and technology are improving our lives and creating the future." This is the world of the fleeting image and the transient sound-bite. It is eminently marketable, but bereft of theory. It arguably detracts from "real" futures work (that is, work with useful social consequences).

- **Problem-oriented work.** This is more serious work. It looks at the ways that societies and organizations are responding, or should respond, to the challenges of the near-term future. So it is largely about social rules and regulations. It emerges most typically in, for example, environmental legislation and organizational innovations, particularly in business – which often gives the impression of being "stranded" at this level.

- **Critical futures studies.** Critical work attempts to "probe beneath the surface" of social life and to discern some of the deeper processes of meaning-making, paradigm formation, and the active influence of obscured worldview commitments (for example, "growth is good" and "nature is merely a set of utilitarian resources"). It utilizes the tools and insights that have emerged within the humanities and which allow us to "interrogate," question, and critique the symbolic foundations of social life and – this is the real point – hence to discern the grounds of new, or renewed, options. Properly understood, the deconstructive and reconstructive aspects of high-quality futures work balance each other in a productive fusion of methods.

- **Epistemological futures work.** Here is where futures studies merges into the foundational areas that feed into the futures enterprise and provide part of its substantive basis. Hence philosophy, ontology, macro-history, the study of time, cosmology, and other disciplines are all relevant at this deep level.

Thus futures studies has developed breadth and depth over the last three decades. It is now a globally distributed meta-discipline taught in a number of universities and which increasingly figures in strategic decision-making, policy debates, and the emergence of social innovations. In essence, it provides interpretative or propositional knowledge about the future, up-dates this regularly, assesses the quality of emerging understandings, and uses them for a range of socially useful purposes. When *Future Shock* was first published such claims could not have been supported, whereas today they are a reality – a fact demonstrated by work such as Wendell Bell's two-volume opus *The Foundations of Futures Studies* (Bell 1996) in *The Knowledge Base of Futures Studies* series (Slaughter 1996).

Drawing on such sources, futurists "study the future" or "construct forward views." In so doing they open out the future as a symbolically and practically significant ream. The discerning of in-depth understanding from these processes enables us to identify options and choices in the present. The point is to move away from a passive or fatalistic acceptance of an "unknown" future to an active and confident participation in creating positively desired futures. Most futurists believe that the future can be shaped by the careful and responsible exercise of human will and effort. Futurists differ in many of their views, but most agree that individuals, organizations, and cultures that attempt to move into the future blindly are taking unnecessary risks. They support the proposition that we need to understand and apply foresight in our private, public, and professional lives.

FROM FUTURE SHOCK TO SOCIAL FORESIGHT

Future Shock was the attempt of one individual to come to terms with change. But building the disciplines of futures studies and strategic foresight is quintessentially a group process involving many actors in many different places. The progression from individual to social capacity is crucial, and there are four "layers of capability" that can lead us there:

- **The brain/mind system.** Human beings are reflexive creatures and their frame of reference embraces past, present, and future. The ability to think ahead is grounded in this biological inheritance. People use foresight informally every day of their lives. This means that we do not have to introduce some new capacity, merely upgrade an existing one.

- **Futures concepts to enable a futures discourse.** Students who take futures courses for the first time find that the language of futures studies and foresight practice opens up the futures domain so that it becomes a symbolically vital realm of understanding and action.

Futures literacy is created by active immersion in the material and the process of considering this with others. From this emerges a distinct futures discourse that enables the forward view.

▪ **Futures methodologies and tools.** But discourse has its limitations. In order to engage with real-world projects, statistics, extended problem analysis (such as the siting of a new road, airport or power utility) a variety of methodologies are used. Chief among these used to be forecasting. Today the focus is more on scenarios and strategic decision-making. These are two of a much wider set of methods that permit us to extend the discourse into new domains and to tackle complex, extended problems.

▪ **Applications of futures work in purpose built organizations or "niches."** But even with all three of these "layers of capability" functioning in concert, still there would be something missing. Futures work would merely be episodic, rising and falling with demand. So the final step is to embed such work in purpose-built organizational niches, or in what I call IOFs or 'institutions of foresight.'

With applied foresight work occurring in this powerfully grounded, way in many social locations and in organizations of all kinds we may reasonably expect to see the beginnings of social foresight. In systems language it would be an "emergent capacity" of all this work (Slaughter 1996a). Thus, 30 years later, future shock has been superseded by more structured and disciplined approaches to the creation of applied social foresight.

FUTURES STUDIES, CYBERCULTURE, AND THE "REAL" FUTURE

Long-term immersion in the futures domain means that the future is no longer seen merely as an empty space or a blank screen upon which the concerns of the present are projected. It takes on positive symbolic content. One way to get a rapid impression of futures in prospect is to pose a number of key questions and then look for high-quality answers. For example:

▪ What are the main continuities?
▪ What are the main trends?
▪ What are the most important change processes?
▪ What are the most serious problems?
▪ What are the new factors in the pipeline?
▪ What are the key sources of inspiration and hope? (Slaughter 1996b).

In-depth answers to such questions begin to reveal the contours of the "real" future from the viewpoint of a particular observer/interpreter. One such reading is as follows. The globalization process is driven by powerful transnational corporations in pursuit of abstract goals such as: growth, innovation, profit, and shareholder value. These processes are not linked with any notion of social need or human value. Indeed, they have many negative human, social, and environmental impacts. They are stimulating the too-rapid development of a series of technological revolutions that threaten to destabilize all human civilizations and societies. Thus, in a nutshell, the most likely future for humankind at this point is a radically diminished one in a world that is mined out, depleted, polluted, and overwhelmed by technologies that we can neither see nor control (See Broderick 1997 for an overview of the latter).

The central challenge of futures studies, foresight work, and, indeed, social and organizational decision-making in general, is to reveal the contours of this unacceptable world and to generate widespread social discussion about feasible alternatives. Alternatively, we can turn the problem on its head and ask: what are the contours of the next level of civilization, the one beyond the mental/egoic, capitalist/late industrial world of the early 21st century? More simply: how might a more positive future be framed?

Remarkably enough, there is a great deal of high-quality material that deals with these generic alternatives. The work of transpersonal synthesist Ken Wilber is paradigmatic. Wilber establishes an interpretive framework that is based on a "four quadrant" view of evolution. The four quadrants cover the following areas (Wilber 1996):

- **The inner individual world.** This is the inner world of reference of each person. It gives access to identity, feelings, emotions, ideas, meaning, and purpose. It provides an overview of individual stages of development.

- **The outer individual world.** This is the human being as known to science. It refers to those aspects that can be directly observed or measured. It is comprised of the structures and functions that enable biological life and awareness.

- **The outer collective world.** This is the rest of the external world as known to science, engineering, architecture, and so on. It embraces the external natural world and the infrastructure of the built environment.

- **The inner collective world.** This is quintessentially the world of reference of stages of social development, of worldviews, languages, professions, and the like.

Through an elegant usage of these distinctions, Wilber has established four profiles of evolutionary development and a method for understanding the likely "deep structure" of civilizations beyond the present one. This is possible because evolution continues in each of the four quadrants along a partly known path. Hence a post-postindustrial world would likely be a world that strove for balance between its different parts. It would be a multi-leveled world, postmaterialist but not anti-technological. Rather, technology would be required to adapt and "fit" clearly articulated human and social values. The main environmental ethic would be a commitment to stewardship. There could also be a re-spiritualization of worldviews and outlooks, as well as a distinct increase in the conscious capabilities of people and organizations.

The foregoing provides a context in which technology, cyberculture, and "the future" can be looked at afresh. Various writers have speculated on the existence of a "discontinuity" ahead as the overlapping developments in biotech, AI, robotics, the miniaturization of computing, and nanotechnology create wave after wave of disorienting change that can't possibly be modeled (Broderick 1997). Some speculate on the "uploading" of human consciousness into enormously sophisticated computer substrates and the potential for machine intelligence to overtake our own redundant biological models (Kurzweil 1999). But from the viewpoint developed here these responses look like technological narcissism, where parts of the human system become split off from the whole and lead to dystopian nightmare futures.

Thus, a central concern for cyberculture is that it skates so perilously close both to technological narcissism on the one hand and to nihilism on the other. The exploration of cyberspace, various "unreal worlds" by Gibson and his successors, would be less of a concern in the context of a shared moral and epistemological framework. But when that framework has been problematized by postmodernism, mass media saturation, and the rise and rise of compulsive global merchandising, it is legitimate to ask if the human race is in peril of undermining itself. That, perhaps, is the core meaning of the highly popular movie *The Matrix* (Slaughter 2001). At least three readings of the film are possible. One is a straight science fiction plot posing standard questions about "the real" and the possibilities of technological domination. Another is an extended homage to the action film/comic genre, albeit with some novel features (such as genuinely meditative scenes and a hero who transcends genre limitations). A third reading is to see *The Matrix* as a compelling postmodern fable, but one whose images, themes, and story lines lack coherence. Which resonates most directly with our hopes and fears for the future? All, or none? *The Matrix* poses many questions but answers few.

CONCLUSION

In the late 19th century a suite of cigarette cards was commissioned in Paris on "life in the year 2000" (Asimov 1986). It explored the universal applications of mechanical lever technology, the leading-edge technology of the time. In the early 21st century levers have all but disappeared. Few can be unaware that it is the computer that is now projected willy-nilly upon the future, along with all its manifestations. Yet we already know that the computer, as such, is about to disappear from our desks into the virtual world of universal information appliances (Norman 1999).

For most of those involved in futures studies and applied foresight, however, the future is not about massive computing, AI, and a vastly expanded Internet. These comprise only a fragment of Wilber's four-quadrant view. The over-identification of technology with the future is a long-standing bias within western culture. A more balanced view suggests that the "real" future is not primarily about technology at all. Rather, it centrally involves a search for new definitions of our shared humanity. In that context Vernor Vinge's "singularity" and Damien Broderick's "spike" provide only a partial view because they only deal with externalities. That is, they primarily focus on the outer collective domain and largely overlook the inner ones that are involved in the constitution of human and cultural significance. Similarly, Toffler's future shock thesis was partly helpful (in proposing social innovations) and partly diversionary (because it focused on external change, and missed the shaping power of self-reference possessed by all human beings).

Since the 1970s people all over the world have become aware of the need to anticipate likely futures, to avoid undesirable ones, and take greater responsibility for the direction of change. Many individuals have contributed to this process. Thus, Toffler's early formulations have been superceded by the development of futures studies as a discipline and the emergence of applied social foresight. The latter may be defined as "the construction and maintenance of high quality forward views and their application in organizationally useful ways."[1] Progress through the four layers of capability previously outlined suggest that social foresight will develop over time. It will be an "emergent capacity" of widespread human effort. The hitherto obscured futures domain will then become an integral part of everyday thinking, life, and work. Society's views of its likely futures will become thoroughly integrated into its present. Instead of falling into dystopia we will see the emergence of "deep design" in every field and a true flowering of human hope and aspiration.

It is for such reasons that shared attempts to develop more highly evolved forms of society and consciousness seem primary: a constructive approach to engaging with the "real" future has more to do with the pursuit of wisdom than the pursuit of, or avoidance of, any technological

---

1 Australian Foresight Institute flyer (Melbourne: Swinburne University, 2000).

capability whatsoever. A possible resolution, however, is that both can be included in a higher order synthesis. Hence the keys to the future lie in seeking a balance between different aspects of humanity in its world: inner, outer, individual, and collective (Wilber 1996).

REFERENCES

Amara, R. 1981. The futures field 1: Searching for definitions and boundaries. *Futurist* 15(1):25-29.

Asimov, I. 1986. *Futuredays: A nineteenth-century vision of the year 2000*. London: Virgin.

Bell, W. 1996. *Foundations of futures studies*. Vol. 1 and 2. New Brunswick: Transaction Publications.

Bezold, C. (ed.). 1978. *Anticipatory democracy*. New York: Vintage.

Broderick, D. 1997. *The spike: Accelerating into an unimaginable future*. Melbourne: Reed Books.

Cornish, E. 1980. Creating a better future: The role of the world future society. *The Futurist* 14(5):15-19.

Cross, N. (ed.). 1974. *Man made futures*. London: Hutchinson.

Hicks, D. and R. Slaughter (eds.). 1998. *Futures education*. World yearbook of education. London: Kogan Page.

Kurzweil, R. 1999. *The age of spiritual machines*. Sydney: Allen & Unwin.

Masini, E. and K. Samset. 1975. Recommendations of the WFSF General Assembly, *WFSF Newsletter* June:15.

Norman, D. 1999. *The invisible computer*. Cambridge, MA: MIT Press.

Platt, J. 1973. What we must do. In *Search for alternatives*, ed. F. Tugwell. Winthrop Publishers, Boston, MA: Winthrop Publishers.

Radnitzky, G. 1972. *Contemporary schools of metascience*. Goteborg: Akademiforlaget.

Slaughter, R. 1988. Futures studies as an intellectual and applied discipline. *American Behavioral Scientist* 42(3):372-385.

——— 1993. Looking for the real megatrends. *Futures* 25(8):827-849.

——— (ed.). 1996. *The knowledge base of futures studies*. Vols 1-3. Melbourne: Futures Study Centre.

——— 1996a. Futures studies: From individual to social capacity. *Futures* 28(8):751-762.

——— 1996b. Mapping the future. *Journal of Futures Studies* 1(1):5-26.

——— 2001. Review of *The Matrix*. *Futures* 33(8): 209-212.

Toffler, A. 1971. *Future shock*. London: Pan Books. Original edition, London: Bodley Head, 1970.

——— (ed.). 1974. *Learning for tomorrow*. New York: Vintage.

Toffler, A. and H. 1994. *Creating a new civilization*. Atlanta: Turner Publishing.

Wilber, K. 1996. *A brief history of everything*. Melbourne: Hill of Content.

Winner, L. 1986. *The whale and the reactor*. Chicago: University of Chicago Press.

RACING TOWARD THE SPIKE

Damien Broderick

I wish I could show you the real future, in detail, just the way it's going to unfold. In fact, I wish I knew its shape myself. But the unreliability of trends is due precisely to *relentless, unpredictable change*, which makes the future interesting but also renders it opaque.

This important notion has been described metaphorically, both in science fiction and in serious essays, as a "technological singularity" by Professor Vernor Vinge, a mathematician in the Department of Mathematical Sciences, San Diego State University (although a few others had anticipated the insight).[1] "The term 'singularity' tied to the notion of radical change is very evocative," Vinge told me, adding, "I used the term 'singularity' in the sense of a place where a model of physical reality fails. I was also attracted to the term by one of the characteristics of many singularities in General Relativity – namely the unknowability of things close to or in the singularity."[2] In mathematics, singularities arise when quantities go infinite; in cosmology, a black hole is the physical, literal expression of that relativistic effect.

For Vinge, accelerating trends in computer sciences will converge somewhere between 2030 and 2100 to form a wall of technological novelties blocking the future from us. However hard we try, we cannot plausibly imagine what lies beyond that wall. "My 'technological singularity' is really quite limited," Vinge stated. "I say that it seems plausible that in the near historical future,

---

1 See Vinge 1987a, 1988; and especially 1987b. An important source is his address to VISION-21 Symposium, March 30-31, 1993, downloadable from http://www-rohan.sdsu.edu/faculty/vinge/misc/singularity.html or
http://www.frc.ri.cmu.edu/~hpm/book98/com.ch1/vinge.singularity.html (accessed Feb. 3, 2001).
For a general survey of this topic in far greater detail than I can provide in this essay, see Broderick 1997, 2001.

2 Private communication, August, 1996. Vinge's own most recent picture of a plausible 2020, cautiously *sans* singularity, emphasizes the role of embedded computer networks so ubiquitous that finally they link into a kind of cyberspace Gaia, even merge with the original Gaia, that geological and biological macro-ecosystem of the globe (Vinge 2000). However, on the last day of 1999, Vinge told me in an email, "The basic argument hasn't changed."

we will cause superhuman intelligences to exist. Prediction beyond that point is qualitatively different from futurisms of the past. I don't necessarily see any vertical asymptotes." So enthusiasts for this perspective (including me) are taking the idea much further than Vinge, because we *do* anticipate the arrival of an asymptote in the rate of change. Humanity, it is argued, will become first "transhuman" and then "posthuman." Under either interpretation, and unlike many currently fashionable debates, Vinge's singularity is an apocalyptic prospect based on testable science rather than religion or New Age millenarianism.

While Vinge first advanced his insight in works of imaginative fiction, he has featured it more rigorously in such formal papers as his address to the VISION-21 Symposium, sponsored by NASA Lewis Research Center and the Ohio Aerospace Institute, March 30-31, 1993. He opened that paper with the following characteristic statement, which can serve as a fair summary of my own starting point:

> The acceleration of technological progress has been the central feature of this century. I argue in this paper that we are on the edge of change comparable to the rise of human life on Earth. The precise cause of this change is the imminent creation by technology of entities with greater than human intelligence (Vinge 1993).

The impact of that distressing but apparently free-floating prediction is much greater than you might imagine. In 1970, Alvin Toffler had already grasped the notion of accelerating change. In *Future Shock* he noted:

> New discoveries, new technologies, new social arrangements in the external world erupt into our lives in the form of increased turn-over rates – shorter and shorter relational durations. They force a faster and faster pace of daily life (Toffler [1970] 1971).

This is the very definition of "future shock."

Thirty years on, we see that this increased pace of change is going to disrupt the nature of humanity as well, due to the emergence of a new kind of mind: AIs (artificial intelligences). With self-bootstrapping minds abruptly arrived in the world, able to enhance and rewrite their own cognitive and affective coding in seconds, science will no longer be restricted to the slow, limited aperture granted by limited human senses (however augmented by wonderful instruments) and brains (however glorious by the standards of other animals). We'll find ourselves, Vinge suggests, in a world where nothing much can reliably be predicted.

On the other hand –

Is that strictly true? There are some negative constraints we can feel fairly confident about. We *can* be fairly sure that Ptolemaic astronomy is not due for a surprise revival (despite the mind-numbing persistence of astrology, based on a model of the heavens known for the last 400 years to be untrue). The sheer reliability and practical effectiveness of quantum theory, and the robust way relativity holds up under strenuous challenge, argues that they will remain at the core of future science – in *some* form, which is rather baffling, since at the deepest levels they disagree with each other about what kind of cosmos we inhabit.[3] In other words, we do already know a great deal, a tremendous amount; corroborated knowledge will not go away.

---

3 For a simplified discussion, see Weinberg 1999.

Meanwhile, what I call the "Spike" (Vernor Vinge's postulate, that we seem to be approaching a technological singularity) apparently looms ahead of us: a horizon of ever-swifter change we can't yet see past. The Spike is a kind of black hole in the future, created by runaway change and accelerating computer power. We can only try to imagine the unimaginable up to a point. That is what scientists and artists (and visionaries and explorers) have always attempted as part of their job description. So let's see if we sketch a number of possible pathways into and beyond the coming technological singularity.

We need to simplify in order to do that. That is, we need to take just one thread at a time and give it priority, treat it as if it were the only big change, modulating everything else that falls under its shadow. It's a risky gambit, since it has never been true in the past and will not strictly be true in the future. The only exception is the dire (and possibly false) prediction that something we do, or something from beyond our control, brings down the curtain, blows the whistle to end the game. So let's call that option:

[A i] *No Spike, because the sky is falling*

In the second half of the 20th century, people feared that nuclear war (especially nuclear winter) might snuff us all out. Later, with the arrival of subtle sensors and global meteorological studies, we worried about ozone holes and industrial pollution and an anthropogenic greenhouse effect combining to blight the biosphere. Later still, the public became aware of the small but assured probability that our world will sooner or later be struck by a "dinosaur-killer" asteroid, which could arrive at any moment. For the longer term, we started to grasp the cosmic reality of the sun's mortality, and hence our planet's fate: solar dynamics will brighten the sun in the next half billion years, roasting the surface of our fair world and killing everything that still lives upon it. Beyond that, the universe as a whole will surely perish one way or another.

Of course, start thinking seriously about posthumans on the far side of the Spike and you have to wonder where their extraterrestrial equivalents are. Maybe they already suffuse the galaxy, as humankind and our descendants surely will do in another million or two years if there are no ETs out there in prior possession of the real estate.

Those are ways in which we have imagined the *end* of life on Earth, or at least the end of humanity. Take a more optimistic view of things. Suppose we survive as a species, and maybe as individuals, at least for the medium term (forget the asteroids and *Independence Day*). That still doesn't mean there must be a Spike, at least in the next century or two. Perhaps artificial intelligence will be far more intractable than Hans Moravec and Ray Kurzweil and other enthusiasts proclaim. Perhaps molecular nanotechnology stalls at the level of micro-scale machines (MEMS) that have an impact but never approach the fecund cornucopia of a true molecular assembler (a "mint," or Anything Box). Perhaps matter compilers will get developed, but the security states of the world agree to suppress them, imprison or kill their inventors, prohibit their use at the cost of extreme penalties. Then we have:

[A ii] *No Spike, steady as she goes*

This obviously forks into a variety of alternative future histories, the two most apparent being:

[A ii a] *Nothing much ever changes ever again*

This is the day-to-day working assumption I suspect most of us default to, unless we force ourselves to think hard. It's that illusion of unaltered identity that preserves us sanely from year to year, decade to decade; that allows us to retain our equilibrium in a lifetime of such smashing disruption that some people alive now went through the whole mind-wrenching transition from agricultural to industrial to knowledge/electronic societies. It's an illusion, and perhaps a comforting one, but I think we can be pretty sure the future is not going to stop happening just because we've arrived in the 21st century.

The clearest alternative to that impossibility is:

[A ii b] *Things change slowly (haven't they always?)*

Well, no, they haven't. This option pretends to acknowledge the century of vast changes just recalled, but insists that, even so, *human nature itself* has not changed. True, racism and homophobia are increasingly despised rather than mandatory. True, warfare is now widely deplored (at least in rich, complacent places) rather than extolled as honorable and glorious. Granted, people who drive fairly safe cars while chatting on the mobile phone live rather strange lives, by the standards of the horse-and-buggy era only a century behind us. Still, once everyone in the world is drawn into the global market, once peasants in India and villagers in Africa also have mobile phones and learn to use the Internet and buy from IKEA, things will ... *settle down.* Nations overburdened by gasping population pressures will pass through the demographic transition, easily or cruelly, and we'll top out at around 10 billion humans living a modest but comfortable, ecologically sustainable existence for the rest of time (or until that big rock arrives).

A bolder variant of this model is:

[A iii] *Increasing computer power will lead to human-scale AI, and then stall*

But why should technology abruptly stall in this fashion? Perhaps there is some technical barrier to improved miniaturization, or connectivity, or dense, elegant coding (but experts argue that there will be ways around such road blocks, and advanced research points to some possibilities: quantum computing, nano-scale processors). Then again, natural selection has not managed to leap to a superintelligent variant of humankind in the last 100,000 years, so maybe there is some structural reason why brains top out at the Murasaki, Einstein, or van Gogh level.

AI research might reach the low-hanging fruit, all the way to human equivalence, and then find it impossible (even with machine aid) to find a path through murky search space to a higher level of mental complexity. Still, using the machines we already have will not leave our world unchanged. Far from it. And even if this story has some likelihood, a grislier variant seems even more plausible:

[A iv] *Things go to hell, and if we don't die we'll wish we had*

This isn't the nuclear winter scenario, or any other kind of doom by weapons of mass destruction – let alone gray nano goo, which by hypothesis never gets invented in this denuded future. But perhaps such a despairing diagnosis of a world deluged by trash is a kind of optical illusion, projecting a refuse pile in a Philippines slum over the clean, comfortable environments most of us inhabit fairly happily.

Still, those benefits demand a toll from the planet's resource base, and our polluted environment. The rich nations, numerically in a minority, notoriously use more energy and materials than the rest, and pour more crap into air and sea. That can change – *must* change, or we are all in bad trouble – but in the short term one can envisage a nightmare decade or two during which the Third World "catches up" with the wealthy consumers, burning cheap, hideously polluting soft coal, running the exhaust of a billion and more extra cars into the biosphere … Some Green activists mock "technical fixes" for these problems, but those seem to me our best last hope.[4] We are moving toward manufacturing and control systems very different from the wasteful, heavy-industrial, pollutive kind that helped drive up the world's surface temperature by 0.4 to 0.8 degrees Celsius (0.7 to 1.4 degrees Fahrenheit) in the 20th century (National Academies 2000).

Pollsters have noted incredulously during the last decade that people overwhelmingly state that their individual lives are quite contented and their prospects good, while agreeing that the nation or the world generally is heading for hell in a hand-basket. It's as if we've forgotten that the vice and brutality of television entertainments do *not* reflect the true state of the world, that it's almost the reverse: we revel in such violent cartoons because, for almost all of us, our lives are comparatively placid, safe, and measured. If you doubt this, go and live for a while in medieval Paris, or paleolithic Egypt (you're not allowed to be a noble).

ROADS FROM HERE AND NOW TO THE SPIKE

I assert that all of these *No Spike* options are of low probability, unless they are brought forcibly into reality by the hand of some Luddite demagogue using our confusions and fears against our own best hopes for local and global prosperity. If I'm right, we are then pretty much on course for an inevitable Spike. We might still ask: what, exactly, is the motor that will propel technological culture up its exponential curve?

Here are seven obvious distinct candidates for paths to the Spike (separate lines of development that in reality will interact, generally hastening but sometimes slowing each other).

[B i] *Increasing computer power will lead to human-scale AI, and then will swiftly self-bootstrap to incomprehensible superintelligence*

This is the "classic" model of the singularity, the path to the ultra-intelligent machine and beyond. But it seems unlikely that there will be an abrupt leap from today's moderately fast machines to a fully functioning artificial mind equal to our own, let alone its self-redesigned kin – although this proviso, too, can be countered, as we'll see. If we can trust Moore's Law – computer power doubling every year – as a guide (and strictly we can't, since it's only a record of the past rather than an oracle), we get the kinds of timelines presented by Ray Kurzweil, Hans Moravec, Michio Kaku, Peter Cochrane, and others. Let's briefly sample those predictions:

▪ **Peter Cochrane.** Several years ago, the British Telecom futures team, led by their guru Cochrane, saw human-level machines as early as 2016. Their remit did not encompass a sufficiently deep range to sight the singularity (British Telecom 1997).

▪ **Ray Kurzweil.** Around 2019, a standard cheap computer has the capacity of a human brain, and some claim to have passed the Turing test (that is, passed as conscious, fully responsive minds). By 2029, such machines are a thousand times more powerful. Machines not only ace the Turing test, they *claim to be conscious*, and are accepted as such. His sketch of 2099 is effectively a Spike: fusion between human and machine, uploads more numerous than the embodied, immortality (Kurzweil 1999).

---

4   See the late economist Julian Simon's readable and optimistic book *The Ultimate Resource* at http://www.inform.umd.edu/EdRes/Colleges/BMGT/.Faculty/JSimon/Ultimate_Resource/

■ Ralph Merkle. While Merkle's special field is nanotech, this plainly has a possible bearing on AI. His is the standard case, although the timeline is still "fuzzy" (personal communication, January 2000): various computing parameters go about as small as we can imagine between 2010 and 2020, if Moore's Law holds up. To get there will require "a manufacturing technology that can arrange individual atoms in the precise structures required for molecular logic elements, connect those logic elements in the complex patterns required for a computer, and do so inexpensively for billions of billions of gates." So the imperatives of the computer hardware industry will create nanoassemblers by 2020 at the latest.[5]

Choose your own timetable for the resulting Spike once both nano and AI are in hand. Merkle still cites the August 1995 *Wired* poll of experts on several aspects of a timeline to working nano: chemistry professor Robert Birge, materials science professor Donald W. Brenner, nanotechnologist Eric Drexler, computer scientist J. Storrs Hall, and Nobel-prize-winning chemist Richard E. Smalley. Here are their now antiquated but interesting estimates:

	Birge	Brenner	Drexler	Storrs Hall	Smalley
Molecular assembler:	2005	2025	2015	2010	2000
Nanocomputer:	2040	2040	2017	2010	2100
Cell repair:	2030	2035	2018	2050	2010
Commercial product:	2002	2000	2015	2005	2000

■ Hans Moravec. Multipurpose "universal" robots by 2010, with "humanlike competence" in cheap computers by around 2039 – a more conservative estimate than Ray Kurzweil's but astonishing nonetheless. Even so, he considers a Vingean singularity as likely within 50 years (Moravek 1998).

■ Michio Kaku. No computer expert, physicist Kaku surveyed some 150 scientists and devised a profile of the next century and beyond. He concludes broadly that from "2020 to 2050, the world of computers may well be dominated by invisible, networked computers which have the power of artificial intelligence: reason, speech recognition, even common sense" (Kaku 1998, 28). In the next century or two, he expects humanity to achieve a Type I Kardeshev civilization, with planetary governance and technology able to control weather but essentially restricted to Earth. Only later, between 800 and 2500 years further on, will humanity pass to Type II, with command of the entire solar system. Once the dream of science fiction, this must now be seen as excessively conservative.

■ Vernor Vinge. His part-playful, part-serious proposal was that a singularity was imminent, at around 2020, marking the end of the human era. Maybe as soon as 2014.

■ Eliezer Yudkowsky. Once we have a human-level AI able to understand and redesign its own architecture, there will be a swift escalation into a Spike. Could be as soon as 2010, with 2005 and 2020 as the outer limits, if the Singularity Institute has anything to do with it (see option [C], below).

---

5 See Merkle's home page at http://merkle.com (accessed Feb. 3, 2001).

[B ii] Increasing computer power will lead to direct augmentation of human intelligence and other abilities

Why build an artificial brain when we each have one already? Well, it is regarded as impolite to delve intrusively into a living brain purely for experimental purposes, whether by drugs or surgery (sometimes dubbed "neurohacking"), except if no other course of treatment for an illness is available. Increasingly subtle scanning machines are now available, allowing us to watch as the human brain does its stuff, and a few brave pioneers are coupling chips to parts of themselves, but few expect us to wire ourselves to machines in the immediate future. That might be mistaken, however. Professor Kevin Warwick of Reading University successfully implanted a sensor-trackable chip into his arm in 1998. A year later, he allowed an implanted chip to monitor his neural and muscular patterns, then had a computer use this information to copy the signals back to his body and cause his limbs to move; he was thus a kind of puppet, driven by the computer signals. He plans experiments where the computer, via similar chips, takes control of his emotions as well as his actions (BBC 1999).

As we gradually learn to read the language of the brain's neural nets more closely, and finally to write directly back to them, we will find ways to expand our senses, directly experience distant sensors and robot bodies (perhaps giving us access to horribly inhospitable environments like the depths of the oceans or the blazing surface of Venus). Instead of hammering keyboards or calculators, we might access chips or the global Net directly via implanted interfaces. Perhaps sensitive monitors will track brainwaves, myoelectricity (muscle signals) and other indices, and even impose patterns on our brains using powerful, directed magnetic fields. Augmentations of this kind, albeit rudimentary, are already seen at the lab level. Perhaps by 2020 we'll see boosted humans able to share their thoughts directly with computers. If so, it is a fair bet that neuroscience and computer science will combine to map the processes and algorithms of the naturally evolved brain, and try to emulate it in machines. Unless there actually is a mysterious non-replicable spiritual component, a soul, we'd then expect to see a rapid transition to self-augmenting machines – and we'd be back to path [B i].

[B iii] Increasing computer power and advances in neuroscience will lead to rapid uploading of human minds

On the other hand, if [B ii] turns out to be easier than [B i], we would open the door to rapid uploading technologies. Once the brain/mind can be put into a parallel circuit with a machine as complex as a human cortex (available, as we've seen, somewhere 2020 and 2040), we might expect a complete, real-time emulation of the scanned brain to be run inside the machine that's copied it. Again, unless the "soul" fails to port over along with the information and topological structure, you'd then find your perfect twin (although grievously short on, ahem, a body) dwelling inside the device.

Your uploaded double would need to be provided with adequate sensors (possibly enhanced, compared with our limited eyes and ears and tastebuds), plus means of acting with ordinary intuitive grace on the world (via physical effectors of some kind – robotic limbs, say, or a robotic telepresence). Or perhaps your upload twin would inhabit a cyberspace reality, less detailed than ours but more conducive to being rewritten closer to your heart's desire. Such VR protocols should lend themselves readily to life as an uploaded personality.

Once personality uploading is shown to be possible and tolerable or, better still, enjoyable, we can expect at least some people to copy themselves into cyberspace. How rapidly this new world is colonized will depend on how expensive it is to port somebody there, and to sustain them. Computer storage and run-time should be far cheaper by then, of course, but still not entirely free. As economist Robin Hanson has argued, the problem is amenable to traditional economic analysis. "I see very little chance that cheap, fast upload copying technology would not be used to cheaply create so many copies that the typical copy would have an income near 'subsistence' level." On the other hand, "If you so choose to limit your copying, you might turn an initial nest egg into fabulous wealth, making your few descendants very rich and able to afford lots of memory."[6]

If an explosion of uploads is due to occur quite quickly after the technology emerges, early adopters would gobble up most of the available computing resources. But this assumes that uploaded personalities would retain the same apparent continuity we fleshly humans prize. Being binary code, after all (however complicated), such people might find it easier to alter themselves – to rewrite their source code, so to speak – and to link themselves directly to other uploaded people, and AIs if there are any around. This looks like a recipe for a Spike to me. How soon? It depends. If true AI-level machines are needed, and perhaps medical nanotechnology to perform neuron-by-neuron, synapse-by synapse brain scanning, we'll wait until both technologies are out of beta-testing and fairly stable. That would be 2040 or 2050, I'd guesstimate.

[B iv] *Increasing connectivity of the Internet will allow individuals or small groups to amplify the effectiveness of their conjoined intelligence*

Routine disseminated software advances will create (or evolve) ever smarter and more useful support systems for thinking, gathering data, writing new programs. The outcome will be an "in-one-bound-Jack-was-free" surge into AI. That is the garage band model of a singularity, and while it has a certain cheesy appeal, I very much doubt that's how it will happen.

But the Internet is growing and complexifying at a tremendous rate. It is possible that one day, as Arthur C. Clarke suggested decades ago of the telephone system, it will just *wake up*. After all, that's what happened to a smart African ape, and unlike computers it and its close genetic cousins weren't already designed to handle language and mathematics.

[B v] *Research and development of microelectromechanical systems (MEMS) and fullerene-based devices will lead to industrial nanoassembly, and thence to "anything boxes"*

Here we have the "classic" molecular nanotechnology pathway, as predicted by Drexler's Foresight Institute and NASA,[7] but also by the mainstream of conservative chemists and adjacent scientists working in MEMS, and funded nanotechnology labs around the world. In a 1995 *Wired* article, Eric Drexler predicted nanotechnology within 20 years. Is 2015 too soon? Not, surely, for the early stage devices under development by Zyvex Corporation in Texas, who hope to have at least preliminary results by 2010, if not sooner.[8] For many years AI was granted huge amounts of research funding, without much result (until recently, with a shift in direction and the wind of Moore's Law at its back). Nano is now starting to catch the research dollars, with substantial

---

6 Personal communication, December 8, 1999.

7 See Nanotechnology Team, Home Page 2000, at http://www.nas.nasa.gov/Groups/SciTech/nano/index.html (accessed Oct. 5, 2000).

8 See Zyvex home page at http://www.zyvex.com (accessed Feb. 2001).

investment from governments (half a billion promised by U.S. President Clinton, in Japan, even Australia) and mega-companies such as IBM. The prospect of successful nanotech is exciting, but should also make you afraid, very afraid. If nano remains (or rather, becomes) a closely guarded national secret, contained by munitions laws, a new balance of terror might take us back to something like the Cold War in international relations – but this would be a polyvalent, fragmented, perhaps tribalist balance. Or building and using nanotech might be like the manufacture of dangerous drugs or nuclear materials: centrally produced by big corporations' mints, under stringent protocols (you hope, fearful visions of Homer Simpson's nuclear plant dancing in the back of your brain), except for those in Colombia and the local bikers' fortress.

Or it might be a Ma & Pa business: a local plant equivalent, perhaps, to a used car yard, with a fair-sized raw materials pool, mass transport to shift raw or partly processed feed stocks in, and finished product out. This level of implementation might resemble a small Internet server, with some hundreds or thousands of customers. One might expect the technology to grow more sophisticated quite quickly, as minting allows the emergence of cheap and amazingly powerful computers. Ultimately, we might find ourselves with the fabled "anything box" in every household, protected against malign uses by an internal AI system as smart as a human, but without human consciousness and distraction. We should be so lucky. But it could happen that way.

A quite different outcome is foreshadowed in a prescient novel by Damon Knight, *A for Anything* (1990, originally published in 1959 as *The People Maker*), in which a "matter duplicator" leads not to utopian prosperity for all but to cruel feudalism, a regression to brutal personal power held by those clever thugs who manage to monopolize the device. A slightly less dystopian future is portrayed in Neal Stephenson's satirical but seriously intended *The Diamond Age* (1995), where tribes and nations and new optional tetherings of people under flags of affinity or convenience tussle for advantage in a world where the basic needs of the many poor are provided free, but with galling drab uniformity, at street corner matter compilers owned by authorities. That is one way to prevent global ruination at the hands of crackers, lunatics, and criminals, but it's not one that especially appeals – if an alternative can be found.

Meanwhile, will nanoassembly allow the rich to get richer – to hug this magic cornucopia to their selfish breasts – while the poor get poorer? Why should it be so? In a world of 10 billion flesh-and-blood humans (ignoring the uploads for now), there is plenty of space for everyone to own decent housing, transport, clothing, arts, music, sporting opportunities … once we grant the ready availability of nano mints. Why would the rich permit the poor to own the machineries of freedom from want? Some optimists adduce benevolence, others prudence. Above all, perhaps, is the basic law of an information/knowledge economy: the more people you have thinking and solving and inventing and finding the bugs and figuring out the patches, the better a nano minting world is for everyone (just as it is for an open-source computing world). Besides, how could they stop us?[9] (Well, by brute force, or in the name of all that's decent, or for our own moral good. None of these methods will long prevail in a world of free-flowing information and cheap material assembly. Even China has trouble keeping dissidents and mystics silenced.)

The big necessary step is the prior development of early nano assemblers, and this will be funded by university and corporate (and military) money for researchers, as well as by increasing numbers of private investors who see the marginal pay-offs in owning a piece of each consecutive

---

9   Some thoughts on the difficulty of containing nanotech (with some comparisons to software piracy "warez"), and the likely evaporation of our current economy, can be found at http://www.cabell.org/Quincy/Documents/Nanotechnology/hello_nanotechnology.html:

improvement in micro- and nano-scale devices. So yes, the rich will get richer – but the poor will get richer too, as by and large they do now, in the developed world at least. Not *as* rich, of course, nor *as* fast. By the time the nano and AI revolutions have attained maturity, these classifications will have shifted ground. Economists insist that rich and poor will still be with us, but the metric will have changed so drastically, so strangely, that we here-and-now can make little sense of it.

[B vi] *Research and development in genomics (the Human Genome Project, etc.) will lead to new "wet" biotechnology, lifespan extension, and ultimately to transhuman enhancements*

This is a rather different approach, and increasingly I see experts arguing that it is the short-cut to mastery of the worlds of the very small and the very complex. "Biology, not computing!" is the slogan. After all, bacteria, ribosomes, viruses – cells for that matter – already operate beautifully at the micro- and even the nano-scales.

Still, even if technology takes a major turn away from mechanosynthesis and "hard" minting, this approach will require a vast armory of traditional and innovative computers and appropriately ingenious software. The IBM petaflop project Blue Gene (doing a quadrillion operations a second) will be a huge system of parallel processors designed to explore protein folding, crucial once the genome projects have compiled their immense catalogue of genes. Knowing a gene's recipe is little value unless you know, as well, how the protein it encodes twists and curls in three-dimensional space. That is the promise of the first couple of decades of the 21st century, and it will surely unlock many secrets and open new pathways.

Exploring those paths will require all the help molecular biologists can get from advanced computers, virtual reality displays, and AI adjuncts. Once again, we can reasonably expect those paths to track right into the foothills of the Spike. Put a date on it? Nobody knows – but recall that DNA was first decoded in 1953, and by around half a century later the whole genome is in the bag. How long until the next transcendent step: complete understanding of all our genes, how they express themselves in tissues and organs and abilities and behavioral bents, how they can be tweaked to improve them dramatically? Cautiously, the same interval: around 2050. More likely (if Moore's Law keeps chugging along), half that time: 2025 or 2030. The usual timetable for the Spike, in other words.

[C] *The singularity happens when we go out and make it happen*

That's Eliezer Yudkowsky's sprightly, in-your-face declaration of intent, which dismisses as jejune and uncomprehending all the querulous cautions about the transition to superintelligence and the singularity on its far side (Yudkowsky 2000). Just getting to human-level AI, this analysis claims, is enough for the final push to a Spike. How so? Don't we need unique competencies to do that? Isn't the emergence of ultra-intelligence, either augmented-human or artificial, the very definition of a Vingean singularity?

Yes, but this is most likely to happen when a system with the innate ability to view and reorganize its own cognitive structure gains the conscious power of a human brain. A machine might have that facility: since its programming is listable, you could literally print it out – in many, many volumes – and check each line. Not so an equivalent human, with our protein spaghetti brains, compiled by gene recipes and chemical gradients rather than exact algorithms; we clearly just can't do that.

So intelligent design turned back upon itself, a cascading multiplier that has no obvious bounds. The primary challenge becomes software, not hardware. The raw petaflop end of the project is chugging along nicely now, mapped by Moore's Law, but even if it tops out, it doesn't matter. A self-improving seed AI could run glacially on a limited machine substrate. The point is, so long as it has the capacity to improve itself, at some point it will do so convulsively, bursting through any architectural bottlenecks to *design* its own improved hardware, maybe even build it (if it's allowed control of tools in a fabrication plant). So what determines the arrival of the singularity is just the amount of effort invested in getting the original seed software written and debugged.

This particular argument is detailed in Yudkowsky's ambitious web documents "Coding a Transhuman AI," "Singularity Analysis," and "The Plan to Singularity." It doesn't matter much, though, whether these specific plans hold up under detailed expert scrutiny; they serve as an accessible model for the process we're discussing. Here we see conventional open-source machine intelligence, starting with industrial AI, leading to a self-rewriting seed AI which runs right into take-off to a singularity. You'd have a machine that combines the brains of a human (maybe literally, in coded format, although that is not part of Yudkowsky's scheme) with the speed and memory of a shockingly fast computer. It won't be like anything we've ever seen on earth. It should be able to optimize its abilities, compress its source code, turn its architecture from a swamp of mud huts into a gleaming, compact, ergonomic office (with a spa and a bar in the penthouse, lest we think this is all grim earnestness).[10] Here is quite a compelling portrait of what it might be like, "human high-level consciousness and AI rapid algorithmic performance combined synergetically," to be such a machine:

> Combining Deep Blue with Kasparov … yields a Kasparov who can wonder 'How can I put a queen here?' and blink out for a fraction of a second while a million moves are automatically examined. At a higher level of integration, Kasparov's conscious perceptions of each consciously examined chess position may incorporate data culled from a million possibilities, and Kasparov's dozen examined positions may not be consciously simulated moves, but 'skips' to the dozen most plausible futures five moves ahead.[11]

Such a machine, we see, is not really human-equivalent after all. If it isn't already transhuman or superhuman, it will be as soon as it has hacked through its own code and revised it (bit by bit, module by module, making mistakes and rebooting and trying again until the whole package comes out right.) If that account has any validity, we also see why the decades-long pauses in the timetables cited earlier are dubious, if not preposterous. Given a human-level AI by 2039, it is not going to wait around biding its time until 2099 before creating a discontinuity in cognitive and technological history. That will happen quite fast, since a self-optimizing machine (or upload, perhaps) will start to function so much faster than its human colleagues that it will simply leave them behind, along with Moore's plodding Law. A key distinguishing feature, if Yudkowsky's analysis is sound, is that we never see HAL, the autonomous AI in the movie *2001*. All we will see is AI specialized to develop software.

Since I don't know the true shape of the future any more than you do, I certainly don't know whether an AI or nano-minted singularity will be brought about (assuming it does actually occur) by careful, effortful design in an Institute with a Spike engraved on its door, by a congeries of industrial and scientific research vectors, or by military ambitions pouring zillions of dollars

---

10 Sorry, that's me again; Yudkowsky didn't say it.

11 See Yudkowsky 1998, at http://www.pobox.com/~sentience/AI_design.temp.html#view_advantage

into a new arena that promises endless power through mayhem, or mayhem threatened. It does strike me as excessively unlikely that we will skid to a stop anytime soon, or even that a conventional utopia minus any runaway singularity sequel (*Star Trek*'s complacent future, say) will roll off the mechano-synthesizing assembly line.[12]

Are there boringly obvious technical obstacles to a Spike? Granted, particular techniques will surely saturate and pass through inflexion points, tapering off their headlong thrust. If the past is any guide, new improved techniques will arrive (or be forced into reality by the lure of profit and sheer curiosity) in time to carry the curves upward at the same acceleration. If not? Well, then, it will take longer to reach the Spike, but it is hard to see why progress in the necessary technologies would simply *stop*.

Well, perhaps some of these options will become technically feasible but remain simply unattractive, and hence bypassed. Dr. Russell Blackford, a lawyer, former industrial advocate and literary theorist who has written interestingly about social resistance to major innovation, notes that manned exploration of Mars has been a technical possibility for the past three decades, yet that challenge has not been taken up (Blackford 1998). Video conferencing is available but few use it (unlike the instant adoption of mobile phones.) While a concerted program involving enough money and with widespread public support could bring us conscious AI by 2050, he argues, it won't happen. Conflicting social priorities will emerge; the task will be difficult and horrendously expensive. Are these objections valid? AI and nano need not be impossibly hard and costly, since they will flow from current work powered by Moore's Law improvements. Missions to Mars, by contrast, have no obvious social or consumer or even scientific benefits beyond their simple feel-good achievement. Profound science can be done by remote vehicles. By contrast, minting and AI or IA will bring immediate and copious benefits to those developing them – and will become less and less expensive, just as desktop computers have.

What of social forces taking up arms against this future? We've seen the start of a new round of protests and civil disruptions aimed at genetically engineered foods and work in cloning and genomics, but not yet targeted at longevity or computing research. It will come, inevitably. We shall see strange bedfellows arrayed against the machineries of major change. The only question is how effective the protests will be.

In 1999, for example, emeritus professor Alan Kerr, winner of the lucrative inaugural Australia Prize for his work in plant pathology, radio-broadcast a heartfelt denunciation of the Greens' adamant opposition to new genetically engineered crops that allow use of insecticide to be cut by half (Kerr 1999). Some aspects of science, though, did concern Dr. Kerr. He admitted he'd been "scared witless" by:

> the thesis is that within a generation or two, science will have conquered death and that humans will become immortal. Have you ever thought of the consequences to society and the environment of such an achievement? If you're anything like me, there might be a few sleep-

---

12 To be fair, the *Star Trek* franchise has always made room for alien civilizations that have passed through a singularity and become as gods. It's just that television's notion of post-Spike entities stops short at mimicry of Greek and Roman mythology (Xena the Warrior Princess goes to the future), spiritualized transformations of humans into a sort of angel (familiar also from *Babylon-5*), down-market cyberpunk collectivity (the Borg), or sardonic whimsy (the entertaining character Q, from the Q dimension, where almost anything can happen and usually does.) It's hard not to wonder why immortality is not assured by the transporter or the replicator, which can obviously encode a whole person as easily as a piping hot cup of Earl Grey tea, or why people age and die despite the future's superb medicine. The reasons, obviously, have nothing to do with plausible extrapolation and everything to do with telling an entertaining tale, using a range of contemporary human actors, that appeals to the largest demographic and ruffles as few feathers as possible while still delivering some faint frisson of difference and future shock.

less nights ahead of you. Why don't the greenies get stuck into this potentially horrifying area of science, instead of attacking genetic engineering with all its promise for agriculture and the environment?[13]

This, I suspect, is a short-sighted and ineffective diversionary tactic. It will arouse confused opposition to life extension and other beneficial ongoing research programs, but will lash back as well against any ill-understood technology.

Cultural objections to AI might emerge, as venomous as yesterday's and today's attacks on contraception and abortion rights, or anti-racist struggles. If opposition to the Spike, or any of its contributing factors, gets attached to one or more influential religions, that might set back or divert the current. Alternatively, careful study of the risks of general assemblers and autonomous artificial intelligence might lead to just the kinds of moratoriums that Greens now urge upon genetically engineered crops and herds. Given the time lag we can expect before a singularity occurs – at least a decade, and far more, probably two or three – there's room for plenty of informed specialist and public debate. Just as the basic technologies of the Spike will depend on design-ahead projects, so too we'll need a kind of think-ahead program to prepare us for changes that might, indeed, scare us witless. And of course, the practical impact of new technologies condition the sorts of social values that emerge; recall the subtle interplay between the oral contraceptive pill and sexual mores, and the swift, easy acceptance of in vitro conception.

Despite these possible impediments to the arrival of the Spike, I suggest that while it might be delayed, it's almost certainly not going to be halted. If anything, the surging advances I see every day coming from labs around the world convince me that we already are racing up the lower slopes of its curve into the incomprehensible. In short, it makes little sense to try to pin down the future. Too many strange changes are occurring already, with more lurking just out of sight, ready to leap from the equations and surprise us. True AI, when it occurs, might rush within days or months to SI (superintelligence), and from there into a realm of Powers whose motives and plans we can't even start to second-guess. Nano minting could go feral or worse, used by crackpots or statesmen to squelch their foes and rapidly smear us all into paste. Or sublime AI Powers might use it to the same end, recycling our atoms into better living through femtotechnology.

The single thing I feel confident of is that one of these trajectories will start its visible run up the right-hand side of the graph within 10 or 20 years, and by 2030 (or 2050 at latest) will have put everything we hold self-evident into question. We will live forever; or we will all perish most horribly; our minds will emigrate to cyberspace, and start the most ferocious overpopulation race ever seen on the planet; or our machines will Transcend and take us with them, or leave us in some peaceful backwater where the meek shall inherit the Earth. Or something else, something far weirder and … *unimaginable*. Don't blame me. That's what I promised you.

---

13 The book that frightened Dr. Kerr was my *The Last Mortal Generation* (Broderick 1999). *The Spike* (1997), by contrast, would surely shock him rigid.

REFERENCES

BBC 1999. Scientist puts computer in control. *BBC News Online* Nov. 3. At
    http://news.bbc.co.uk/hi/english/sci/tech/newsid_503000/503552.stm (accessed Oct. 5, 2000).

British Telecom. 1997. *Technology calendar.* At http://btlabs1.labs.bt.com/library/on-line/calendar/index.htm (accessed Feb. 7, 2001).

Blackford, R. 1998. Singlularity shadow. *Quadrant* 347 (June): 24-29.

Broderick, D. 1997. *The spike: Accelerating into the unimaginable future.* Melbourne: Reed Books/New Holland.

———— 1999. *The last mortal generation.* Sydney: New Holland.

———— 2001. *The spike: Accelerating into the unimaginable future.* (Revised and updated edition.) New York: Forge.

Kaku, Michio. 1998. *Visions: How science will revolutionize the 21st century and beyond.* Oxford: Oxford University Press

Kerr, A. 1999. Genetically engineering crops. Interview with Robyn Williams, *Ockham's Razor.* ABC Radio National Sept 26.
    http://abc.net.au/rn/science/ockham/stories/s54399.htm (accessed Oct. 5, 2000).

Knight, Damon. 1990. *A for Anything.* New York: Tor Books.

Kurzweil, R. 1999. *The age of spiritual machines: When computers exceed human intelligence.* Sydney: Allen & Unwin.

Moravek, H. 1998. *ROBOT: Mere machine to transcendent mind.* Oxford: Oxford University Press. Also at
    http://www.frc.ri.cmu.edu/~hpm/book98/

Nanotechnology Team, Home Page 2000. Numerical Aerospace Simulation Systems Division (NAS) of the National Aeronautics
    and Space Administration (NASA). At http://www.nas.nasa.gov/Groups/SciTech/nano/index.html (accessed Oct. 5, 2000).

National Academies 2000. New evidence helps reconcile global warming discrepancies (media release). Jan. 16. At
    http://www4.nationalacademies.org/news.nsf/isbn/0309068916?OpenDocument (accessed Oct. 5, 2000).

Simon, J. 1998. *The ultimate resource.* College of Business and Management, University of Maryland. At
    http://www.inform.umd.edu/EdRes/Colleges/BMGT/.Faculty/JSimon/Ultimate_Resource/ (accessed Oct. 5, 2000).

Stephenson, N. 1995. *The Diamond Age.* New York: Bantam Books.

Toffler, A. 1971. *Future Shock.* London: Pan Books. Original edition, London: Bodley Head, 1970.

Vinge, Vernor. 1987a. *True names ... and other dangers.* New York: Baen Books

———— 1987b. *Marooned in Realtime.* London: Pan Books

———— 1988. *Threats ... and other promises.* New York: Baen Books

———— 1993. Address to VISION-21 Symposium, March 30-31. Downloadable from
    http://www-rohan.sdsu.edu/faculty/vinge/misc/singularity.html (accessed Feb. 3, 2001)
    http://www.frc.ri.cmu.edu/~hpm/book98/com.ch1/vinge.singularity.html (accessed Feb 3, 2001).

———— 2000. The digital Gaia. *Wired* (January):74-78.

Weinberg, S. 1999. A unified physics by 2050? *Scientific American* (December):36-43.

Yudkowsky, E. 1998. Coding a transhuman AI. At http://www.pobox.com/~sentience/AI_design.temp.html (accessed Oct. 5, 2000).
    Updated 2000 (Coding a transhuman AI 2. At http://singinst.org/CaTAI.html [accessed Feb. 3, 2001]).

———— 2000. Staring into the singularity. At http://pobox.com/~sentience/singularity.html or
    http://sysopmind.com/singularity.html (accessed Feb. 3, 2001).

CODA

MEMORIES OF THE FUTURE: EXCAVATING THE JET AGE AT THE TWA TERMINAL

MEMORIES OF THE FUTURE :
EXCAVATING THE JET AGE AT THE TWA TERMINAL

"One thing I miss is the time when America had big dreams about the future. Every year there'd
be a new scientific discovery or new invention and everyone would think the future was going
to be even more wonderful than ever. Now it seems like nobody has big hopes for the future.
We all seem to think that it's going to be just like it is now, only worse"
ANDY WARHOL, AMERICA

"It is the year 2000 and I don't see any flying cars! I was promised flying cars. I don't see any
flying cars! *Why? Why? Why?*"
MAN IN AN IBM COMMERCIAL

Mark Dery

ROCKET GARDEN

It's 2002; do you know where your future is?

The future is obsolete, as quaintly archaic as the mothballed spacecraft in the Kennedy
Space Center's Rocket Garden – relics of a more optimistic decade, when the romance of boldly
going where none had gone before, immortalized in footage of Apollo rockets thundering heaven-
ward to the crackling chatter of mission control, held the mass imagination captive. Now, in the
wake of Three-Mile Island and Love Canal, Bhopal and Chernobyl, we're losing our religion.
Broken faith in the forward march of big science and technological progress has dampened
American enthusiasm for extraterrestrial adventures, especially at a time when the space program's annual operating expenses have mounted into the billions (Wade 1993, 34). In the United
States, people are voting with their remote controls. The viewing audience – American for "citi-
zenry" – can't be bothered to watch the latest shuttle maneuvers, weary of what the science writer
Nicholas Wade calls the "weightless and often witless antics" of astronauts (34).

We are, as the SF writer J. G. Ballard says, looking back on the Space Age, and the heady
visions of empire, inexhaustible resources, and push-button solutions to social ills it once sym-
bolized. In "Memories of the Space Age," Ballard imagines a post-apocalyptic Florida of the
future, where the gantries of Cape Canaveral stand rusting in the sun, cenotaphs to the dream of
escape velocity: "A threatening aura emanated from these ancient towers, as old in their way as
the great temple columns of Karnak, bearers of a different cosmic order, symbols of a view of the
universe that had been abandoned ..." (Ballard 1991, 137).

Yet, even as the Space Age recedes in history's rearview mirror, its fantasies of liberation from
gravity, limits to growth, and even mortality live on, in our cyberspace age, in the bubble-brained
visions, recently burst, of instant millions spun from thin air by profitless dot.coms; in the cold-

warrior dream of a missile-defense system that would seal Fortress America in an impregnable bubble; and in futurists' premonitions of nanotech alchemy that turns garbage into gold and cryogenic necromancy that guarantees immortality – the techno-fundamentalist gamble of freezing your head, as the wags put it, to save your ass.

Not that our resemblance to things past makes us any less affectionately condescending toward the obsolete tomorrows of earlier, dorkier decades. The turn-of-the-millennium vogue for the space-Pop aesthetic of the 1960s is soaked through with the been-there, done-that smugness of our post-everything culture. At the same time, an earnest yearning hides behind the ironic GenX/Boomer fetish for Egg chairs, shag carpets, space-age bachelor-pad music, and other retro-futurist kitsch: a closeted desire for the future that never was. "Design from the 1960s … expressed the collective optimism … Americans felt about our race to … the moon," says Scott Reilly, founder of the "Internet supersite for 20th century design," retromodern.com. "It was an opportunity to see a glimpse of what our future might look like within a generation – but to have it now" (quoted in Frauenfelder 1998). From *Wallpaper* magazine to the Austin Powers movies, Eero Aarnio's 1967 Ball chair on the cover of *Harper's Bazaar* to Arne Jacobsen's iconic chair in a J. Crew ad, nostalgia for the future, at once deeply sincere and deeply ironic, is an essential part of our post-millennial hangover. A campy elegy to the Death of the Future, the midcentury-modernism craze is borne of the longing articulated by the New York deejay and vintage electronics collector Frankie Inglese, who lamented in 1998, "It's nearly the year 2000. We have great technology, like cell phones, the Internet, and digital cameras. But we're still sitting on eyesores by Ikea. This is not what the future was supposed to look like" (quoted in Chaplin 1998, 2). Quintessentially, the future was supposed to look like Eero Saarinen's TWA Terminal, endlessly celebrated in swooning design-magazine spreads and paid winking homage by Bo Welch in his design for the intergalactic command center in the movie *Men in Black*. More than any other touchstone of Space-Age style, the TWA Terminal has been embraced as a symbol of things to come that never came.

TERMINAL CONDITION

The sad state of the actual building, which every day looks more and more like Ballard's brooding ruins of futures past, is a leading cultural indicator of the fact that the future just isn't what it used to be.

Design freaks who know the swooping, gull-winged icon only from Ezra Stoller's period photos in the reverent monograph, *The TWA Terminal*, are in for a rude reality check if they make the pilgrimage to New York's famously inconvenient JFK. Taken after the building's completion in 1962, Stoller's elegant black and white studies make it look at once ancient and futuristic: Brancusi's "Bird in Space" re-imagined as a Pharaonic monument. The information desk rises up out of the floor, all sinuous curves, its elliptical display of arrival and departure information looking like a television for aliens or a deep-space obelisk. Stoller's skinny-tied businessmen and white-gloved women stroll beneath soaring arches and a vaulted ceiling that give the terminal a decidedly churchlike air – an effect that is surely intentional, given the architect's debt to Le Corbusier's Chapel at Ronchamp.

(The church I attended as a boy, growing up in suburban Southern California, reversed Saarinen's architectural metaphor. A pyramid-shaped artifact of the Space Age, St. Mark's Lutheran Church in Chula Vista resembles a redwood treehouse idling on a concrete launchpad: house of worship as rustic spaceship, a visual trope that reconciled the era's crunchy-granola vibe with its star-trekking pioneer spirit. Designed in 1966 by Robert Des Lauriers, St. Mark's serves two masters, aesthetically speaking: the Rousseauian back-to-naturism of the counterculture and the rocket-finned futurism of the space race and mass air travel – culturequakes whose tremors were deeply felt in Southern California, where the aeronautics industry was a major employer. At the time he designed St. Mark's, the architect recalls, "Everybody was thinking of doing space things." Des Lauriers went on to make a name for himself as the creator of strikingly modernist churches throughout Southern California. His flirtations with the aesthetic reached their dizzy apogee in the Carlton Hills Lutheran Church in Santee, a Jetsonian traffic-stopper whose "flying effects," says Des Lauriers, exploit the hyperbolic paraboloid.)

In its heyday, the TWA Terminal was a Jet Age cathedral, consecrated to speed and – unintentionally – to the double-edged sword of technological utopianism, which staked its faith in a future whose chrome was never supposed to flake. After his last visit to the construction site, Saarinen (who died shortly before the building was finished) remarked, "If anything happened and they had to stop work right now and just leave it in this state, I think it would make a beautiful ruin, like the Baths of Caracalla"(quoted in Lamster 1999, 5).

His dream is coming true, though with far less sublime melancholy and far more bus-station seediness than he could have imagined. Once upon a time, the terminal was "a destination point in itself," according to the architectural critic Mark Lamster. "Its chic dining facilities, clubs, and lounges (the gallery-level restaurant and cafeteria were actually designed by the office of industrial designer Raymond Loewy) offered locals spectacular views and the *frisson* of mingling with the new 'jet set' in high style" (Lamster 1999, 4). Now, despite its landmark status, the terminal is in a terminal state, its concrete walls scabbed by slapdash patch jobs, its postage-stamp-sized "midget" floor tiles crumbling, its once-stylish Paris Café blighted by tacky updates, the upholstery in its Lisbon Lounge splitting and leaking stuffing.

Ironically, Saarinen's masterwork was ready for the wrecking ball almost from day one. Designed with propeller aircraft in mind, it was too small for the human traffic it was soon forced to accommodate, a flood unleashed by the advent of commercial jet aircraft like the Boeing 707 and the Douglas DC-8. Aerodynamic archetype of life lived at Mach 1, the TWA Terminal was anachronistic from the very beginning.

CLASS CONSCIOUSNESS

Time and irony have transformed Saarinen's masterwork into an elegy to the Jet Age, when the world above cloud cover was reserved for sharp-dressed executives and smiling stewardesses while the antlike masses traveled via earthbound modes more appropriate to their station. A seductive blend of upper-class elegance and ultramodern chic, equal parts Cunard liner and moonliner, once prevailed at cruising altitude.

In a January 2001 Opinion Editorial on the passing of the bankrupt TWA, whose assets will be acquired by American Airlines, the former TWA flight attendant Ann Hood reflects wistfully on the bygone era when white-gloved attendants like herself carved chateaubriand, mixed the perfect martini, and "dressed lamb chops in tiny gold foil booties" – worldly skills acquired at the company's training academy, "a Jetson-style office complex with sunken living rooms and modular furniture" (Hood 2001). To a small-town girl like Hood, the glamorous airline "represented the future." She mourns the end of the age "when TWA was all first class," a decline and fall marked, in her eyes, by the transition from cream pitchers spirited away on silver trays to Styrofoam cups collected by dreary attendants dragging garbage bags down the aisles.

The 70s witnessed the deregulation of international air travel, enabling cheap fares; also, Boeing introduced its wide-bodied 747 "jumbo" jet, which carried twice as many passengers as any previous plane. Thus was the jet set democratized – to a minus effect, in the minds of some. By 1978, a United spokesperson was already complaining, in Kenneth Hudson and Julian Pettifer's social history of air travel, *Diamonds in the Sky*, "Air travel is not the big thing it used to be. It has become the bus service of the nation" (Hudson and Pettifer 1979, 199). As the authors note, "Many people, especially those in positions of power and influence, dislike travelling by bus" (199). And, it might be added, travelling with the class of people who usually travel by bus.

The well-rewarded should count their blessings. For the masses, air travel means inedible food, cattle-car conditions, and frightful displays of bad breeding back in steerage. In his neogothic novel, *Hannibal*, Thomas Harris evokes the horrors of air travel among the lower orders when his antihero, Hannibal Lecter, has to fly (heaven forefend!) *Economy Class*. Lecter is a high-born cannibal whose penchant for Mozart, Steuben chandeliers, and eviscerating the "free-range rude" places him squarely on the George Will side of the culture wars. But lo, how the mighty are fallen: on the lam, Lecter is forced to travel incognito with the package-tour untouchables. Harris sets the scene:

> Shoulder room is 20 inches. Hip room between armrests is 20 inches. This is two inches more space than a slave had on the Middle Passage. The passengers are being slopped with freezing-cold sandwiches of slippery meat and processed cheese food, and are rebreathing the farts and exhalations of others in economically reprocessed air, a variation on the ditch-liquor principle established by cattle and pig merchants in the 1950s (Harris 1999, 247).

It's understandable, given such conditions, that "air rage," if not serial cannibalism, is on the rise. The term entered the media lexicon in 1995, when a drunken customer filed a complaint against the crew's refusal to serve him another martini by yanking his pants down and taking a dump on the drink cart. Since then, flight attendants have been punched, peed on, hurled into bulkheads and bashed with bottles. Clearly, the barbarians are at the boarding gates.

In keeping with the time-honored, middlebrow tradition of policing the border between the rabble and the power elite, some experts have hinted that "air rage" is the inevitable result of the erosion of traditional values, specifically our unquestioned respect for a Man in a Uniform. "Passengers have lost respect for the pilot, the flight attendants, and each other, and this is what happens," admonishes Dr. Arnold Nerenberg, a Whittier, California-based authority on "road rage" (quoted in Elliott 1998a). The square-jawed guy with the right stuff – every man's fantasy of grace under pressure and every woman's dream of rocket-jock hunkiness – has gone the way of *Airport* and *Coffee, Tea or Me?*.

Hudson and Pettifer (1979) offer a dryly funny Freudian reading of the pilot as paterfamilias:

> Throughout the 1920s and 1930s, and certainly during the war years pilots had almost godlike status, both for what they did and for what they were … Psychologists have gone further and seen the pilot as an essential father-figure. As such, his strengthening, encouraging influence should be all-pervasive in the aircraft. If his first task is to fly the plane safely, his second is to reassure passengers that they are in good hands … Only the word of the captain contains the assurance needed to reduce passenger's anxiety to the fullest possible extent. If information had to be relayed instead of coming directly in the voice of the father-figure, the communication should reach passengers in some such form as 'Captain McGregor says that we will be in

Paris in about an hour and that the sun is shining there.' The ritualistic use of his name is essential if the message is to have its maximum effect. Or so the theory goes. (156-157).

Come to think of it, the captain sounds a lot like another unseen, omniscient patriarch: the Judeo-Christian sky god. As Ronald Reagan's eulogy to the crew of the ill-fated *Challenger* ("They touched the face of God") suggests, fans of Bronze Age belief systems still think of the Big Guy as, you know, somebody *up there*. Hence, it's understandable that pilots, at home in the heavens, should have "almost godlike status."

Inversely, it also stands to reason that Our Heavenly Father should be conceived of as a pilot. Remember God Is My Co-Pilot? And isn't the biblical Cap'n Yahweh always issuing prophetic bulletins over some sort of cosmic PA system or through uniformed emissaries who return from that exalted flight deck, the Holy of Holies, with messages validated by "the ritualistic use of His name"? A hardcore fan of Xtreme exegesis (pushing the envelope of overinterpretation to rocket-sled extremes), I can't resist noting that the traditionally female flight attendant stands in for the Holy Mother, in our Jet-Age theology. As Hudson and Pettifer (1979) point out:

> If the pilot is father, the stewardess is mother, waiting on her children hand and foot, listening and watching to see if one is crying or unhappy, always at hand with comforting words and an understanding smile (157).

Nevertheless, in the same way that midcentury visions of terminals as modish "jetports" have given way to the drab reality of our overcrowded Greyhound stations of the skies, pilots have been demoted from deities or dashing starship captains to sleep-deprived Airbus drivers. As the comedian George Carlin puts it in his routine, "Airline Announcements" (on *Jammin' in New York*), "Who made this man a *captain*, might I ask? Did I sleep through some sort of armed-forces swearing-in ceremony or something? He's a fucking *pilot*, and let him be happy with that!" To add insult to injury, computer systems inch closer, every day, to the science-fiction goal of HAL at the helm; nowhere does the shadow of impending human obsolescence fall as long and dark as it does in the cockpit of the jet fighter or the commercial airliner, where autopilot and flight-director systems may one day put human pilots out of a job.

CABIN FEVER

Then again, as common sense (and Hannibal Lecter) would seem to suggest, the recent nosedive in civility in the air surely has less to do with our plummeting respect for authority than with the simmering class war that rises to a boil when one too many proles are penned up in Economy Class, breathing recycled farts and choking down McGlop, while the scent of First-Class cuisine wafts tauntingly down the aisle, hinting at the privileges of rank.

(This is *not* an apologia for escapees from Darwin's waiting room like the former Stone Roses singer Ian Brown, who provided unscheduled in-flight entertainment on a 1998 British Airways flight. After threatening to cut off the hands of a flight attendant, Brown pummeled the cockpit door while the pilots attempted to land the plane. Obviously, sociopaths like Brown should fly alongside the tranquilized Rottweiler in the cargo hold, if at all.)

There's no denying that the industry's clueless pursuit of profit maximization has turned airliners into social and psychological pressure-cookers. A typical example of the calculus of greed: the introduction of "thinline" seats, whose composite materials and ergonomic styling reduce seatback thickness by as much as 40 percent, has exacerbated the sardine-can overcrowding in steerage. Carriers could pass the legroom freed up by slimmer seatbacks on to their passengers; instead, they're using the newly available space for – you guessed it – more seats.[1]

In these times of mounting income disparity, the have-nots have one thing to burn: serious attitude. Air travel, a magnifier of class differences since the days of the jet set, is an unpleasant reminder to the "classless" American masses that the socioeconomic chasm between them and the new, new rich is a veritable Grand Canyon. The cultural critic Tom Vanderbilt reads commercial cabin design as "a microcosm of the rising global wealth imbalance ... rarely is the divide between have and have-not under 'turbo-charged capitalism' so nakedly, if politely, expressed"(Vanderbilt 2000, 73). For every digeratus contemplating that Lear-jet time-share advertised in *Wired*, for every geek genius flying high in his super-cool helicopter like Jim Clark in *The New New Thing*, for every dot.com mogul like Mark Cuban buying a $40 million Gulfstream V jet "because I can," there are countless lumpen, crammed into airborne cattle cars. Muddle their oxygen-starved brains with booze and rub their noses in the ample legroom and complimentary cocktails on the other side of the curtain, in Virgin Atlantic's "Upper Class" or Delta's "Elite," and you've got the makings of class war at 35,000 feet.

Despite his hilarious protestations that he is neither a communist nor a Democrat, ABC-NEWS.com's "Crabby Traveler" columnist Christopher Elliott shows alarming signs of (Economy) class consciousness when he recounts the experience of a reader named Julie. Sitting in First Class, watching other, less fortunate souls shuffle by, Julie experienced a sudden spasm of solidarity with her comrades in the cheap seats. "As they filed past her," writes Elliott, quoting Julie:

> a little boy – about 3 – said to his mother, who was holding an infant, 'Mommy, let's sit in these seats. They look more comfortable.' Of course, they headed on back. 'Somehow,' Julie concludes, 'it just does not feel right' (Elliott 1998b).

The First Class/Economy Class divide offers a marvelous civics lesson, dramatizing the fact that, in an ever more privatized nation, the gains of the wealthy can be the lower classes' losses, purchased at the price of a smaller, meaner public sphere. Acknowledging that space is a scarce resource on commercial airliners, Elliott proposes a radical solution: jettison First Class. "Get rid of those oversized seats [and] space the Economy Class seats further apart," he argues, in a column subtitled "Few Luxuriate at Expense of Many" (Elliott 1998c). (Not that he's a *communist* or anything!) According to Elliott, "airlines know that it's the right thing to do, but they're hiding behind a number of tired excuses. They claim their frequent fliers wouldn't tolerate the elimination of First Class and that they couldn't turn a profit with a one-class configuration. (In fact, more than half of First-Class seats are filled with passengers using upgrades, which erodes the profit argument" (Elliott 1998c).)

The stratification of airline seating by class has less to do with market logic, it turns out, than perpetuating the myth that the Jet Age never died – that air travel is still about gracious living

---

1 As long as we're on the subject of greed, the moral fungus that thrives on the underside of turbo-capitalist cultures, I can't resist mentioning that greed is a factor in airline safety as well as cabin design. In *Deadly departure: Why the experts failed to prevent the TWA flight 800 disaster and how it could happen again*, Christine Negroni (2000) argues convincingly that the explosion that ripped apart Flight 800 was caused by Boeing's lunatic decision to use the plane's fuel tanks as heat absorbers for its airconditioning units. As every high school science nerd knows, fuel becomes more flammable when heated. Combine that fact with the intriguing bit of trivia that the faulty insulation on Boeing's wiring has been known to cause sparks, and you have the makings of an explosive exposé. What makes the story even more explosive is that Boeing likely knew about the dangers of its fuel tank design as early as the late 60s and, weighing the cost of less flammable – but more expensive – systems against the cost of the number of fatal accidents expected to happen in the next decade, opted for the laissez-faire approach (in both senses of the word). But I digress ...

and creature comforts and smiling servants who come when the call button is pressed. First class seating is to commercial air travel what the Horatio Alger myth is to American capitalism: the juicy morsel on the end of the hook. The myth of the self-made man and the level playing field keeps working stiffs punching those time cards. It ensures, too, that they'll endure the inequities of a system rigged in favor of those born with the right skins, the right chromosomes, and the right zip codes.

Similarly, the fantasy of one day landing in the tax bracket that comes with a First-Class seat or of being suddenly upgraded, Cinderella-like, to upper-class status helps the unfortunates in budget class make it through the flight. "For semifrequent travelers, First Class can be like the lottery," says Margaret Frey, of Jericho, Vermont. "There's always the hope that you might get bumped into one of those seats" (quoted in Elliott 1998b). Drowsing bolt upright, at the inhuman angle only a cabin designer could call "reclined," suffering the skull-busting headache that comes with cutbacks in air rations, processing the after-effects of that slippery meat and processed cheese food, we can dream, can't we?

DEBRIS FIELD

Suffice it to say, then, that the Jet Age incarnated by the TWA Terminal is well and truly gone, and with it the belief that we are cleared for takeoff to a brighter tomorrow, master-planned by social engineers and watched over by technocrats who will ensure that the monorails run on time.

The gathering clouds of our collective loss of faith in technology and teleology are gradually assuming a discernible shape, and it is the shape of a jetliner hurtling toward certain doom. Ironically, the growing fear of flying coalesces at a time when the triumph over time and space allegedly made possible by email and teleconferencing was supposed to have rendered jet travel obsolete. Early rhapsodists of virtual reality, such as Timothy Leary, bemoaned the bothersome, environmentally unfriendly necessity of hauling our bodies from one time zone to another, and looked forward to the day when business travel would mean rolling out of bed and into a virtual-reality pod. Like the paperless office, the teleconferencing cybernaut, jacked into his virtual corporation from his electronic cottage, is still largely a wish-fulfillment fantasy. Even *Homo cyber* can't seem to shake his vestigial need to interact in meatspace, in real time – an evolutionary hangover underscored by telling expressions such as "face to face" and "press the flesh."

Statistically, commercial flight has never been safer: in the United States, for example, there were 394 aviation-related deaths in 1996; by comparison, 41,907 Americans died in car crashes that year. Even so, the increase in the sheer number of flights ensures, in the words of a 1996 report, that if "the current accident rate remains steady and air travel continues to grow rapidly, a passenger jet may crash as often as once a week by the year 2010" (quoted in MacPherson 1998, x).

Anecdotal evidence suggests that fear of flying is on the rise, fanned by tabloid-gothic accounts of aviation disasters. Reconstructed from black box recordings of frantic cockpit exchanges, these reports inevitably feature a second-by-second countdown to impact. They offer front-row seats to someone else's real-life disaster movie, complete with the heart-rending last words of crewmembers who know they're about to die ("Amy, I love you" or "Oh, Mama, gonna miss those baseball games") and the clamor of whooping alarms and robotic voices yelling, "Terrain! Terrain! Pull up! Pull up!"

The heavens seem to be raining planes, these days, from the mysterious explosion of TWA Flight 800, whose entrails are still being examined by conspiracy theorists; to EgyptAir Flight 990's nose-dive into oblivion, fertile ground for anti-Arab caricatures of a Muslim pilot on a suicide mission from God; to the crash of a Concorde, that endangered machine species from a supersonic future that never came, as sweetly sad in its own way as the last passenger pigeon. Our mental screens flicker with images of planes bellyflopping and bursting into flames, of National Transportation Safety Board investigators picking grimly through "debris fields," of aircraft reconstructed on hangar floors, their fragments tracing their outlines, like the bones of beached whales. The debris field is the graveyard of Jet-Age futurism, the frenzied chatter of the black box its unsettling epitaph.

What is it, these days, that draws us to websites like www.PlaneCrashInfo.com or plays like *Charlie Victor Romeo*, a gut-lurching dramatization of aviation disasters based on cockpit voice recordings? Is their appeal just a throttled-up version of what Ballard called the "erotic power of disaster"? Is our attraction to them part of the same psychopathology of everyday life that makes us rubberneck at multiple-car pileups or watch tabloid-TV shockumentaries? There's no denying the voyeuristic subtext of PlaneCrashInfo.com, a site for aviation buffs that feels, at times, like a disaster-porn peepshow, where the uncontrolled dive is foreplay and the moment of impact the money shot. There, on-line rubberneckers can ogle accident photos, browse grim statistics ("Mortality Risk by Type of Scheduled Service"), thrill to actual samples of automated cockpit warnings ("Pull Up!," "Windshear!," "Don't Sink!"), and, creepiest of all, relive the last seconds of doomed airliners through black box recordings ("Caution, may be disturbing").

The Black Box: All-New Cockpit Voice Recorder Accounts of In-Flight Accidents, by *Newsweek* correspondent Malcolm MacPherson (1998), excavates the same psychic debris field. A pillow book for wreck-chasers, *The Black Box* is also a thrill ride from hell, a wrenching memorial to anonymous casualties of information overload, and an existential comedy of errors, as when the Chinese pilot of a Boeing jet, lost in fog and baffled by the ground-proximity warning system telling him, in English, to "Pull up! Pull up!" asks, "What means, 'Pull up'?" (MacPherson 1998, xiv).

But what is this morbid little book's deeper fascination, beyond the obvious, undying appeal of (someone else's) death? Perhaps it has something to do with the catharsis of vicariously living out the worst nightmares of a high-tech society that seems to be running increasingly on auto-pilot. In our age of complex systems and autonomous technology, a software bug, a hardware failure or the fatal mistakes of humans trying to multitask at machine speed can mean certain disaster for the rest of us. Cowering, heads down, in the aft cabin, we wonder if the end will come in fiery agony or instantaneously, the summary judgement of "multiple anatomical separations secondary to velocity impact," in the antiseptic poetry of a coroner's report quoted in *The Black Box* (112). Nonetheless, we're being asked, more and more, to trust in technologies whose speed and complexity are leaving humans in the dust, even as that trust is eroded by firsthand acquaintance with the fickle, glitchy, virus-infested reality of computers and, at the societal level, by tabloid tales of planes falling from the sky.

Reassuring statistics about the safety of commercial flight notwithstanding, *The Black Box* draws back the curtain on what some of us melodramatically imagine are the backstage dramas taking place behind our society's cool façade of technocratic expertise and infallible technology. Sometimes, it is precisely our over-reliance on technology – our fundamentalist faith in the omniscience of our intelligent machines – that is our undoing. In the case of Aeroperu Flight 603 (included in *The Black Box*), the altimeter, airspeed indicator, and other instruments malfunctioned, giving readings the pilots knew couldn't possibly be accurate. As a commercial pilot quoted in a *Wall Street Journal* article put it, "The brain of the autopilot and auto-guidance system had a stroke. All the crew had to do was turn it off and fly the plane" (quoted in Lawrence 2000).

Tragically, the pilot couldn't bring himself to shut down the computer and rely on his own expertise, despite his mounting frustration with the bewildering data the computer was sending him. Drowning in information and forced to make life-or-death decisions at machine speed, the pilot allowed his deep-rooted faith in artificial omniscience to trump his confidence in his own abilities, with disastrous results.

Sometimes, of course, bad things happen to dumb people. One obvious candidate for the Darwin Award, in *The Black Box*, is the anonymous captain doofus who is so "relaxed," in the deadpan understatement of a subsequent NTSB report, that he doesn't notice ice buildup on the wings or the fact that he has selected the wrong autopilot setting, causing the plane to stall and plunge 12,000 feet before the panicked crew regains control (MacPherson 1998, chap. 19). More tragically, there's the ground crew that forgets to remove masking tape from a Boeing 757's "static ports" after washing them, effectively blinding the plane; unable to read air speed or altitude, the crew flies into the sea, killing all aboard (MacPherson 1998, chap. 2).

But the cliché that human error is the weakest link in any cyborgian chain is the moral of only some of MacPherson's Information-Age fables. In others, machines go haywire, doing almost willfully perverse things. On May 26, 1991, Lauda Air Flight 004 was 4,000 feet over Thailand when for some unknown reason – maybe an isolation valve failed, who knows? – one of the jet's thrust reversers deployed. Thrust reversers are used to brake an airplane on the ground. "A deployment in flight of the thrust reverser is by definition catastrophic," notes MacPherson. "It is almost unthinkable" (MacPherson 1998, 27). The plane disintegrated in mid-air.

Read as postindustrial myth, these tales of humans huddled in a pressurized metal cylinder, streaking across the sky at supersonic speeds with a machine (and two or three technoliterate human attendants) at the helm provide a concise metaphor for our predicament as a society, in the year 2001.

Increasingly, those of us who don't speak fluent code feel as if we're along for the ride, and our nagging fears aren't likely to be calmed by the queasy realization that the real answer to the question we dare not ask – "Who's flying this thing?" – is "No one." As Air Force Captain John Varljen, who works at the Pentagon, told a newspaper reporter, "We're beginning to lean toward thinking of the aircraft as another member of the crew. [A]s aircraft become smarter and smarter, it's more likely that the plane will go off and do things that you may not expect" (quoted in Lawrence 2000). Have a nice flight.

Ironically, in the unlikely event that you or I become one of the ill-fated few who die in a commercial airline accident, the last voice we hear may be an artificial one, like the automated announcement that coolly informed the passengers of Korean Air Flight 007, as they plummeted toward the ocean after a Soviet MIG blew them out of the air, "Put out your cigarette. This is an emergency descent. Put out your cigarette. This is an emergency descent. Put the mask over your nose and mouth and adjust the headband. Put the mask over your nose and mouth …" (MacPherson 1998, 35). Death, in the computer age, sometimes comes in the guise of a bad Laurie Anderson routine.

DEPARTURE LOUNGE

One gray winter day in late 2000, I drove out to the TWA Terminal for a firsthand look at Saarinen's shrine to Jet-Age futurism. Exploring its empty, echoing lobby (I went on a day when not many flights were scheduled), bluffing my way into its Ambassador Club and VIP Lounge, I decided that the terminal had something of the art critic Ralph Rugoff's "forensic aesthetic" about it. In *Scene of the Crime*, Rugoff argues that avant-garde artists from the 70s on have co-opted the semiotics of crime-scene and disaster-zone investigation to frame their cryptic installations and staged psychodramas. This forensic aesthetic, says Rugoff, is characterized by "a strong sense of aftermath" and the requirement that "the viewer arrive at an interpretation by examining traces and marks and reading them as clues" (Rugoff 1997, 62).

The missing link between the funereal and the futuristic, the TWA Terminal is haunted by what the SF novelist William Gibson would call the "semiotic ghosts" of yesterday's tomorrows. Like Oscar Niemeyer's Brasilia, the Brazilian city that looks like a moonbase by Le Corbusier, or Arcosanti, Paolo Soleri's half-built science-fiction technopolis decaying in the Arizona desert, the terminal is one of those cultural wormholes where the future as we knew it in the 20th century, disappeared into some alternate universe – gone without a trace, like the abducted traveler in the creepy thriller, *The Vanishing*. In that sense, Saarinen's building *is* the scene of a crime.

The terminal's second-floor gallery is lined with glass cases displaying the vintage uniforms of "air hostesses" long gone. I half-expected to find a mummified flight attendant in one of these glass sarcophagi, still natty in the decaying remains of her "Jet Age" uniform, designed for 1960 crews by Loewy, the apostle of streamlining. One exhibit struck an off-key note of 90s knowing-ness. A peppy article on the attaché case of the future (i.e., 2001 A.D.) from a 1969 issue of TWA's in-flight magazine, *Ambassador*, it was obviously included as a wasn't-the-future-wonderful smirk at the clunky earnestness of 60s futurism. After extolling the features of the briefcase of the future ("a portable communications and computing center which even has space for extra socks … [and] ample room for futuristic lunch"), the un-by-lined writer exhorts his 60s reader to "zip up your spacesuit, pack your pills, pick up your attaché case and count-down. The world of the future will be here sooner than you think."

So soon, I think, alone in this mausoleum of outworn and stillborn tomorrows, that it often overtakes our visions of things to come, rendering them obsolete before they arrive.

REFERENCES

Ballard, J. G. 1991. Memories of the space age. In *War fever*. New York: Farrar, Straus, Giroux.

Chaplin, J. 1998. Generation wallpaper. *New York Times*, Sept. 6, Sunday Styles section.

Elliott, C. 1998a. Cabin fever rages: Flight attendants take the heat. *The crabby traveler* (abcnews.com).
    At http://www.abcnews.go.com/sections/travel/Crabby/cabinfev.html (accessed Feb. 6, 2001).

————— 1998b. Class conflict: Is banning first class a political statement? *The crabby traveler* (abcnews.com).
    At http://www.abcnews.go.com/sections/travel/Crabby/firstclass1.html (accessed Feb. 6, 2001).

————— 1998c. Can first class: Make room for the masses. *The crabby traveler* (abcnews.com).
    At http://www.abcnews.go.com/sections/travel/Crabby/firstclass.html (accessed Feb. 6, 2001).

Frauenfelder, M. 1998. The plastic fantastics. *Wired News*.
    At http://www.wirednews.com/news/print/0,1294,10542,00.html (accessed Feb. 6, 2001).

Harris, T. 1999. *Hannibal*. New York: Delacorte Press.

Hood, A. 2001. When TWA was all first class. *New York Times*, Jan. 13: A13.

Hudson, K. and J. Pettifer. 1979. *Diamonds in the sky: A social history of air travel*.
    London: The Bodley Head/British Broadcasting Corporation.

Lamster, M. 1999. Introduction. In *The TWA terminal*, by Ezra Stoller. New York: Princeton Architectural Press.

Lawrence, L. 2000. *Charlie Victor Romeo* offers a peek inside the black box. Review. Wall Street Journal (no date given).
    At http://www.collective.weird.org/Calendar/shows/cvr/wsj.htm (accessed Feb. 6, 2001).

MacPherson, M. (ed.). 1998. *The black box: All-new cockpit voice recorder accounts of in-flight accidents*. New York: Quill.

Negroni, C. 2000. *Deadly departure: Why the experts failed to prevent the TWA Flight 800 disaster and how it could happen again*.
    New York: Cliff Street Books.

Rugoff, R. 1997. More than meets the eye. In *Scene of the crime*, ed. R. Rugoff. Cambridge, MA: MIT Press.

Vanderbilt, T. 2000. Class wars. *Artbyte* (October).

Wade, N. 1993. Method and madness: No space for NASA. The *New York Times* magazine, Nov. 14: 34.

NOTES ON CONTRIBUTORS
AND INDEX

NOTES ON THE EDITORS

SENIOR EDITOR

Dr Darren Tofts is Chair of Media and Communications at Swinburne University of Technology in Melbourne. He is an internationally recognized writer on cyberculture, new media arts, critical theory and hypertextual poetics. His work has appeared in publications such as *Mediamatic, Senses of Cinema, UTS Review, Mesh* and *Social Semiotics*. He is the author, with artist Murray McKeich, of *Memory Trade. A Prehistory of Cyberculture* (Sydney: Interface Books 1998) and *Parallax: Essays on Art, Culture and Technology* (Sydney: Interface Books 1998). Formerly an editorial correspondent for *21C* and *World Art* magazines, he is on the editorial boards of *Postmodern Culture, Hypermedia Joyce Studies* and *Continuum. The Australian Journal of Media and Culture* and is a media arts editor for *RealTime*. He is currently working with an international consortium of scholars on the first hypermedia edition of James Joyce's *Ulysses*.

dtofts@swin.edu.au

EDITORS

Annemarie Jonson teaches in the Arts Informatics program at the University of Sydney. Her book on Artificial Life is due to be published by Routledge in 2003. She has published and edited widely around technology and the media arts, and has written for publications such as *Art&Text, UTS Review* and *Australian Feminist Studies*. She was joint coordinating editor of *OnScreen* in *RealTime*, and co-editor of *Essays in Sound* and worked extensively on the international Artificial Life conference held at Sydney's Powerhouse Museum in 2000.

annemarie.jonson@ozemail.net.au

Alessio Cavallaro is Producer and Curator of New Media Projects at the Australian Centre for the Moving Image, Federation Square, Melbourne, Australia. He was Director of dLux Media Arts, Sydney, and was an inaugural member of the New Media Arts Board of the Australia Council for the Arts. In this role Alessio conceptualized and convened the annual *futureScreen* new media arts forums, exhibitions, and screening programs held in Sydney from 1997. He is consulting editor and former joint coordinating editor of *OnScreen*, and has previously co-edited the journal *Essays in Sound*, the catalog of the Third International Symposium of Electronic Arts (Sydney 1992) and subedited *Media International Australia* (1996-1997).

alessio@acmi.net.au

NOTES ON CONTRIBUTORS

Russell Blackford is an Australian writer who writes mainly about philosophy, science and science fiction. He has published articles, reviews and stories in many journals, magazines, anthologies and reference works in Australia, North America and Europe, dealing with such topics as the prospects of computer super-intelligence; cyberspace; cloning and genetic engineering; radical life extension; and the relationship between religion and science. He has won the William Atheling, Jr Award for Criticism or Review on three occasions and his own fiction has won the Aurealis Award and the Ditmar Award.

Damien Broderick holds a PhD from Deakin University in the discursive systems of the sciences and the humanities. He is a Senior Fellow in the Department of English and Cultural Studies at the University of Melbourne, Australia. His futurist studies include *The Last Mortal Generation: How Science Will Alter our Lives in the 21st Century* (1999) and *The Spike: How Our Lives are being Transformed by Rapidly Advancing Technologies* (1997; revised US edition 2001). Theoretical studies include *The Architecture of Babel: Discourses of Literature and Science* (1994), *Reading by Starlight: Postmodern Science Fiction* (1995), *Theory and its Discontents* (1997) and *Transrealist Fiction: Writing in the Slipstream of Science* (2000). He has also published a large body of science fiction. See http://www.thespike.addr.com

Justine Cooper is a New York based artist whose interests lie in how medical technologies mediate our conceptions of identity, time, and space, and more recently in systems of translation, pattern and randomness. Cooper's work has been shown internationally in over 30 shows in 12 countries across five continents, and is held in both public and private collections including the Metropolitan Museum of Art. She has recently completed an artist-in-residency at The American Museum of Natural History in New York (funded by the Australian Network for Art and Technology), and until recently was an artist in residency at the World Trade Center (through the Lower Manhattan Cultural Council's *World Views* program).

Char Davies' immersive virtual environments are internationally recognized as landmarks in the field of interactive media art, and are known for challenging conventional ways of thinking about virtual reality. More than 20,000 "immersants" have breathed/floated through her works *Osmose* (1995) and *Ephémère* (1998) around the world, including at the National Gallery of Canada, the Barbican Centre (London), the Museum of Monterrey (Mexico) and the San Francisco Museum of Modern Art. Davies' work has been written about widely in the international press, including *New Scientist*, *Art in America*, and *World Art*, and has been the subject of numerous scholarly essays and documentary films. She has lectured at venues ranging from the Museum of Modern Art, NYC, to the University of Cambridge (2001). Davies has also written in depth about virtual space, with texts forthcoming from MIT Press and University of Cambridge Press. Formerly a painter, Davies became involved with digital media in the mid-1980s as a founding director of the software company Softimage. In 1998, she founded her own company, Immersence, for producing new work. She has received many awards, most recently, in 2002, an Honorary Doctorate in Fine Arts from the University of Victoria, British Columbia.

Erik Davis is a San Franciso-based writer, culture critic, and independent scholar. His book *TechGnosis: Myth, Magic, and Mysticism in the Age of Information* was released by Harmony Books in the fall of 1998. Davis is currently a contributing editor for *Wired*, and writes "The Posthuman Condition" column for the online magazine *Feed*. Davis has also contributed articles and essays to *ArtByte*, *Spin*, *Mediamatic*, the sadly departed *Gnosis* and *21C*, and the *Village Voice*. He has contributed to a number of cyberculture anthologies, including *ReadMe! ASCII Culture and the Revenge of Knowledge* and *Flame Wars*. Davis appeared prominently in Craig Baldwin's recent film, *Specters of the Spectrum*, and has lectured internationally at media arts conferences on topics relating to cyberculture, contemporary electronic music, and spirituality in the postmodern world. Davis was one of the organizers of Planetwork, a conference on information technology and global ecology held in San Francisco. Some of his work can be accessed at http://www.levity.com/figment, and he can be reached at figment@sirius.com.

Mark Dery (markdery@mindspring.com) is a cultural critic. He wrote *Escape Velocity: Cyberculture at the End of the Century* and edited *Flame Wars: The Discourse of Cyberculture*. His essay on guerrilla media activism and disinformation warfare, "Culture Jamming: Hacking, Slashing, and Sniping in the Empire of the Signs," is a bona fide Net.classic, endlessly recirculated on the Internet. His most recent book is the essay collection, *The Pyrotechnic Insanitarium: American Culture on the Brink* (www.levity.com/markdery/). His byline has appeared in *Salon*, *Feed*, *Suck*, *Rolling Stone*, *The New York Times Magazine*, *Lingua Franca*, *Wired*, and the online version of *The Atlantic Monthly*, to name a few. A frequent lecturer in the U.S. and Europe on new media, fringe art, and unpopular culture, he was appointed Chancellor's Distinguished Fellow at UC Irvine, for January 2000.

Troy Innocent combines computer animation, generative systems, multimedia design and interactivity to create virtual worlds which explore the "language of computers." He has exhibited widely at international events including SIGGRAPH, the International Symposium of Electronic Arts (ISEA), and Digital Salon (New York). Innocent is Senior Lecturer in Multimedia and Digital Arts at Monash University, Melbourne, Australia, and is currently producing *Evolglyph*, an interactive virtual world commissioned by the Australian Centre for the Moving Image at Federation Square, Melbourne.

Stephen Jones has been producing *The Brain Project*, on aspects of the nature of consciousness, since August 1996 and is an active and regular participant in international academic conferences and forums on artificial intelligence, consciousness and complexity. He also builds physical immersion installations based on the incunabula of computing and works as an electronic engineer. Jones is an internationally recognized video artist, working with the electronic music band Severed Heads in the 1980s.

Evelyn Fox Keller received her Ph.D. in theoretical physics at Harvard University, worked for a number of years at the interface of physics and biology, and is now Professor of History and Philosophy of Science in the Program in Science, Technology and Society at MIT. She is the author of *A Feeling for the Organism: The Life and Work of Barbara McClintock*; *Reflections on Gender and Science*; *Secrets of Life, Secrets of Death: Essays on Language, Gender and Science*; *Refiguring Life: Metaphors of Twentieth Century Biology* (Columbia University Press 1995) and *The Century of the Gene* (Harvard University Press 2000). Her most recent book is *Making Sense of Life: Explaining Biological Development with Models, Metaphors, and Machines* (Harvard University Press, 2002).

Bruce Mazlish is Professor of History, MIT. His areas of interest and expertise are Western intellectual and cultural history, with a special nod to history of science and technology, the culture of capitalism, and history of the social sciences. He is also an authority in the interdisciplinary field of psychohistory as well as historical methodology; most recently he has spearheaded an effort to conceptualize global history (editing a volume by that name which appeared in 1993). His most recent publications are *The Uncertain Sciences*; *The Fourth Discontinuity. The Co-Evolution of Humans and Machines*; and *A New Science: The Breakdown of Connections and the Birth of Sociology*. He is a Fellow of the American Academy of Arts and Sciences. In 1986 he was awarded the Toynbee Prize, an international award in social science.

Jon McCormack is an Australian researcher, computer scientist and electronic media artist. His artwork explores the dialogues of art and science and has been extensively exhibited internationally at, for example, the Centre Georges Pompidou, Paris, the Museum of Modern Art, New York, the ACM SIGGRAPH Electronic Theater, Los Angeles, ARTEC, International Biennale, Nagoya, International Symposium on Electronic Art (ISEA), Montréal, and the London Film Festival. He heads the Centre for Electronic Media Art at Monash University, Melbourne, Australia.

Scott McQuire wrote the essay is this book while holding an Australian Postdoctoral Fellowship in the School of Social Inquiry at Deakin University, Australia, where his research project concerned interactions between mass media, architecture, and urban space. He is the author *Visions of Modernity: Representation, Memory, Time and Space in the Age of the Camera* (London: Sage 1998), and *The Look of Love* (Masterthief 1998), a limited edition artist's book produced with photomonteur Peter Lyssiotis. Other recent publications include two major research reports on the film industry, *Crossing the Digital Threshold* (1997) and *Maximum Vision: Large Format and Special Venue Cinema* (1999), both commissioned by the Communications Law Centre and published by the Australian Film Commission and the Australian Key Centre for Media and Cultural Policy. Scott McQuire is currently a Senior Lecturer in the Media and Communications Program at the University of Melbourne.

Simon Penny is an Australian artist, theorist and teacher in the field of interactive media art, and Professor of Arts and Engineering at University of California Irvine (a joint appointment between the Henry Samuel School of Engineering and the Claire Trevor School of Arts). His art practice consists of interactive and robotic installations, which have been exhibited in the U.S., Australia and Europe. Recent awards include a grant from the Langlois foundation (with Bill Vorn), first prize in the Cyberstar 98 awards (GMD/WDR, Germany) and a residency at the Institut fur Bildmedien, ZKM Karlsruhe, spring 1997. He is a widely published and translated essayist and he edited the anthology *Critical Issues in Electronic Media* (SUNY Press 1995). Penny is currently working on a book on embodied interaction and procedural aesthetics for MIT Press. He is also establishing a new graduate program in Arts, Computation and Engineering and is Layer Leader for the Arts in the California Institute for Telecommunications and Information Technology, CAL(IT) 2.

Patricia Piccinini is an artist living and working in Melbourne. Her work explores contemporary life, and therefore makes frequent reference to technologies and their social and cultural effects. 2002 exhibitions include the Sydney and Liverpool Biennales and a major solo exhibition at the Museum of Contemporary Art, Sydney. Past exhibitions include the Berlin, Kwangju and Melbourne Biennales, Plasticology (NTTICC, Tokyo) as well as numerous other shows in Australia, Japan, Netherlands, Peru, Spain, Germany, France, China and elsewhere.

John Potts is Senior Lecturer in Media and Communication at Macquarie University, Sydney. He is the author of *Radio In Australia* (Sydney: UNSW Press 1989), and co-author with Andrew Murphie of *Culture and Technology* (London: Macmillan 2002). He has published numerous articles and book chapters on media studies and digital culture. He is also a radio producer and sound artist who has produced several audio arts works for ABC Radio, Australia.

Francesca da Rimini (GashGirl, dollyoko, LiQuiD_Nation) has been working with new media since 1984 as an artist, writer, filmmaker, curator, corporate CADCAM geisha girl, VNS Matrix cyberfeminist collective member, puppet mistress and ghost. Squatting the screens at System-X and The Thing, her projects have been widely exhibited and accessed.

Richard A. Slaughter is Foundation Professor of Foresight at Swinburne University of Technology and a Director of Foresight International, Brisbane. He is a fellow and the current president of the World Futures Studies Federation (WFSF) and a professional member of the World Future Society. In 1997 he was elected to the executive council of the WFSF. He a consulting editor to *Futures* (Oxford, UK), and on the editorial boards of the *Journal of Futures Studies* (Tamkang University, Taiwan), *On the Horizon* (University of N. Carolina), *Foresight* (Manchester, UK) and series editor of *The Knowledge Base of Futures Studies* (FSC Melbourne). He is co-author of *Education for the 21st Century* (Routledge 1993), author of *The Foresight Principle – Cultural Recovery in the 21st Century* (Praeger 1995), editor of *New Thinking for a New Millennium* (Routledge 1996) and co-editor of the *World Yearbook of Education 1998: Futures Education* (London: Kogan Page 1998). His most recent books are *Futures for the Third Millennium – Enabling the Forward View* (Sydney: Prospect 1999) and *Gone Today, Here Tomorrow: Millennium Previews* (Sydney: Prospect 2000).

Stelarc has performed extensively in Japan, Europe and the USA, using medical instruments, prosthetics, robotics, Virtual Reality systems and the Internet to explore alternate, intimate and involuntary interfaces with the body. In 1997 he was appointed Honorary Professor of Art and Robotics at Carnegie Mellon University. He is currently Principal Research Fellow in the Performance Arts Digital Research Unit at The Nottingham Trent University, UK. Current projects include EXTRA EAR, a surgically constructed ear that, coupled with a modem and a wearable computer, will act as an internet antenna, able to hear RealAudio sounds.

Zoë Sofoulis (who also publishes as Zoë Sofia) is Senior Lecturer in Feminist and Cultural Studies at the University of Western Sydney and has a long-standing interest in the irrational and corporeal dimensions of our relations to high technology. Her publications include *Whose Second Self? Gender and (Ir)rationality in Computer Culture* (Geelong: Deakin University Press 1993), *Planet Diana: Cultural Studies and Global Mourning* (Research Centre for Intercommunal Studies at UWS Nepean 1997), for which she was a contributing co-editor, and various papers on feminism, technology, and contemporary art. She is currently researching and writing a book about different kinds of container technologies.

John Sutton teaches Philosophy at Macquarie University in Sydney. He works on theories of memory, dreams, and embodied cognition in the philosophy of psychology and the history of science. He is the author of *Philosophy and Memory Traces: Descartes to Connectionism* (Cambridge 1998), and (with Stephen Gaukroger and John Schuster) he co-edited *Descartes' Natural Philosophy* (Routledge 2000). A regular contributor to the Philosophy Nights series of performances in Sydney, he wrote and directed *Empedocles: Love and Strife* in the 1998 series, and *Kenelm Digby and the Liquid Empire* in 2000. He currently presents a weekly community radio show, *Ghost in the Machine*, discussing minds and bodies, machines and memory, on Eastside Radio (89.7FM) in Sydney.

Donald F. Theall, University Professor and President (Emeritus) at Trent University, Peterborough, Ontario was previously Molson Professor, Chair of English and Founding Director to the Graduate Communication Programme at McGill University. His recent publications include *Beyond the Word: Reconstructing Sense in the Joyce Era of Technology, Culture and Communication* (University of Toronto Press 1995), *James Joyce's Techno-poetics* (University of Toronto Press 1997) and *The Virtual Marshall McLuhan* (McGill-Queen's University Press 2001) as well as numerous articles including a seminal article "Beyond the Orality/Literacy Principle: James Joyce and the Pre-history of Cyberspace." He has written articles on communication theory and history, aesthetic theory, science fiction, film, media, modernist poetics and issues of freedom of communication and dissemination on the Internet. Currently he is pursuing a research project on *Radical Modernism and its Aftermath: Cultural Production and the Pre-history of the Digital Infomatrix and Cyberculture* under a grant from the Social Science and Humanities Research Council of Canada.

Gregory L. Ulmer, Professor of English and Media Studies at the University of Florida, is the author of *Heuretics: The Logic of Invention* (Johns Hopkins 1994); *Teletheory: Grammatology in the Age of Video* (Routledge 1989); and *Applied Grammatology: Post(e)-Pedagogy from Jacques Derrida to Joseph Beuys* (Johns Hopkins 1985). In addition to two other monographs and a textbook for writing about literature, Ulmer has authored some 50 articles and chapters exploring the shift in the apparatus of language from literacy to electracy. His media work includes two video tapes in distribution (one with Paper Tiger Television, the other with the Critical Art Ensemble). He has given invited addresses at international media arts conferences in Helsinki, Sydney, Hamburg, Nottingham, and Halifax as well as at many sites in the United States, and is on the faculty of the European Graduate School. Ulmer's Internet experiments are organized around the problematic of electronic monumentality – concerned with the mutation of the public sphere in electracy and the consequences for national identity. His current work is a collaboration with the Florida Research Ensemble on the project for a new consultancy – the EmerAgency – as described in their coauthored book, *Miami Miautre: Mapping the Virtual City*.

Samuel J. Umland is Professor of English and Film Studies at the University of Nebraska at Kearney, where in addition to film he teaches courses in media theory. He edited a collection of essays on Philip K. Dick, *Philip K. Dick: Contemporary Critical Interpretations* (Westport, CT: Greenwood Press 1995). With Rebecca A. Umland he co-authored a critical biography of the late British film director Donald Cammell, forthcoming from FAB Press (UK).

VNS Matrix (Josephine Starrs, Francesca da Rimini, Julianne Pierce, and Virginia Barratt [until 1996]) emerged from the cyberswamp as VNS Matrix during a steamy southern Australian summer, circa 1991. From 1991-2001 the group made new media, photography, sound works, video installations, computer games and public art works in Australia and overseas exploring ideas around technology, sexuality, and control in technoculture from a feminist perspective. Together with Sadie Plant, VNS coined the term "cyberfeminism" in the early 1990s.

Catherine Waldby is Reader in Sociology and Communications and the Director of the Centre for Research in Innovation, Culture and Technology at Brunel University, London. She is also Adjunct Associate Professor at the National Centre in HIV Social Research, University of New South Wales, Sydney. She is the author of *AIDS and the Body Politic: Biomedicine and Sexual Difference* (Routledge 1996), *The Visible Human Project: Informatic Bodies and Posthuman Medicine* (Routledge 2000), and numerous articles about science, technology and the body. She is currently researching social aspects of tissue transfer, with a focus on blood, embryos and human stem cell technologies.

McKenzie Wark is the author of three books, including *Virtual Geography* (Indiana University Press) and *Celebrities, Culture and Cyberspace* (Sydney: Pluto Press). He is co-author of the collaborative anti-hypertext work *Speed Factory* (Fremantle Arts Centre Press) and with Brad Miller co-authored the multimedia work *Planet of Noise* (Australian Film Commission). He is Senior Lecturer in Media Studies at Macquarie University in Sydney, Australia, and a Visiting Professor in Comparative Literature at the State University of New York, Binghamton.

http://www.mcs.mq.edu.au/~mwark

Margaret Wertheim is a science writer and commentator currently living in Los Angeles. She is the author of *The Pearly Gates of Cyberspace: A History of Space from Dante to the Internet* (New York: W. W. Norton 1999), and *Pythagoras' Trousers* (New York: W. W. Norton 1997) a history of the relationship between physics and religion. She writes about the intersection between science and the wider cultural landscape for many publications including the *Los Angeles Times, LA Weekly, New Scientist, The Guardian, T.L.S.* and *Die Zeit*. She is a columnist for *The Age* newspaper in Melbourne, and has written a dozen television science documentaries. With her husband Cameron Allan, she produced and directed the documentary film *It's Jim's World … we just live in it* on visionary "outsider physicist" James Carter and is currently writing a book on the role of "aberrant" creativity in the evolution of scientific theories, to be titled *Imagining the World*. Wertheim lectures frequently at colleges and universities across the USA. She is a Research Associate at the American Museum of Natural History in New York and a fellow of the Los Angeles Institute for the Humanities. She is founding director of the recently established Institute For Figuring.

Karl Wessel is a Los Angeles writer who has written on Philip K. Dick. While at UCLA he studied mathematical logic under Alonzo Church, the father of the Church-Turing Thesis and the grandfather of LISP, the language of artificial intelligence.

Elizabeth A. Wilson is a Research Fellow in the Research Institute for Humanities and Social Sciences, University of Sydney. She is the author of *Neural Geographies: Feminism and the Microstructure of Cognition* (Routledge 1998) and guest editor of a special issue of *Australian Feminist Studies* (1999) on science studies. She is currently completing a book on neuroscience, affects and evolution.

INDEX

Harris, Thomas 297

Hayles, N. Katherine 14, 19, 41 n.6

Hebb, Donald 58

Heidegger, Martin 174

Helmholtz, Hermann von 54

Heraclitus 68, 77

Hermelin, Beatte 77

heterogeneity 9, 41, 90, 151

    and body's open-ended becoming 9

    and Frankenstein's monster 9

Hooke, Robert 135

Hudson, Kenneth 297-298

human xiii, xiv, 3, 8, 9, 17, 24, 29, 30, 33, 35,
    36, 37, 41 n.6, 42, 42 n.9, 48, 50, 71, 72,
    73, 74, 79, 83, 87, 89, 91, 103, 140, 159,
    160, 172, 182, 184, 185, 186, 196, 198, 200,
    211, 213, 222, 228, 231, 243, 246, 248, 252,
    253, 257, 258, 259, 261, 264, 265, 266, 267,
    268, 269, 270, 272, 273, 274, 275, 276, 279,
    281, 282, 283, 284, 286, 287, 288, 289, 289
    n.12, 291, 298, 301, 302, 312

    as machinic 9, 33, 42 n.9, 46, 147

    becoming informatic 8

human nature 3, 11, 107, 212, 252, 257, 262, 281

    as technologized 9

    post-dualistic approach to 9

See also cyborg, posthuman

Huxley, Aldous 64

Huxley, Thomas 235, 236, 237, 238, 239

hybridity 33, 37, 88, 91, 92, 96, 97, 100

    and cyborg 11, 33, 85, 96, 184

    and non-oppositional thinking 8

intelligence 11, 36, 38, 41, 41 n.6, 45 n.14, 46,

    47, 48, 49, 50, 56, 86, 118, 119, 120, 125,
    132, 136, 140, 186, 187, 194, 200, 204, 238,
    239, 240, 248, 253, 261, 262, 279, 284,
    285, 291

    and Turing 4, 9, 10, 38, 39, 40, 41, 44
    n.13, 46

    as posthuman 260

intelligent machines xiv, 2, 10, 41 n.6, 47, 50,
    93, 102, 129, 185, 196, 261, 265, 282, 301

    and Turing xiv, 10

interface xiii, 5, 11, 32, 46, 91, 103, 106, 107,
    108, 113, 114, 115, 116, 119, 121, 122, 123,
    127, 129, 132, 138, 145, 166, 175, 183, 186,
    192, 200, 204, 222, 284, 309, 311

    of humans and computer technology 2, 4,
    9, 88, 96, 101, 111, 112, 129, 149, 184, 187,
    204

Irigaray, Luce 93

Jacob, François 52

Jantsch, Erich 64

Jencks, Charles 247

Joyce, James 108, 142, 144, 145, 145 n.2, 146,
    147, 148, 149, 150, 151, 152

Kaku, Michio 283

Kant, Immanuel 53, 54, 55, 57, 58, 62

Kauffman, Stuart 81

Kelly, Kevin 222, 223

Kerr, Alan 289-290

Knight, Damon 286

Koolhaas, Rem 174

Kosko, Bart 66, 67

See also artificial intelligence

Minsky, Marvin 14, 61, 62, 62 n.11

Mitchell, William 219

molecular biology 10, 52, 56, 58, 61, 63
  and cybernetics 57

monstrosity 29, 35, 36, 37, 96

Moravec, Hans 13, 280, 283

More, Max 248

More, Thomas 211, 217, 218, 220, 221, 224,
  225, 240, 243, 244

Morgan, T.H. 55-56

Moylan, Tom 259, 260

Muller, Nat 99-100

Mulvey, Laura 84

Mumford, Lewis 258, 265

Munster, Anna 96

nanotechnology 249, 275, 280, 281, 283, 285,
  286, 287, 289

Negroponte, Nicholas 219, 220, 222

Neuromancer 18, 97, 98, 108, 166, 167, 168,
  171-172, 172 n.4, 173, 253

non-oppositional thinking
  and hybridity 8

organic machine 30, 33, 184, 187
  body as 9, 31

organization 3, 10, 31, 53, 54, 55, 56, 57, 58,
  59, 60, 186, 233, 243, 265, 267, 271, 272,
  273, 274, 275
  as defining characteristic of organism 53,
  54, 59, 60, 61, 63, 64
  biological concept of 52, 53, 60, 63

Papert, Seymour 62

Pask, Gordon 61

Pettifer, Julian 297-298

Phaedrus 112, 113, 117, 127, 128

Piaget, Jean 40

Plant, Sadie 93, 99, 100

Plato 107, 110, 111, 112, 113, 117, 118, 119, 125,
  126, 132, 220, 244, 249

Platt, John 269

Popper, Karl 72, 76, 80

post-dualism
  and cyberculture 9

posthuman 3, 5, 8, 11, 14, 27, 50, 91, 100, 102,
  107, 147, 168, 211, 212, 213, 214, 248, 249,
  252, 253, 258, 261, 262, 279, 280, 308, 312
  as beyond binary opposition 8
  as beyond essentialist thinking 8
  cyborg as paragon of 8
  intelligence 260

See also cyborg

postmodern 11, 78, 79, 87, 91, 95, 176, 247, 275

Presley, Elvis 107, 125, 126, 127, 128

Probyn, Elspeth 159

prosthetics 8, 121, 128, 131, 136, 137, 175,
  184, 186, 211, 231, 311

Rabelais, François 247

Radnitzky, Gerard 269

Regis, Ed 252, 253

Reilly, Scott 295

Rheingold, Howard 174, 219

Ricardo, David 79

robotics 8, 13, 41 n.6, 185, 275, 284, 310, 311

320

321

322